INTERMEDIATE CLASSICAL MECHANICS

INTERMEDIATE CLASSICAL MECHANICS

Joseph Norwood, Jr.
East Carolina University

Prentice-Hall, Inc., Englewood Cliffs, N.J.   07632

Library of Congress Cataloging in Publication Data

NORWOOD, JOSEPH.
    Intermediate classical mechanics.

    Includes index,
    1. Mechanics. I. Title.
QA805.N77   531   78-15683
ISBN 0-13-469635-2

Editorial/production supervision and interior design by:
Cathy Van Yperen
Cover Design by: Edsal Enterprise
Manufacturing buyer: Gordon Osbourne

Printed in the United States of America

10 9 8 7 6 5 4 3 2 1

PRENTICE-HALL INTERNATIONAL, INC., *London*
PRENTICE-HALL OF AUSTRALIA PTY. LIMITED, *Sydney*
PRENTICE-HALL OF CANADA, LTD., *Toronto*
PRENTICE-HALL OF INDIA PRIVATE LIMITED, *New Delhi*
PRENTICE-HALL OF JAPAN, INC., *Tokyo*
PRENTICE-HALL OF SOUTHEAST ASIA PTE. LTD., *Singapore*
WHITEHALL BOOKS LIMITED, *Wellington, New Zealand*

CONTENTS

v

PREFACE

My reason for writing a book on classical mechanics at the intermediate (undergraduate) level when so many good texts are already on the market has to do with the conviction that classical mechanics should be taken to include not only particle mechanics and a bit on rigid bodies, but also many-body systems often neglected in texts at this level. Consequently, the present text treats elastic bodies, and also many-body systems from a fluid and statistical point of view. There is a considerable amount of material on hydromagnetic and plasma systems including a discussion of waves in hydromagnetic and hydrodynamic media from the elegant vantage point afforded by the theory of characteristics introduced early in the text. Other features worth noting include a comprehensive discussion of vectors, a treatment of adiabatic invariance, a thorough discussion of rocket motion, Lagrangian and Hamiltonian formulations via the calculus of variations, a chapter on stability, and a modern treatment of relativity including the general theory.

Appropriate problems have been included at the end of each section rather than lumped at the end of the chapter. The book as a whole is suitable for either a one or two-semester course with the first six chapters comprising the first semester. The level of mathematics is such that the two-semester course of elementary calculus is prerequisite and the course of ordinary differential equations is corequisite.

The author would like to thank his typist who also doubles as his wife and the mother of four tolerant children, for her hard work and support during the preparation of this book. Special thanks go to Professor Carl Adler of East Carolina University for a critical reading of the manuscript. Appreciation is also due to the reviewers and my editor Logan Campbell for many helpful suggestions and comments.

Joseph Norwood, Jr.

xi

PRINCIPLES
OF NEWTONIAN MECHANICS

1-1 The Nature of Mechanics

Mechanics, the study of the reaction of massive bodies to forces imposed on them, was the first exact science developed by man. The early Greeks dealt with questions concerning mechanics, but the answers obtained were based, for the most part, on esthetic arguments rather than experimental evidence. Clearly, the success of any theory generated in this way can only be qualitative at best, and even this marginal triumph requires luck. The chance of an exact agreement to within the limits of experimental error is vanishingly small. It was not until the time of Aristotle that the necessity for interplay between theory and experiment began to become apparent. In his *De Generatione et Corruptione* Aristotle notes:[1]

> Lack of experience diminishes our power of taking a comprehensive view of the admitted facts. Hence, those who dwell in intimate association with nature and its phenomena grow more and more able to formulate, as the foundation of their theories, principles such as to admit of a wide and coherent development; while those whom devotion to abstract discussions has rendered unobservant of the facts are too ready to dogmatize on the basis of a few observations.

Unfortunately, Aristotle did not always practice what he preached in this regard, and it was not until the time of Galileo that the *scientific method* finally became established.

---

[1] Aristotle, *De Generatione et Corruptione*, book I, cap. II, H. H. Joachim (tr.). Oxford: Clarendon Press, 1922.

The history of mechanics is closely associated with man's interest in, and ideas about, the motion of the heavenly bodies. Greek science had developed a geocentric cosmology wherein the sun, stars, and planets all revolved in circular orbits with the earth stationary and at the center of the universe. This cosmology became dogmatized by the Church and was not seriously challenged until Copernicus published his heliocentric theory 17 centuries after the time of Aristotle. Galileo, whose career began about 40 years after the publication of Copernicus' theory, was an outspoken critic of Aristotle's doctrines. The first to utilize the telescope for astronomical observations, Galileo began to amass evidence against the geocentric cosmology of Aristotle. The adverse reaction of the Church, however, caused him to become cautious in his later work and to publish in the form of dialogues. These dialogues presented both sides of the argument in such a way that the fallacies of the conventional view would be obvious to an openminded reader without the necessity for Galileo to commit himself directly in print. He wrote two of these dialogues: *Dialogues on the Ptolemaic and Copernican Systems* published in 1632 and *Dialogues on Two New Sciences* (motion and cohesion) published in 1636. The dialogues on motion report Galileo's experiments on accelerating bodies and contain his mature deliberations on their significance. He states, for example, that all bodies fall at the same speed if the resistance of the medium can be neglected, and he establishes that the path of a projectile is parabolic under certain conditions.[2] His work on mechanics exerted a profound influence on Newton, born in 1642, the same year that Galileo died.

Although apparently a slow starter as a student, Newton's genius asserted itself shortly after he entered Cambridge in 1661; he discovered the binomial theorem, developed properties of infinite series, and was one of the inventors of differential calculus while still a student. The plague that swept Europe in the 1660s caused the university to close its doors at various periods. According to Newton, it was during this time, which was spent at home on his mother's estate at Woolsthorpe, that the idea of universal gravitation was developed.[3] Newton had read the work of Kepler, a contemporary of Galileo, while at Cambridge. Kepler was a theoretician who had taken the precise planetary observations of his mentor, Tycho Brahe, and had deduced three empirical laws of planetary motion from these data. He recognized in a qualitative way the presence of a universal binding force for planetary systems but was unable to take the critical step to formulate a theory. Newton sought to find a law of attraction between two massive bodies—for instance, the sun and a planet —such that Kepler's third law, which states that the square of the period of rotation is proportional to the cube of the mean distance separating the two

---

[2] Galileo understood that forces serve as mechanical agents, but he did not manage to establish a quantitative connection between force and motion.

[3] The "falling apple" incident supposedly occurred during this period.

bodies, is obtained as a result. Newton found that a gravitational attraction that varies as the inverse square of the separation distance would produce this relation. He attempted to test this inverse-square law by calculating the measurable ratio of the acceleration of the moon toward the earth to the acceleration of falling bodies at the earth's surface. It was not until Newton managed to prove that homogeneous spheres attract one another as though all their mass were concentrated at the centers that this test was successful. Through his law of universal gravitation, Newton could now derive all three of Kepler's empirical laws.

In 1687 the Royal Society published Newton's *Principia*, in which he expounded his ideas on mechanics. The first two of the three parts of this work are concerned with establishing the foundations of Newton's mechanics; the third part presents a detailed treatment of planetary motion. The three laws of motion on which classical mechanics are based appear for the first time in this book. In formulating these laws, Newton had to break considerable new ground; he was the first, for instance, to distinguish between weight and mass.

A broad range of phenomena can be described directly on the basis of Newton's laws. This textbook is concerned primarily with such phenomena. Certain types of problems are more amenable to treatment by the use of alternative formulations of Newton's theory that are due to Lagrange and Hamilton. If the speeds attained by the bodies of interest are allowed to approach the speed of light or if very large masses or distances are contemplated, then the theory of relativity developed by Einstein in the years 1904 through 1916 must be used. These departures from pure Newtonian mechanics are presented in the closing chapters as introductory material to a more advanced treatment that is beyond the scope of this book.

Kinematics is the branch of mechanics that deals with the classification and description of the types of motions experienced by massive bodies. In many of the problems with which we will deal, the net force acting on the body vanishes. Such a condition is referred to as a state of *equilibrium* and can be further classified as *static* if the velocity of the body is zero or *stationary* if the velocity is constant but nonzero. The question of the *stability* of equilibria will also be considered. In other cases, we shall be interested in bodies that move under the action of forces; this branch of mechanics is called *dynamics*. The bodies whose mechanical state we will examine include point masses (particles), systems of particles, systems of many particles (gases), fluids, elastic bodies, and rigid bodies.

1-2 Scalars, Vectors, and Tensors

The concept of a *field* is almost universal in physics. A field can be regarded as a mathematical idealization of some physical phenomenon involving the notion of extension. The simplest fields can be specified by assigning a single *measure number* at each point of space, as, for example, the temperature distribution in an extended body. Fields of this type are known as *scalars*. Other fields descriptive of common phenomena require three measure numbers at each point of space and are called *vector* fields.[4] Scalars and vectors are special cases of an elegant class of functions known as *tensors*. Scalars are tensors of zero *rank* and vectors are tensors of the first rank.

The special property of tensors that makes them so uniquely useful in formulating physics is their invariance under coordinate transformation. Clearly, it is desirable for a physical theory to be formulated in such a way as to be independent in its results of the choice of coordinate system. This choice should be made purely on the basis of convenience. For instance, a temperature distribution may be written in one coordinate system as $T(x, y, z)$ and in terms of another coordinate system as $T(x', y', z')$. The property that distinguishes a scalar like T from a nonscalar function of the coordinates is that for scalars

$$T(x, y, z) = T(x', y', z'). \tag{1-1}$$

A vector is similarly distinguished from a nonvector function by its invariance under coordinate transformation. In order to ascertain the mathematical criteria for this invariance, let us examine the changes in the coordinates of a point as the coordinate system is rotated about an axis passing through its origin. In Fig. 1-1 we show a point P from the point of view of

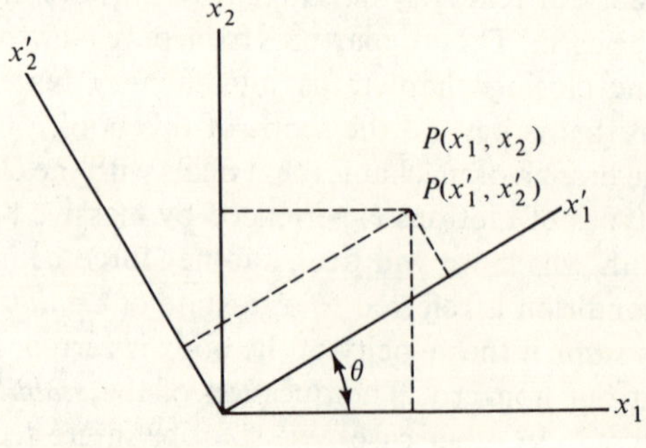

Fig. 1-1

two coordinate systems rotated through an angle θ with respect to one another. The axes are labeled x_1, x_2 instead of x, y in order to facilitate the use of summation notation. By simple trigonometry, we see that the primed

---

[4] In spaces having a dimensionality higher than three, more measure numbers at each point are required, the number being equal to the dimensionality of the space in question. In the present discussion and throughout most of the book we shall only be concerned with the three-dimensional Euclidean space.

coordinates of P are given in terms of the unprimed coordinates and the angle θ by

$$x_1' = x_1 \cos \theta + x_2 \sin \theta,$$
$$x_2' = -x_1 \sin \theta + x_2 \cos \theta. \tag{1-2}$$

Let us introduce

$$\lambda_{ij} = \cos (x_i', x_j) \tag{1-3}$$

for the cosine of the angle between the x_i' and x_j axes. From Fig. 1-1 we see that

$$\lambda_{11} = \cos (x_1', x_1) = \cos \theta,$$
$$\lambda_{12} = \cos (x_1', x_2) = \cos \left(\frac{\pi}{2} - \theta\right) = \sin \theta,$$
$$\lambda_{21} = \cos (x_2', x_1) = \cos \left(\frac{\pi}{2} + \theta\right) = -\sin \theta,$$
$$\lambda_{22} = \cos (x_2', x_2) = \cos \theta, \tag{1-4}$$

in terms of which Eq. (1-2) can be written

$$x_1' = \lambda_{11} x_1 + \lambda_{12} x_2,$$
$$x_2' = \lambda_{21} x_1 + \lambda_{22} x_2. \tag{1-5}$$

The extension to three dimensions is perfectly straightforward:

$$x_1' = \lambda_{11} x_1 + \lambda_{12} x_2 + \lambda_{13} x_3,$$
$$x_2' = \lambda_{21} x_1 + \lambda_{22} x_2 + \lambda_{23} x_3,$$
$$x_3' = \lambda_{31} x_1 + \lambda_{32} x_2 + \lambda_{33} x_3. \tag{1-6}$$

Equation (1-6) can be written much more compactly in summation notation as

$$x_i' = \sum_{j=1}^{3} \lambda_{ij} x_j, \qquad i = 1, 2, 3. \tag{1-7}$$

The symmetry of this equation allows the inverse transformation to be written by inspection:

$$x_i = \sum_{j=1}^{3} \lambda_{ji} x_j', \qquad i = 1, 2, 3. \tag{1-8}$$

The λ_{ij} are often written as a square array called a *matrix*, where λ denotes the matrix

$$\lambda = \begin{pmatrix} \lambda_{11} & \lambda_{12} & \lambda_{13} \\ \lambda_{21} & \lambda_{22} & \lambda_{23} \\ \lambda_{31} & \lambda_{32} & \lambda_{33} \end{pmatrix}. \tag{1-9}$$

Let us examine the properties of this matrix. In Fig. 1-2 we have a line of length R drawn in an arbitrary direction from the origin of a rectangular coordinate system to the point P. The direction of OP is described by the angles α, β, γ between OP and the x_1, x_2, x_3 axes, respectively. If the coordinate axes are mutually orthogonal, then $R_{x_1}^2 + R_{x_2}^2 + R_{x_3}^2 = R^2$ by virtue of the theorem of Pythagorus. Since $R_{x_1} = R \cos \alpha$ and similarly for the x_2 and x_3 components, the *direction cosines* must be related according to

$$\cos^2 \alpha + \cos^2 \beta + \cos^2 \gamma = 1. \qquad (1\text{-}10)$$

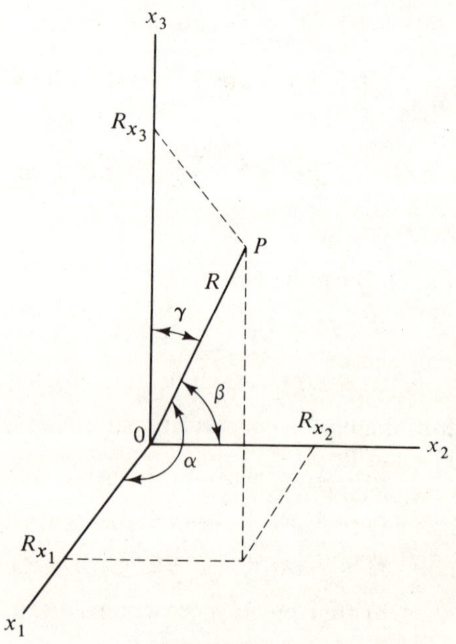

Fig. 1-2

If we take OP to be, say, the x_1' axis in the (x_1, x_2, x_3) coordinate system, we see that

$$\lambda_{11}^2 + \lambda_{12}^2 + \lambda_{13}^2 = 1$$

and similarly for the x_2' and x_3' axes. This result can be expressed in summation notation as

$$\sum_j \lambda_{ij}\lambda_{kj} = 1, \qquad i = k, \qquad (1\text{-}11)$$

which amounts to three relations between the nine elements of the λ matrix.

Next, consider two lines drawn from the origin of a rectangular coordinate system such that the lines are characterized by direction cosines ($\cos \alpha_1$,

$\cos \beta_1, \cos \gamma_1)$ and $(\cos \alpha_2, \cos \beta_2, \cos \gamma_2)$, respectively. The cosine of the angle θ between these lines is then given by

$$\cos \theta = \cos \alpha_1 \cos \alpha_2 + \cos \beta_1 \cos \beta_2 + \cos \gamma_1 \cos \gamma_2. \qquad (1\text{-}12)$$

In terms of, say, the x_1' and x_2' axes in the (x_1, x_2, x_3) coordinate system, Eq. (1-12) can be expressed in terms of the matrix elements as

$$\cos \frac{\pi}{2} = 0 = \lambda_{11}\lambda_{21} + \lambda_{12}\lambda_{22} + \lambda_{13}\lambda_{23}$$

and we are led to a second general relation

$$\sum_j \lambda_{ij}\lambda_{kj} = 0, \qquad i \neq k. \qquad (1\text{-}13)$$

Equations (1-11) and (1-13) can be combined into a single summation equation

$$\sum_j \lambda_{ij}\lambda_{kj} = \delta_{ik}, \qquad (1\text{-}14)$$

where δ_{ik} is the *Kronecker delta* symbol, defined by

$$\delta_{ik} = \begin{cases} 0, & i \neq k, \\ 1, & i = k. \end{cases} \qquad (1\text{-}15)$$

The six relations represented by Eq. (1-14), which is called the *orthogonality condition*, are based on the fact that the coordinate axes in each of the coordinate systems are mutually perpendicular.

The λ matrix given in Eq. (1-9) is a *square matrix*—that is, it has an equal number of rows and columns. Such is not true of all matrices, however. For instance, we can write the three measure numbers that define the coordinates of a point as either a *column matrix*

$$\mathbf{x} = \begin{pmatrix} x_1 \\ x_2 \\ x_3 \end{pmatrix} \qquad (1\text{-}16)$$

or a *row matrix*

$$\mathbf{x} = (x_1 \quad x_2 \quad x_3). \qquad (1\text{-}17)$$

Next, we need to establish the rules describing the multiplication of two matrices. These rules have, in effect, already been established by the requirement that they be consistent with Eq. (1-7) when x_i and x_i' are written in matrix form. Thus we find that

$$\mathbf{x}' = \lambda\mathbf{x} = \begin{pmatrix} x_1' \\ x_2' \\ x_3' \end{pmatrix} = \begin{pmatrix} \lambda_{11} & \lambda_{12} & \lambda_{13} \\ \lambda_{21} & \lambda_{22} & \lambda_{23} \\ \lambda_{31} & \lambda_{32} & \lambda_{33} \end{pmatrix} \begin{pmatrix} x_1 \\ x_2 \\ x_3 \end{pmatrix} \qquad (1\text{-}18)$$

must be equivalent to

$$x_1' = \lambda_{11}x_1 + \lambda_{12}x_2 + \lambda_{13}x_3,$$
$$x_2' = \lambda_{21}x_1 + \lambda_{22}x_2 + \lambda_{23}x_3, \qquad (1\text{-}19)$$
$$x_3' = \lambda_{31}x_1 + \lambda_{32}x_2 + \lambda_{33}x_3.$$

This establishes the multiplication rule for the special case of the product of a 3×3 matrix and a 3×1 matrix. For the more general case where the number of rows of the first matrix and the number of columns of the second matrix are arbitrary, the product rule is

$$R_{ij} = \sum_k P_{ik}Q_{kj} \qquad (1\text{-}20)$$

for the product $R = PQ$. We see that in order for the product QP to be defined, Q must have the same number of columns as P has rows. A particularly interesting aspect of matrix multiplication is that the product is, in general, noncommutative. That is,

$$PQ \neq QP. \qquad (1\text{-}21)$$

Indeed, unless the two matrices are both square, QP is not even defined.

The *transpose* of a matrix P is accomplished by interchanging rows and columns—that is,

$$\tilde{P}_{ij} = P_{ji}, \qquad (1\text{-}22)$$

where $\tilde{P}$ denotes the transpose of the matrix P. If we consider two matrices, P of order $(n \times h)$ and Q of order $(h \times m)$, then the product PQ is defined and $R = PQ$ is of order $(n \times m)$. Consider the transposed matrices. The transpose of P is of order $(h \times n)$ and $\tilde{Q}$ is of order $(m \times h)$. Thus $\tilde{P}\tilde{Q}$ is undefined and we find for $\tilde{R}$

$$\tilde{R} = \tilde{Q}\tilde{P}. \qquad (1\text{-}23)$$

In general, if $F = ABCD\ldots X$, then $\tilde{F} = \tilde{X}\ldots\tilde{D}\tilde{C}\tilde{B}\tilde{A}$.

The *unit matrix* or *identity matrix* is given by

$$E = \delta_{ij}, \qquad (1\text{-}24)$$

where δ_{ij} is the Kronecker delta. So it is a matrix, all of whose off-diagonal elements are zero, with the diagonal elements all equal to one. This matrix leaves any matrix with which it is multiplied in either order unchanged:

$$EP = PE = P. \qquad (1\text{-}25)$$

The λ matrix with which we have been concerned in this section enjoys the property that

$$\lambda = \tilde{\lambda}^{-1} \qquad (1\text{-}26)$$

or

$$\lambda\tilde{\lambda} = \tilde{\lambda}\lambda = E \qquad (1\text{-}27)$$

and is known as an *orthogonal matrix.* The special property of an *orthogonal transformation* like Eq. (1-18) that recommends it to us is that the *magnitude* of a vector remains invariant under such a transformation. The magnitude of a vector x, written $|x|$ or x, is defined as the positive square root of $\tilde{x}x$:

$$x^2 = \tilde{x}x = (x_1 x_2 x_3) \begin{pmatrix} x_1 \\ x_2 \\ x_3 \end{pmatrix} = \sum_j x_j^2. \tag{1-28}$$

Thus if

$$x' = \lambda x, \tag{1-29}$$

which has as its transpose

$$\tilde{x}' = \tilde{x}\tilde{\lambda}, \tag{1-30}$$

then

$$\tilde{x}'x' = \tilde{x}\tilde{\lambda}\lambda x = \tilde{x}Ex = \tilde{x}x. \tag{1-31}$$

Consequently, $x' = x$ if the vectors are associated through an orthogonal transform.

EXERCISES

1-2.1. Consider the case where the coordinate axes are rotated *CCW* through an angle of 90° about the x_3 axis. Calculate the (3×3) transformation matrix λ_1 that describes this rotation.

1-2.2. Determine the orthogonal matrix λ_2 that describes rotation of the coordinate axes through an angle of 90° about the x_1 axis.

1-2.3. Show that the rotation described in problem 1-2.1 followed by the rotation described in problem 1-2.2 can be written as a single orthogonal transformation

$$x'' = \lambda_3 x,$$

where

$$\lambda_3 = \lambda_1 \lambda_2.$$

1-2.4. Show both by evaluating the matrix product and by sketching the sequence of coordinate axis rotations that

$$\lambda_1 \lambda_2 = \lambda_3 \neq \lambda_4 = \lambda_2 \lambda_1.$$

1-2.5. Derive the orthogonal matrix λ_5 that describes the reflection of all three coordinate axes through the origin—that is,

$$x_1' = -x_1,$$

$$x_2' = -x_2,$$

$$x_3' = -x_3.$$

This is called an *inversion transformation.*

1-2.6. Evaluate the determinant of each of the orthogonal matrices $\lambda_1, \ldots, \lambda_5$ derived in the preceding exercises. Note that all transformations resulting from a sequence of rotations starting from the original set of axes have the same value of the determinant and are known as *proper rotations*.

1-3 Vector Algebra

Any quantity $\phi(x, y, z)$ that is unaffected by an orthogonal coordinate transformation such that $\phi(x, y, z) = \phi(x', y', z')$ is a scalar. Any set of three quantities (P_1, P_2, P_3) that transform like a line segment under an orthogonal transformation

$$P_i' = \sum_j \lambda_{ij} P_j \tag{1-32}$$

is a vector. We shall represent vector quantities by a boldface italic letter; the same letter in ordinary italics will denote the magnitude of the vector as defined by Eq. (1-28). An italic Greek letter is used to denote a scalar.

The addition of vectors is defined operationally as

$$\begin{aligned} \boldsymbol{P} + \boldsymbol{Q} &= (P_1, P_2, P_3) + (Q_1, Q_2, Q_3) \\ &= (P_1 + Q_1, P_2 + Q_2, P_3 + Q_3), \end{aligned} \tag{1-33}$$

that is, by adding each corresponding component in $\boldsymbol{P}$ and $\boldsymbol{Q}$ to obtain the components of $\boldsymbol{P} + \boldsymbol{Q}$. As a result of this rule, it is clear that vector addition obeys a commutative law

$$\boldsymbol{P} + \boldsymbol{Q} = \boldsymbol{Q} + \boldsymbol{P} \tag{1-34}$$

and an associative law

$$\boldsymbol{P} + (\boldsymbol{Q} + \boldsymbol{R}) = (\boldsymbol{P} + \boldsymbol{Q}) + \boldsymbol{R}. \tag{1-35}$$

The same is true of scalars:

$$\phi + \psi = \psi + \phi, \tag{1-36}$$

$$\phi + (\psi + \theta) = (\phi + \psi) + \theta. \tag{1-37}$$

Note that the addition of a vector and a scalar is undefined; only tensors of equal rank can be added.

Multiplication is somewhat more involved. The simplest case is the multiplication of a vector by a scalar. This product, $\phi\boldsymbol{P}$, is a vector. It has a magnitude given by $|\phi|P$ and a direction parallel to $\boldsymbol{P}$ if $\phi > 0$ and anti-parallel to $\boldsymbol{P}$ if $\phi < 0$. Clearly, this product commutes

$$\phi\boldsymbol{P} = (\phi P_1, \phi P_2, \phi P_3) = (P_1\phi, P_2\phi, P_3\phi) = \boldsymbol{P}\phi, \tag{1-38}$$

is associative in the case of two scalars multiplying a vector

$$\phi(\psi\boldsymbol{P}) = (\phi\psi)\boldsymbol{P}, \tag{1-39}$$

and obeys a distributive law in the case of a scalar multiplying the sum of two vectors

$$\phi(P + Q) = \phi P + \phi Q \tag{1-40}$$

or a vector multiplying the sum of two scalars

$$(\phi + \psi)P = \phi P + \psi P. \tag{1-41}$$

Before going on to describe products of vectors, we must see how magnitude and direction can be represented algebraically. In Fig. 1-3 we have

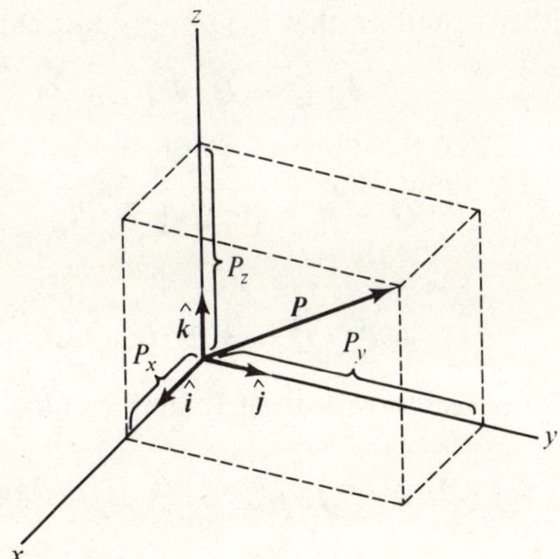

Fig. 1-3 Components of a vector.

drawn a vector whose origin is coincident with the origin of a cartesian coordinate system. The *components* of the vector P are defined as its projections P_x, P_y, P_z onto the x, y, and z axes. These components may be positive or negative, depending on whether the projection along a particular axis lies in the positive or negative portion of the axis. In Fig. 1-3 we have also drawn three vectors $\hat{i}, \hat{j}, \hat{k}$ of unit length. These vectors lie along the x, y, and z axes, respectively. The caret over a vector will be used to denote unit magnitude, and we shall call them *unit vectors*. It follows from our definition of scalar-vector multiplication and vector addition that we can also write the vector P as

$$P = P_x\hat{i} + P_y\hat{j} + P_z\hat{k}. \tag{1-42}$$

In terms of this definition, the sum of two vectors can be written

$$P + Q = (P_x + Q_x)\hat{i} + (P_y + Q_y)\hat{j} + (P_z + Q_z)\hat{k}, \tag{1-43}$$

and the product of a scalar and a vector as

$$\phi P = \phi P_x\hat{i} + \phi P_y\hat{j} + \phi P_z\hat{k}. \tag{1-44}$$

In considering the multiplication of a vector by a vector, it is useful to define two types of multiplication operations. The first is known as the *scalar* or *dot product* and can be defined as

$$\boldsymbol{P} \cdot \boldsymbol{Q} = PQ \cos \theta, \tag{1-45}$$

where θ is the angle between the vectors $\boldsymbol{P}$ and $\boldsymbol{Q}$, or, alternatively, as an operational definition

$$\boldsymbol{P} \cdot \boldsymbol{Q} = \sum_{i=1}^{3} P_i Q_i = P_x Q_x + P_y Q_y + P_z Q_z. \tag{1-46}$$

It follows from either definition that the dot product commutes

$$\boldsymbol{P} \cdot \boldsymbol{Q} = \boldsymbol{Q} \cdot \boldsymbol{P}, \tag{1-47}$$

is distributive

$$\boldsymbol{P} \cdot \boldsymbol{Q} + \boldsymbol{R} = \boldsymbol{P} \cdot \boldsymbol{Q} + \boldsymbol{P} \cdot \boldsymbol{R}, \tag{1-48}$$

and associative

$$(\phi \boldsymbol{P}) \cdot \boldsymbol{Q} = \phi(\boldsymbol{P} \cdot \boldsymbol{Q}). \tag{1-49}$$

The dot product of a vector with itself is simply equal to the square of its magnitude

$$\boldsymbol{P} \cdot \boldsymbol{P} = P^2. \tag{1-50}$$

Since the dot product of two vectors is proportional to the cosine of the angle between them, we can use this product to determine if two vectors are mutually perpendicular, since $\boldsymbol{P} \cdot \boldsymbol{Q} = 0$ if $\boldsymbol{P} \perp \boldsymbol{Q}$.

The other type of product involving two vectors is known as the *vector* or *cross product*. It can be defined as

$$\boldsymbol{P} \times \boldsymbol{Q} = PQ \sin \theta \hat{\boldsymbol{n}}, \tag{1-51}$$

where θ is the angle between the vectors as before and $\hat{\boldsymbol{n}}$ is a unit vector perpendicular to the plane defined by $\boldsymbol{P}$ and $\boldsymbol{Q}$ and directed in the sense of a *right-handed* rotation of $\boldsymbol{P}$ into $\boldsymbol{Q}$ (see Fig. 1-4). It is clear from this definition that the cross product anticommutes

$$\boldsymbol{P} \times \boldsymbol{Q} = -\boldsymbol{Q} \times \boldsymbol{P}. \tag{1-52}$$

The distributive

$$\boldsymbol{P} \times (\boldsymbol{Q} + \boldsymbol{R}) = \boldsymbol{P} \times \boldsymbol{Q} + \boldsymbol{P} \times \boldsymbol{R} \tag{1-53}$$

and associative laws apply

$$(\phi \boldsymbol{P}) \times \boldsymbol{Q} = \phi(\boldsymbol{P} \times \boldsymbol{Q}). \tag{1-54}$$

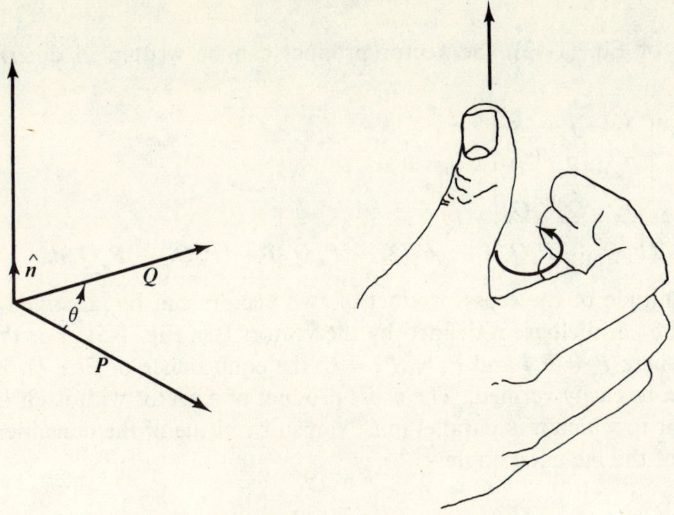

Fig. 1-4 Definition of the cross product.

An operational definition in correspondence with Eq. (1-46) can be given as

$$(\boldsymbol{P} \times \boldsymbol{Q})_i = \sum_{j,k} \epsilon_{ijk} P_j Q_k, \tag{1-55}$$

where ϵ_{ijk} is the *Levi-Civita symbol*, given by

$$\epsilon_{ijk} = \begin{cases} 0 & \text{for any two indices equal} \\ +1 & \text{for } i, j, k \text{ an even permutation of } 1, 2, 3 \\ -1 & \text{for } i, j, k \text{ an odd permutation of } 1, 2, 3. \end{cases} \tag{1-56}$$

In other words,

$$\epsilon_{122} = \epsilon_{313} = \epsilon_{211} = 0$$

$$\epsilon_{123} = \epsilon_{231} = \epsilon_{312} = +1$$

$$\epsilon_{132} = \epsilon_{213} = \epsilon_{321} = -1.$$

Setting $i = 1$, let us evaluate the first component of $\boldsymbol{P} \times \boldsymbol{Q}$ by using the prescription of Eq. (1-55). The only nonzero ϵ_{ijk}'s are ϵ_{123} and ϵ_{132}. Thus

$$(\boldsymbol{P} \times \boldsymbol{Q})_1 = \sum_{j,k} \epsilon_{1jk} P_j Q_k = \epsilon_{123} P_2 Q_3 + \epsilon_{132} P_3 Q_2$$

$$= P_2 Q_3 - P_3 Q_2 \tag{1-57}$$

and, similarly,

$$(\boldsymbol{P} \times \boldsymbol{Q})_2 = P_3 Q_1 - P_1 Q_3, \tag{1-58}$$

$$(\boldsymbol{P} \times \boldsymbol{Q})_3 = P_1 Q_2 - P_2 Q_1. \tag{1-59}$$

13

In terms of Eq. (1-42), the vector product can be written in determinant form as

$$P \times Q = \begin{vmatrix} \hat{i} & \hat{j} & \hat{k} \\ P_x & P_y & P_z \\ Q_x & Q_y & Q_z \end{vmatrix}$$

$$= (P_y Q_z - P_z Q_y)\hat{i} + (P_z Q_x - P_x Q_z)\hat{j} + (P_x Q_y - P_y Q_x)\hat{k}. \quad (1\text{-}60)$$

The magnitude of the cross product of two vectors can be identified as the area of the parallelogram defined by the vectors (see Fig. 1-5). For the case shown, where $P = P_x \hat{i}$ and $P_y = P_z = 0$, the equivalence of Eqs. (1-60) and (1-51) can be easily verified. The cross product of a vector with itself or with any vector to which it is parallel must vanish by virtue of the dependence on the sine of the included angle.

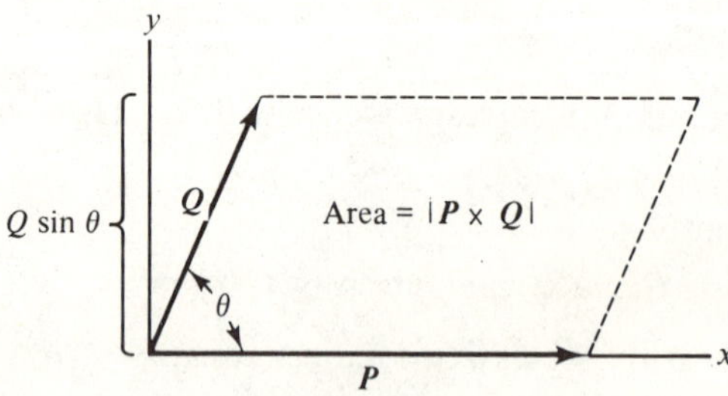

Fig. 1-5 A geometrical interpretation of the cross product.

The vector formed by the cross product of two vectors is peculiar in the sense that the inversion of two vectors give as their vector product a result that is parallel to the cross product of the uninverted vectors. This situation is shown in Fig. 1-6. The vectors P and Q whose cross product $P \times Q$ is directed into the page are inverted about the origin to produce two new vectors P' and Q' of opposite handedness:

$$P' = (P'_x, P'_y, P'_x) = (-P_x, -P_y, -P_z) = -P$$

$$Q' = (Q'_x, Q'_y, Q'_z) = (-Q_x, -Q_y, -Q_z) = -Q. \quad (1\text{-}61)$$

The primed vectors form a left-handed set in the sense that the cross product $P' \times Q' = -P \times -Q = P \times Q$. As we saw in problem 1-2.6 the inversion of all three axes constitutes an *improper rotation*. In the cross product, however, the product of two inverted vectors remains uninverted and so corresponds to a proper rotation. Consequently, the product enjoys the required vector transformation properties. This is a quirk of three-dimensional space. In spaces of higher dimensionality, the resulting array of measure numbers would not transform like a vector.

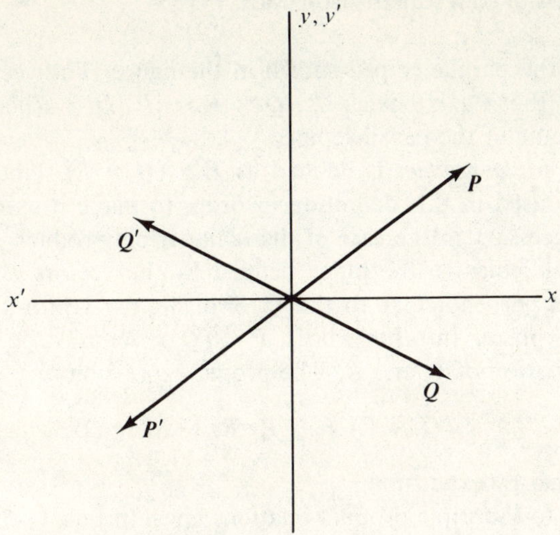

Fig. 1-6 The inversion of vectors about the origin.

Let us examine two other products known as *triple products*, formed by using different combinations of dot and cross products to associate three vectors. The *scalar triple product* of the vectors P, Q, and R is defined as $P \cdot Q \times R$. This product is invariant to cyclic permutation

$$P \cdot Q \times R = Q \cdot R \times P = R \cdot P \times Q, \qquad (1\text{-}62)$$

as can be seen by writing the product in determinant form

$$P \cdot Q \times R = \begin{vmatrix} P_x & P_y & P_z \\ Q_x & Q_y & Q_z \\ R_x & R_y & R_z \end{vmatrix}. \qquad (1\text{-}63)$$

A geometrical interpretation of the scalar triple product can be made with the help of Fig. 1-7. The quantity $|Q \times R| = QR\sin\theta$ is equal to the area

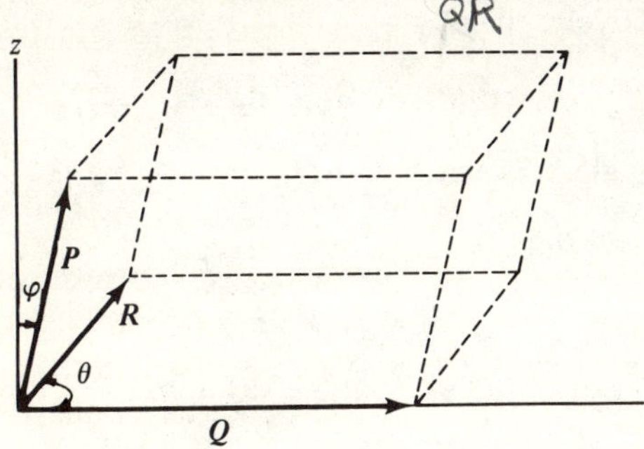

Fig. 1-7 A geometrical interpretation of the scalar triple product.

15

of the base of the parallelepiped shown in the figure. The vector $Q \times R$ is directed along the z axis; hence $P \cdot Q \times R = |P||Q \times R| \cos \varphi$, which is simply the volume of the parallelepiped.

The *vector triple product* is defined as $P \times (Q \times R)$. Note that parentheses must be used in this definition in order to make it explicit, whereas they were unnecessary in the case of the scalar triple product. This product must be perpendicular to the plane defined by the vectors P and $Q \times R$; since $Q \times R$ is perpendicular to the Q, R plane, the vector triple product must lie in this plane. In other words, $P \times (Q \times R)$ must be expressible as a linear combination of Q and R. The proper expression is

$$P \times (Q \times R) = Q(P \cdot R) - R(P \cdot Q), \tag{1-64}$$

as can be verified by expansion.

In addition to the triple product relations given in Eqs. (1-63) and (1-64), the following vector identities involving the product of four vectors may prove useful:

$$P \times Q \cdot R \times S = (P \cdot R)(Q \cdot S) - (P \cdot S)(Q \cdot R), \tag{1-65}$$

$$(P \times Q) \times (R \times S) = (P \times Q \cdot S)R - (P \times Q \cdot R)S. \tag{1-66}$$

EXERCISES

1-3.1. Given the sum $P + Q$ and difference $P - Q$ of two vectors, determine the vectors P and Q.

1-3.2. The vector R has a magnitude $R = 8$ and is known to make equal angles with the three coordinate axes. Find the components.

1-3.3. Show that if two vectors P and Q have direction cosines $\alpha_1, \beta_1, \gamma_1$ and $\alpha_2, \beta_2, \gamma_2$, respectively, then

$$\cos \theta = \alpha_1 \alpha_2 + \beta_1 \beta_2 + \gamma_1 \gamma_2,$$

where θ is the angle between P and Q.

1-3.4. Given the vectors

$$P = 3\hat{\imath} + 2\hat{\jmath} - \hat{k}$$

$$Q = -6\hat{\imath} - 4\hat{\jmath} + 2\hat{k}$$

$$R = \hat{\imath} - 2\hat{\jmath} - \hat{k},$$

find the ones that are mutually perpendicular and the ones that are parallel or antiparallel.

1-3.5. Determine the unit vector $\hat{A}$ perpendicular to the two vectors

$$P = 2\hat{\imath} + \hat{\jmath} - \hat{k}$$

and
$$Q = \hat{\imath} - \hat{\jmath} + \hat{k}.$$

1-3.6. Show that

$$P \times (Q \times R) + Q \times (R \times P) + R \times (P \times Q) = 0.$$

1-3.7. Verify the relation

$$(P \times Q)^2 = (P \times Q) \cdot (P \times Q) = P^2 Q^2 - (P \cdot Q)^2.$$

1-3.8. Write a unit vector that lies in the plane of the vectors $\hat{\imath} + \hat{\jmath}$ and $\hat{\jmath} + \hat{k}$ and is perpendicular to the vector $\hat{\imath} + \hat{\jmath} + \hat{k}$.

1-3.9. Show that

$$(P - Q) \times (P + Q) = 2P \times Q.$$

1-3.10. If P, Q, and R are coplanar and have magnitudes $P = 1$, $Q = 2$, $R = 3$ and directions related by $(P, Q) = 45°$, $(P, R) = 90°$, $(Q, R) = 45°$, where the parenthetical expressions denote the angles separating the indicated vectors, write an expression for each of these vectors as a linear function of the other two. Evaluate the coefficients numerically to three significant figures.

1-3.11. Show that if we rotate our cartesian coordinate axes to new directions x', y', z', then

$$P \cdot Q = P'_x Q'_x + P'_y Q'_y + P'_z Q'_z$$

and

$$P \times Q = \begin{vmatrix} \hat{\imath}' & \hat{\jmath}' & \hat{k}' \\ P'_x & P'_y & P'_z \\ Q'_x & Q'_y & Q'_z \end{vmatrix}.$$

1-3.12. Construct a unit vector that is everywhere tangent to the circumference of a circle of unit radius. Can this problem be extended to three dimensions to find a vector that is everywhere tangent to the surface of a sphere?

1-3.13. Verify Eq. (1-64).

1-3.14. Verify Eq. (1-65).

1-3.15. Verify Eq. (1-66).

1-4 Vector Calculus

Mechanics is normally formulated in terms of vector calculus—specifically, vector differential equations. The extension of our knowledge of the calculus of scalars to embrace vectors is quite straight-forward. The basic rule for the differentiation of a vector by a scalar t is simply

$$\frac{d\boldsymbol{P}}{dt} = \lim_{\Delta t \to 0} \frac{\boldsymbol{P}(t + \Delta t) - \boldsymbol{P}(t)}{\Delta t}. \tag{1-67}$$

The rule governing the differentiation of a product of two variables is directly extended to apply to vectors:

$$\frac{d}{dt}(\phi \boldsymbol{P}) = \phi \frac{d\boldsymbol{P}}{dt} + \boldsymbol{P} \frac{d\phi}{dt}, \tag{1-68}$$

$$\frac{d}{dt}(\boldsymbol{P} \cdot \boldsymbol{Q}) = \boldsymbol{P} \cdot \frac{d\boldsymbol{Q}}{dt} + \boldsymbol{Q} \cdot \frac{d\boldsymbol{P}}{dt}, \tag{1-69}$$

$$\frac{d}{dt}(\boldsymbol{P} \times \boldsymbol{Q}) = \boldsymbol{P} \times \frac{d\boldsymbol{Q}}{dt} + \frac{d\boldsymbol{P}}{dt} \times \boldsymbol{Q}. \tag{1-70}$$

The time derivative of a vector $\boldsymbol{P}$ having a constant magnitude but changing direction is a vector perpendicular to $\boldsymbol{P}$. This fact can easily be shown. If P is constant, then $\boldsymbol{P} \cdot \boldsymbol{P} = P^2$ is constant and we can write

$$\frac{dP^2}{dt} = \frac{d}{dt}(\boldsymbol{P} \cdot \boldsymbol{P}) = 2\boldsymbol{P} \cdot \frac{d\boldsymbol{P}}{dt} = 0. \tag{1-71}$$

If neither P nor dP/dt vanish, then $\boldsymbol{P}$ and dP/dt must be mutually perpendicular by virtue of the definition of the dot product. The time derivative of a vector $\boldsymbol{P}$ that is constant in direction but variable in magnitude is a vector in the direction (but not necessarily the sense) of $\boldsymbol{P}$.

It is helpful to define several vector differential operators, the use of which provides the most convenient and powerful extension of calculus to solve practical vector problems. Let $\phi(x, y, z)$ be a scalar function that we will assume to have continuous first spatial derivatives. In Fig. 1-8 we have sketched a surface specified by $\phi = \text{const}$. A radius vector r is drawn from the point A to the point B in the surface. If we also define a point C in the surface, then the distance from B to C can be specified either by the distance Δs in the surface or by the vector Δr. It can easily be seen that

$$\lim_{\Delta s \to 0} \frac{\Delta r}{\Delta s} = \frac{dr}{ds} = \hat{\boldsymbol{\imath}} \tag{1-72}$$

serves as the definition of a unit vector tangent to the surface $\phi = \text{const}$ at the point $B \to C$. If A is taken as the origin of our cartesian coordinate system, the radius vector will be given by

$$r = x\hat{\boldsymbol{\imath}} + y\hat{\boldsymbol{\jmath}} + z\hat{\boldsymbol{k}} \tag{1-73}$$

and
$$dr = dx\,\hat{\boldsymbol{\imath}} + dy\,\hat{\boldsymbol{\jmath}} + dz\,\hat{\boldsymbol{k}}. \tag{1-74}$$

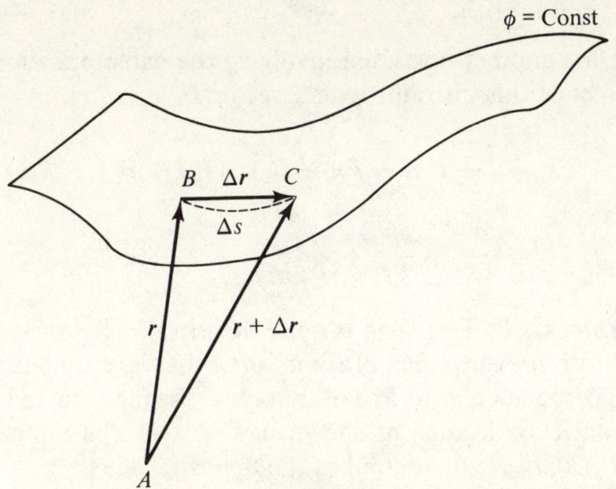

Fig. 1-8

The differential $d\phi$ is

$$d\phi = \frac{\partial \phi}{\partial x}\, dx + \frac{\partial \phi}{\partial y}\, dy + \frac{\partial \phi}{\partial z}\, dz.$$

It may also be written as the dot product of two vectors

$$d\phi = \left(\frac{\partial \phi}{\partial x}\,\hat{\imath} + \frac{\partial \phi}{\partial y}\,\hat{\jmath} + \frac{\partial \phi}{\partial z}\,\hat{k}\right) \cdot (dx\,\hat{\imath} + dy\,\hat{\jmath} + dz\,\hat{k})$$

$$= \nabla\phi \cdot d\boldsymbol{r}, \tag{1-75}$$

where we have defined a vector operator

$$\nabla \equiv \frac{\partial}{\partial x}\,\hat{\imath} + \frac{\partial}{\partial y}\,\hat{\jmath} + \frac{\partial}{\partial z}\,\hat{k}, \tag{1-76}$$

which is known as the *nabla* or *del* operator. The vector quantity $\nabla\phi$ is called the *gradient* of ϕ. Let us see what its properties are. We write the derivative of ϕ with respect to s, the length variable in the surface $\phi = $ const, as

$$\frac{d\phi}{ds} = \frac{\partial \phi}{\partial x}\frac{dx}{ds} + \frac{\partial \phi}{\partial y}\frac{dy}{ds} + \frac{\partial \phi}{\partial z}\frac{dz}{ds} = \nabla\phi \cdot \frac{d\boldsymbol{r}}{ds} = \nabla\phi \cdot \hat{\boldsymbol{t}} = 0, \tag{1-77}$$

where $\hat{\boldsymbol{t}}$ is the unit tangent vector defined above. The function ϕ naturally does not vary within a surface defined by the spatial constancy of ϕ; consequently, $\nabla\phi$ must be perpendicular to the surface and thus represents the direction of the greatest spatial rate of change of the function $\phi(x, y, z)$.

19

Let us define another operation involving the nabla operator. We define the dot product of this operator with a vector $\boldsymbol{P}$

$$\boldsymbol{\nabla} \cdot \boldsymbol{P} = \left(\frac{\partial}{\partial x}\hat{\imath} + \frac{\partial}{\partial y}\hat{\jmath} + \frac{\partial}{\partial z}\hat{k}\right) \cdot (P_x\hat{\imath} + P_y\hat{\jmath} + P_z\hat{k})$$

$$= \frac{\partial P_x}{\partial x} + \frac{\partial P_y}{\partial y} + \frac{\partial P_z}{\partial z} \tag{1-78}$$

as the *divergence* of $\boldsymbol{P}$. This term is quite descriptive. Suppose that we take the vector $\boldsymbol{P}$ to represent a flux of some sort—that is, a quantity per second per unit area perpendicular to $\boldsymbol{P}$. Let us look at the input to and output from an infinitesimal cube located as shown in Fig. 1-9. The input through the face at x is $P_x(x)\,dy\,dz$; the output at the opposite face is

$$P_x(x + dx)\,dy\,dz = \left(P_x + \frac{\partial P_x}{\partial x}\,dx\right)dy\,dz.$$

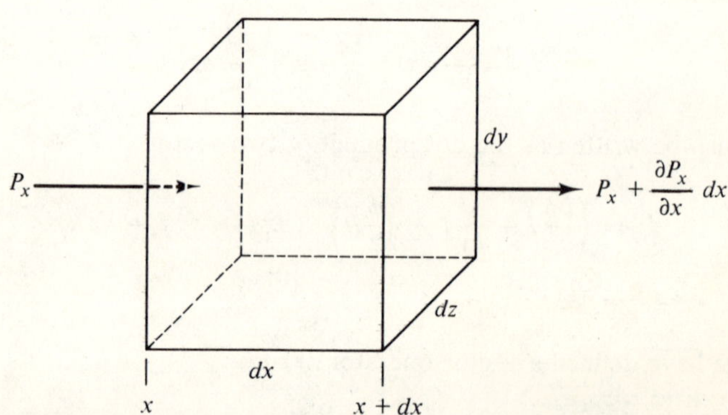

Fig. 1-9 Geometrical definition of the divergence of a vector P.

So the net output in the x direction is merely the difference

$$\left(P_x + \frac{\partial P_x}{\partial x}\,dx - P_x\right)dy\,dz = \frac{\partial P_x}{\partial x}\,dx\,dy\,dz, \tag{1-79}$$

and similarly for the other faces. The net flow per second out of the cube is

$$\left(\frac{\partial P_x}{\partial x} + \frac{\partial P_y}{\partial y} + \frac{\partial P_z}{\partial z}\right)dx\,dy\,dz = \boldsymbol{\nabla} \cdot \boldsymbol{P}\,d\tau, \tag{1-80}$$

where $d\tau$ is the volume of the cube. This operation serves to delineate the presence of *sources* or *sinks* in a vector field. The divergence of a vector is therefore a scalar quantity, as the dot product implies.

The possibility obviously exists to operate with the nabla operator on a vector by means of the cross product. Let us see what the significance of this operation might be. We define the *curl* of P to be

$$\nabla \times P = \begin{vmatrix} i & j & \hat{k} \\ \dfrac{\partial}{\partial x} & \dfrac{\partial}{\partial y} & \dfrac{\partial}{\partial z} \\ P_x & P_y & P_z \end{vmatrix}$$

$$= \left(\frac{\partial P_z}{\partial y} - \frac{\partial P_y}{\partial z}\right)i + \left(\frac{\partial P_x}{\partial z} - \frac{\partial P_z}{\partial x}\right)j + \left(\frac{\partial P_y}{\partial x} - \frac{\partial P_x}{\partial y}\right)\hat{k}, \quad (1\text{-}81)$$

which is a vector. In order to gain some insight into the physical significance of this operation, consider a vector v that we shall take to be the linear velocity of some fluid. Let us specify this velocity in such a way that

$$v_x = v_x(0, y, 0),$$

$$v_y = v_y(x, 0, 0), \quad (1\text{-}82)$$

$$v_z = 0.$$

The behavior of the x component of this velocity is shown in Fig. 1-10. A flow with *shear*—that is, one that varies monotonically perpendicular to the direction of the flow vector as shown—will result in fluid rotation about the z axis. If $\partial v_x/\partial y < 0$ and $\partial v_y/\partial x > 0$, the direction is positive. The total rotation for the vector specified in Eq. (1-79) is simply

$$\nabla \times v = (\nabla \times v)_z \hat{k} = \left(\frac{\partial v_y}{\partial x} - \frac{\partial v_x}{\partial y}\right)\hat{k}. \quad (1\text{-}83)$$

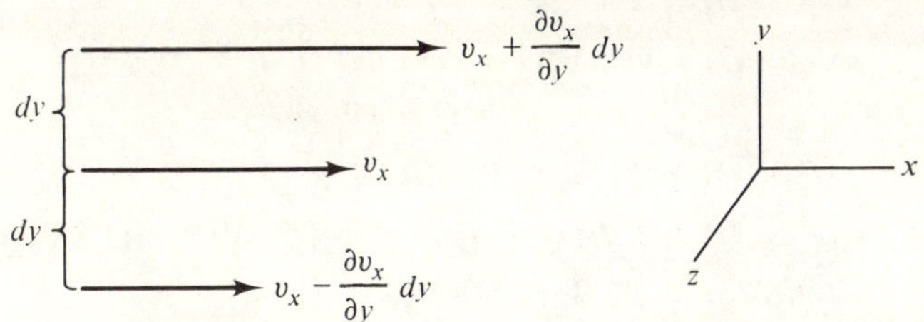

Fig. 1-10 Geometrical interpretation of the curl of a vector.

It should be remembered that the nabla operator defined in Eq. (1-76) is a vector operator and *not* a vector in its own right; that is, it has no inherent magnitude or direction. So when we take the cross product of the nabla operator with some vector P, thereby generating a vector $Q = \nabla \times P$, we find that Q and P are not necessarily perpendicular, as our experience with the cross product might have led us to believe. There is, in fact, a large and interesting class of vector fields for which the curl of the vector is parallel (or antiparallel) to the vector itself:

$$\nabla \times P = \alpha P. \tag{1-84}$$

Such fields are called *Beltrami* fields.

Let us define one last differential operator involving the nabla. This operator is written ∇^2 (or Δ in European textbooks), is defined as

$$\nabla^2 \equiv \nabla \cdot \nabla = \frac{\partial^2}{\partial x^2} + \frac{\partial^2}{\partial y^2} + \frac{\partial^2}{\partial z^2}, \tag{1-85}$$

and is known as the *Laplacian* operator. This operator is ordinarily considered to operate on a scalar. The definition can be extended to define a vector Laplacian operator

$$\nabla^2 \equiv \nabla(\nabla \cdot \ ) - \nabla \times \nabla \times \tag{1-86}$$

It is left as an exercise for the reader to show that the vector and scalar Laplacian operators are identical for the choice of cartesian coordinates. In general, this statement is not true for other coordinate systems.

Before leaving differential vector calculus and considering the integration of vectors, it is useful to collect for reference a number of identities of vector differential calculus:

$$\nabla(\phi + \psi) = \nabla\phi + \nabla\psi \tag{1-87}$$

$$\nabla(\phi\psi) = \phi\nabla\psi + \psi\nabla\phi \tag{1-88}$$

$$\nabla \cdot (P + Q) = \nabla \cdot P + \nabla \cdot Q \tag{1-89}$$

$$\nabla \times (P + Q) = \nabla \times P + \nabla \times Q \tag{1-90}$$

$$\nabla \cdot (\phi P) = P \cdot \nabla\phi + \phi\nabla \cdot P \tag{1-91}$$

$$\nabla \times (\phi P) = \nabla\phi \times P + \phi\nabla \times P \tag{1-92}$$

$$\nabla(P \cdot Q) = (P \cdot \nabla)Q + (Q \cdot \nabla)P + P \times (\nabla \times Q) \\ + Q \times (\nabla \times P) \tag{1-93}$$

$$\nabla \cdot (P \times Q) = Q \cdot \nabla \times P - P \cdot \nabla \times Q \tag{1-94}$$

$$\nabla \times (P \times Q) = P\nabla \cdot Q - Q\nabla \cdot P + (Q \cdot \nabla)P - (P \cdot \nabla)Q \tag{1-95}$$

$$\nabla \cdot (\phi\nabla\psi) = \phi\nabla^2\psi + \nabla\phi \cdot \nabla\psi \tag{1-96}$$

$$\nabla \times \nabla \times P = \nabla\nabla \cdot P - \nabla^2 P \tag{1-97}$$

$$\nabla \times \nabla\phi = 0 \tag{1-98}$$

$$\nabla \cdot \nabla \times P = 0 \tag{1-99}$$

$$\nabla \cdot (\nabla\phi \times \nabla\psi) = 0 \tag{1-100}$$

$$\nabla \cdot r = 3 \tag{1-101}$$

$$\nabla \times r = 0. \tag{1-102}$$

The only one of the operation sequences that needs clarification is $(P \cdot \nabla)Q$:

$$(P \cdot \nabla)Q = \left(P_x \frac{\partial}{\partial x} + P_y \frac{\partial}{\partial y} + P_z \frac{\partial}{\partial z}\right)Q. \tag{1-103}$$

The integrals of practical interest in vector analysis can be classified as *line integrals* or *surface integrals*. In the first category can be found integrals of the following types:

$$\int_{\mathscr{C}} \phi \, dr, \tag{1-104}$$

$$\int_{\mathscr{C}} P \cdot dr, \tag{1-105}$$

$$\int_{\mathscr{C}} P \times dr. \tag{1-106}$$

The object of our approach to any vector integral is to reduce the problem to a collection of scalar integrals. The reduction of Eq. (1-104) is trivial.

$$\int_{\mathscr{C}} \phi(x, y, z) \, dr = i \int_{\mathscr{C}} \phi(x, y, z) \, dx + j \int_{\mathscr{C}} \phi(x, y, z) \, dy + \hat{k} \int_{\mathscr{C}} \phi(x, y, z) \, dz. \tag{1-107}$$

To effect the solution of these scalar integrals, the path of integration $\mathscr{C}$ must be specified in order to cast the integrand of each integral in terms of the variable of integration. In general, the value of the integral is path dependent.

The integration of the parallel component of a vector along a specified path is the line integral most commonly encountered in mechanics. Let us define a vector function

$$P = u(x, y, z)\hat{i} + v(x, y, z)\hat{j} + w(x, y, z)\hat{k}. \tag{1-108}$$

The line integral of the parallel component of this vector along a curve specified by $\mathscr{C}$ is

$$\int_{\mathscr{C}} P \cdot dr = \int_{\mathscr{C}} u \, dx + \int_{\mathscr{C}} v \, dy + \int_{\mathscr{C}} w \, dz. \tag{1-109}$$

In general, as in the case of integrals of the type given in Eq. (1-104), the path must be specified. An important exceptional case must be noted. Suppose that

$$P \cdot dr = u \, dx + v \, dy + w \, dz = d\phi, \tag{1-110}$$

an exact differential. As we saw in Eq. (1-75), $d\phi$ can also be expressed as $\nabla \phi \cdot dr$. This fact implies that vector fields for which the condition of Eq. (1-110) applies can be expressed as the gradient of some scalar function

$$P = \nabla \phi. \tag{1-111}$$

The line integral of P between the points q_2 and q_1 is

$$\int_{q_1}^{q_2} P \cdot dr = \int_{q_1}^{q_2} d\phi = \phi(q_2) - \phi(q_1), \tag{1-112}$$

which depends *only* on the value of the endpoints and not on the path of integration between them. The integral of such a function around a closed path obviously vanishes. Such integrals are called *contour integrals* and are denoted by a loop across the integral sign

$$\oint P \cdot dr = 0. \tag{1-113}$$

It can be seen from Eq. (1-98) that if $P = \nabla\phi$, then the curl of P must vanish. Vector fields that have this property of being derivable from a *scalar potential* function are called *conservative* or *irrotational* fields. A test for determining whether $P \cdot dr$ is an exact differential is easily made. Since the coefficients of P can be identified with the components of $\nabla\phi$

$$\frac{\partial \phi}{\partial x} = u, \qquad \frac{\partial \phi}{\partial y} = v, \qquad \frac{\partial \phi}{\partial z} = w,$$

it follows that

$$\frac{\partial^2 \phi}{\partial x\, \partial y} = \frac{\partial u}{\partial y}, \qquad \frac{\partial^2 \phi}{\partial y\, \partial x} = \frac{\partial v}{\partial x}.$$

If the order of differentiation is immaterial, as is the case, then

$$\frac{\partial u}{\partial y} = \frac{\partial v}{\partial x}, \tag{1-114}$$

and, similarly,

$$\frac{\partial v}{\partial z} = \frac{\partial w}{\partial y}, \qquad \frac{\partial u}{\partial z} = \frac{\partial w}{\partial x} \tag{1-115}$$

must obtain if $P = P(u, v, w)$ is an irrotational vector.

Let us examine vector surface integrals

$$\iint_{\mathscr{S}} P \cdot d\sigma, \tag{1-116}$$

$$\iint_{\mathscr{S}} P \times d\sigma. \tag{1-117}$$

To see how the vector element of area $d\sigma$ is defined, we consider a surface specified by $\phi(x, y, z) = $ const. The gradient $\nabla\phi$ is a vector normal to the surface. A *unit normal* vector can therefore be specified as

$$\hat{n} = \frac{\nabla\phi}{|\nabla\phi|}. \tag{1-118}$$

The vector element of area

$$d\boldsymbol{\sigma} = \hat{n}\, d\sigma \tag{1-119}$$

is shown in Fig. 1-11. The direction of the vector $d\boldsymbol{\sigma}$ is the direction of the *outward* normal to the surface. Referring to the figure, we can define an element of area $dA = \hat{k}\, dx\, dy$, where $\hat{k}$ is the unit vector in the z direction.

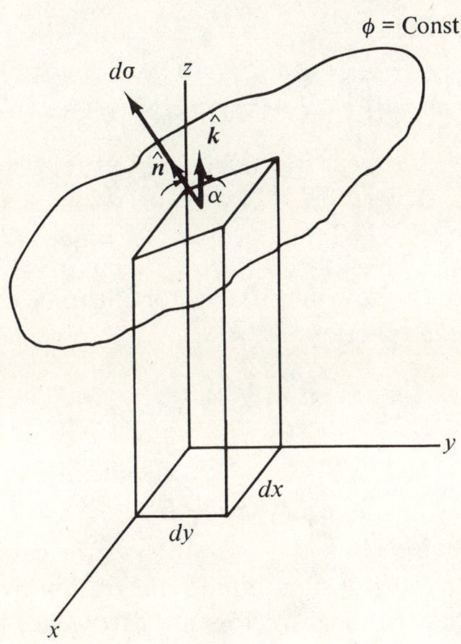

Fig. 1-11 Geometrical definition of the vector element of surface area.

The element of surface area $d\boldsymbol{\sigma}$ can then be written

$$d\boldsymbol{\sigma} \cdot \hat{k} = d\sigma\, \hat{n} \cdot \hat{k} = d\sigma \cos \alpha = dx\, dy;$$

so

$$d\sigma = dx\, dy \sec \alpha, \tag{1-120}$$

where

$$\sec \alpha = (\hat{n} \cdot \hat{k})^{-1} = \frac{|\nabla\phi|}{\nabla\phi \cdot \hat{k}} = \frac{\sqrt{(\partial\phi/\partial x)^2 + (\partial\phi/\partial y)^2 + (\partial\phi/\partial z)^2}}{\partial\phi/\partial z}. \tag{1-121}$$

As an example of this type of surface integral, let us calculate the integral of the normal component of $\boldsymbol{P} = x\hat{i} + y\hat{j}$ over the unit sphere above the xy plane. First, we calculate the unit normal. Since

$$\phi = x^2 + y^2 + z^2 = 1,$$

$$\nabla\phi = 2(x\hat{i} + y\hat{j} + z\hat{k}),$$

and

$$|\nabla\phi| = 2(x^2 + y^2 + z^2) = 2,$$

then

$$\hat{n} = x\hat{\imath} + y\hat{\jmath} + z\hat{k}. \tag{1-122}$$

With

$$\sec\alpha = (\hat{n}\cdot\hat{k})^{-1} = z^{-1},$$

our integral can be written

$$\iint_{\mathscr{S}} \boldsymbol{P}\cdot d\boldsymbol{\sigma} = \iint_{\mathscr{S}} (x\hat{\imath} + y\hat{\jmath})\cdot(x\hat{\imath} + y\hat{\jmath} + z\hat{k})\frac{dx\,dy}{z}$$

$$= \int_0^1 \int_0^{\sqrt{1-x^2}} \frac{(x^2 + y^2)\,dx\,dy}{\sqrt{1 - x^2 - y^2}} = \frac{4}{3}\pi.$$

It is left as an exercise to show that if a vector $\boldsymbol{P}$ can be expressed as the curl of some *vector potential* function

$$\boldsymbol{P} = \nabla \times \boldsymbol{Q}, \tag{1-123}$$

then

$$\oiint \boldsymbol{P}\cdot d\boldsymbol{\sigma} = 0, \tag{1-124}$$

where the loop on the integral sign implies integration over a *closed* surface. Equation (1-99) implies that such vectors are divergence free:

$$\nabla \cdot \boldsymbol{P} = 0. \tag{1-125}$$

We refer to such a vector as a *solenoidal* vector.

There are two integral theorems of vector calculus with which the reader should become familiar. The first, known as the *Gauss divergence theorem*, is essentially an integral version of the interpretation that we gave to the divergence operation:

$$\iiint_{\mathscr{V}} \nabla \cdot \boldsymbol{P}\,d\tau = \oiint_{\mathscr{S}} \boldsymbol{P}\cdot d\boldsymbol{\sigma}, \tag{1-126}$$

where the closed surface $\mathscr{S}$ defined the boundary of the volume $\mathscr{V}$. This result follows from our earlier discussions without formal proof. The second integral theorem, known as *Stokes' theorem*, follows from the definition of the curl operation

$$\iint_{\mathscr{S}} \nabla \times \boldsymbol{P}\cdot d\boldsymbol{\sigma} = \oint_{\mathscr{C}} \boldsymbol{P}\cdot d\boldsymbol{r}. \tag{1-127}$$

Imagine many adjacent infinitesimal circulations as suggested in Fig. 1-12. As the figure shows, the circulations interior to the open surface $\mathscr{S}$ cancel, and the only component that contributes to the net curl of P integrated over $\mathscr{S}$ is the tangential component of P on the boundary.

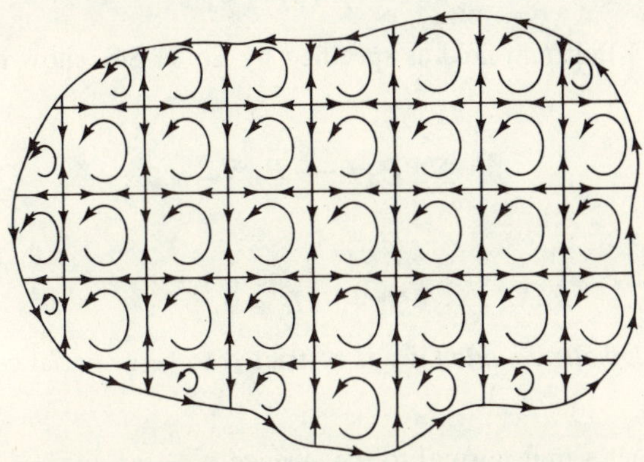

Fig. 1-12 Geometrical basis for Stokes' theorem.

In addition to Eqs. (1-126) and (1-127), we list a few other useful integral vector identities:

$$\iiint_{\mathscr{V}} (\phi\, \nabla^2 \psi - \psi\, \nabla^2 \phi)\, d\tau = \oiint_{\mathscr{S}} (\phi\, \nabla\psi - \psi\, \nabla\phi) \cdot d\boldsymbol{\sigma}, \qquad (1\text{-}128)$$

which is known as *Green's theorem*, and

$$\iiint_{\mathscr{V}} \nabla\phi\, d\tau = \oiint_{\mathscr{S}} \phi\, d\boldsymbol{\sigma}, \qquad (1\text{-}129)$$

$$\iiint_{\mathscr{V}} \nabla \times P\, d\tau = \oiint_{\mathscr{S}} d\boldsymbol{\sigma} \times P, \qquad (1\text{-}130)$$

$$\iint_{\mathscr{S}} d\boldsymbol{\sigma} \times \nabla\phi = \oint_{\mathscr{C}} \phi\, d\boldsymbol{r}. \qquad (1\text{-}131)$$

EXERCISES

1-4.1. Show that

$$\nabla \times P \times P = (P \cdot \nabla)P - \tfrac{1}{2}\nabla P^2. \qquad (1\text{-}132)$$

1-4.2. Show that if the vectors P and Q are constant in space, then

$$\nabla(P \cdot Q \times r) = P \times Q.$$

1-4.3. Show by explicit expansion that in cartesian coordinates the definition of the vector Laplacian operator given in Eq. (1-86) is equivalent to that of the scalar Laplacian operator, Eq. (1-85).

1-4.4. Prove that

$$\mathbf{\nabla} \times (\phi\,\mathbf{\nabla}\phi) = 0.$$

1-4.5. If $\mathbf{P}$ is a Beltrami field as specified by Eq. (1-84), show that

(a)

$$\mathbf{\nabla} \cdot \mathbf{P} = -\frac{1}{\alpha}\,\mathbf{P} \cdot \mathbf{\nabla}\alpha,$$

and

(b)
$$\nabla^2 \mathbf{P} + \alpha^2 \mathbf{P} = 0,$$

the vector *Helmholtz equation* if α is restricted to be a spatial constant in the latter case.

1-4.6. Construct a unit normal to the surface $x^2 + y^2 + z^2 = c$ at the point x_0, y_0, z_0 in the surface. Find the equation of a plane tangent to the surface at (x_0, y_0, z_0).

1-4.7. If a vector $\mathbf{P}$ is known to be a function of space and time, show that the differential of $\mathbf{P}$ can be written

$$d\mathbf{P} = \frac{\partial \mathbf{P}}{\partial t}\,dt + (d\mathbf{r} \cdot \mathbf{\nabla})\mathbf{P}. \tag{1-133}$$

1-4.8. Prove Eq. (1-94).

1-4.9. Given the vector $\mathbf{P} = -x\mathbf{i} - y\mathbf{j}$, evaluate the line integral between the endpoints (α, α) and (β, β) by each of the paths

(a) $\alpha, \alpha \rightarrow \beta, \alpha \rightarrow \beta, \beta$

(b) $\alpha, \alpha \rightarrow \alpha, \beta \rightarrow \beta, \beta$

(c) $\alpha, \alpha \rightarrow \beta, \beta$ along $x = y$.

1-4.10. Prove Eq. (1-100). Doing so amounts to proving that if two vectors are each irrotational, their cross product is solenoidal.

1-4.11. Devise a vector that is *both* solenoidal and irrotational.

1-4.12. Show that

$$\oiint_{\mathscr{S}} \mathbf{\nabla} \times \mathbf{Q} \cdot d\boldsymbol{\sigma} = 0.$$

1-4.13. Show that

(a) $$\nabla\phi = \lim_{\iiint d\tau \to 0} \frac{\oiint_{\mathscr{S}} \phi \, d\boldsymbol{\sigma}}{\iiint d\tau},$$

(b) $$\nabla \cdot \boldsymbol{P} = \lim_{\iiint d\tau \to 0} \frac{\oiint_{\mathscr{S}} \boldsymbol{P} \cdot d\boldsymbol{\sigma}}{\iiint d\tau},$$

(c) $$\nabla \times \boldsymbol{P} = \lim_{\iiint d\tau \to 0} \frac{\oiint_{\mathscr{S}} d\boldsymbol{\sigma} \times \boldsymbol{P}}{\iiint d\tau},$$

(d) $$\nabla^2\phi = \lim_{\iiint d\tau \to 0} \frac{\oiint_{\mathscr{S}} \nabla\phi \cdot d\boldsymbol{\sigma}}{\iiint d\tau}.$$

1-4.14. If $\boldsymbol{P} = \nabla \times \boldsymbol{Q}$, then by Stokes' theorem

$$\iint_{\mathscr{S}} \boldsymbol{P} \cdot d\boldsymbol{\sigma} = \oint_{\mathscr{C}} \boldsymbol{Q} \cdot d\boldsymbol{r}.$$

Show that both sides of this equation remain invariant under the *gage transformation*

$$\boldsymbol{Q} \to \boldsymbol{Q} + \nabla\psi.$$

1-4.15. Prove Eq. (1-128).

1-5 Orthogonal Coordinate Systems

      All the vector analysis presented so far has been limited to cartesian or rectangular coordinates. Many problems in mechanics possess symmetries that imply that the problem would best be formulated in terms of a coordinate system in which the symmetry can most conveniently be exploited to simplify the solution.

      In the cartesian description we defined three mutually orthogonal (perpendicular) families of planes—$x = $ const, $y = $ const, $z = $ const—and described a point in terms of its intersects (x, y, z). Here we would like to superimpose on this picture a new set of coordinate surfaces, not necessarily planar: $q_1 = $ const, $q_2 = $ const, $q_3 = $ const, in terms of which the point (x, y, z) can be alternately identified as (q_1, q_2, q_3). In principle, there must exist a set of transformation equations

$$\begin{aligned} x &= x(q_1, q_2, q_3) \\ y &= y(q_1, q_2, q_3) \\ z &= z(q_1, q_2, q_3) \end{aligned} \tag{1-134}$$

and inverse relations

$$q_1 = q_1(x, y, z)$$

$$q_2 = q_2(x, y, z) \tag{1-135}$$

$$q_3 = q_3(x, y, z).$$

Each of the surfaces $q_i = $ const has associated with it a unit vector $\hat{e}_i$ normal to the surface and directed toward increasing q_i.

The distance between two neighboring points is specified in the most general way by

$$ds^2 = dx^2 + dy^2 + dz^2 = \sum_{i,j} h_{ij}^2 \, dq_i \, dq_j, \tag{1-136}$$

where the h_{ij} are *metric coefficients* that specify the geometrical nature of the coordinate system. For notational convenience, let us set $x = x_1$, $y = x_2$, $z = x_3$ and write the differential of Eq. (1-134):

$$dx_k = \sum_i \frac{\partial x_k}{\partial q_i} \, dq_i. \tag{1-137}$$

Squaring

$$dx_k^2 = \sum_i \sum_j \frac{\partial x_k}{\partial q_i} \frac{\partial x_k}{\partial q_j} \, dq_i \, dq_j$$

and substituting into Eq. (1-136), we find for the metric

$$h_{ij}^2 = \sum_k \frac{\partial x_k}{\partial q_i} \frac{\partial x_k}{\partial q_j}. \tag{1-138}$$

At this point we wish to limit ourselves to orthogonal systems. This requirement can be satisfied if

$$h_{ij} = h_{ii}\delta_{ij}, \tag{1-139}$$

that is, only $h_{ii} = h_i$, $h_{jj} = h_j$, and $h_{kk} = h_k$ are nonzero. For a general orthogonal coordinate system, the element of length becomes

$$ds^2 = \sum_i h_i^2 \, dq_i^2. \tag{1-140}$$

The quantity $h_i \, dq_i = ds_i$ must have the units of length, although the coordinate q_i need not. The area and volume elements are

$$d\sigma_{ij} = ds_i \, ds_j = h_i h_j \, dq_i \, dq_j, \tag{1-141}$$

$$d\tau = ds_1 \, ds_2 \, ds_3 = h_1 h_2 h_3 \, dq_1 \, dq_2 \, dq_3. \tag{1-142}$$

In terms of the generalized system that we have defined, the differential operations of vector calculus can be written[5]

$$\nabla\phi = \sum_i \frac{1}{h_i}\frac{\partial\phi}{\partial q_i}\,\hat{e}_i,$$ (1-143)

$$\nabla\cdot\boldsymbol{P} = \sum \frac{1}{h_i h_j h_k}\frac{\partial}{\partial q_i}\,(h_j h_k P_i),$$ (1-144)

$$\nabla\times\boldsymbol{P} = \frac{1}{h_1 h_2 h_3}\begin{vmatrix} h_1\hat{e}_1 & h_2\hat{e}_2 & h_3\hat{e}_3 \\ \dfrac{\partial}{\partial q_1} & \dfrac{\partial}{\partial q_2} & \dfrac{\partial}{\partial q_3} \\ h_1 P_1 & h_2 P_2 & h_3 P_3 \end{vmatrix},$$ (1-145)

$$\nabla^2\phi = \sum \frac{1}{h_i h_j h_k}\frac{\partial}{\partial q_i}\left(\frac{h_j h_k}{h_i}\frac{\partial\phi}{\partial q_i}\right),$$ (1-146)

where the summation in Eqs. (1-144) and (1-146) is cyclical.

Besides the cartesian system, two other coordinate systems are useful for many mechanical (and electrical) problems. They are the cylindrical and spherical systems shown in Figs. 1-13 and 1-14, respectively.

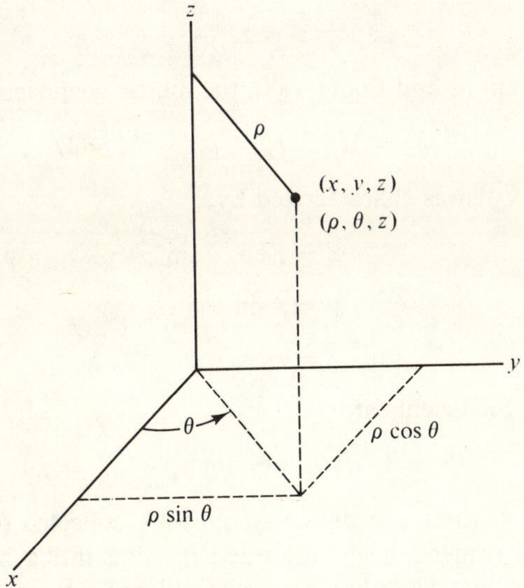

Fig. 1-13 Cylindrical coordinates.

---

[5] For the derivation of these forms, see any good treatment of orthogonal coordinate systems, for instance, F. B. Hildebrand, *Advanced Calculus for Applications*. Englewood Cliffs, N.J.: Prentice-Hall, Inc., 1962.

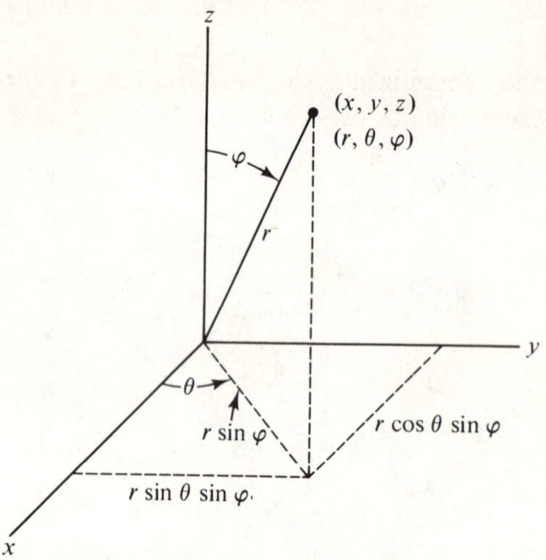

Fig. 1-14 Spherical coordinates.

For the cylindrical system, the transformation equations can be seen from Fig. 1-13 to be

$$x = \rho \cos \theta$$
$$y = \rho \sin \theta \tag{1-147}$$
$$z = z.$$

Using these relations and Eq. (1-138), the metric coefficients are found to be

$$h_\rho = 1, \qquad h_\theta = \rho, \qquad h_z = 1. \tag{1-148}$$

The spherical system is characterized by

$$x = r \cos \theta \sin \varphi$$
$$y = r \sin \theta \sin \varphi \tag{1-149}$$
$$z = r \cos \varphi$$

and the metric coefficients are

$$h_r = 1, \qquad h_\theta = r \sin \varphi, \qquad h_\varphi = r. \tag{1-150}$$

The properties of these coordinate systems are collected for convenience in Table 1-1. The reader who is interested in acquainting himself with other orthogonal curvilinear coordinate systems will find a comprehensive reference in George Arfken, *Mathematical Methods for Physicists*. New York: Academic Press, 1970, Chapter 2.

32

Table 1-1 VECTOR RELATIONS IN COMMON COORDINATE SYSTEMS

	Cartesian	Cylindrical	Spherical
Transformation from cartesian		$x = \rho \cos \theta$ $y = \rho \sin \theta$ $z = z$	$x = r \cos \theta \sin \varphi$ $y = r \sin \theta \sin \varphi$ $z = r \cos \varphi$
	$h_x = 1;\ h_y = 1;\ h_z = 1$	$h_\rho = 1;\ h_\theta = \rho;\ h_z = 1$	$h_r = 1;\ h_\theta = r \sin \varphi;\ h_\varphi = r$
Line element	$ds = \sqrt{dx^2 + dy^2 + dz^2}$	$ds = \sqrt{d\rho^2 + \rho^2\, d\theta^2 + dz^2}$	$ds = \sqrt{dr^2 + r^2 \sin^2 \varphi\, d\theta^2 + r^2\, d\varphi^2}$
Volume element	$d\tau = dx\, dy\, dz$	$d\tau = \rho\, d\rho\, d\theta\, dz$	$d\tau = r^2 \sin \varphi\, dr\, d\theta\, d\varphi$
Surface elements	$d\sigma_x = dy\, dz$ $d\sigma_y = dx\, dz$ $d\sigma_z = dx\, dy$	$d\sigma_\rho = \rho\, d\theta\, dz$ $d\sigma_\theta = d\rho\, dz$ $d\sigma_z = \rho\, d\rho\, d\theta$	$d\sigma_r = r^2 \sin \varphi\, d\theta\, d\varphi$ $d\sigma_\theta = r\, dr\, d\varphi$ $d\sigma_\varphi = r \sin \varphi\, dr\, d\theta$
Gradient	$\nabla\phi = \dfrac{\partial\phi}{\partial x}\,\hat{\imath} + \dfrac{\partial\phi}{\partial y}\,\hat{\jmath} + \dfrac{\partial\phi}{\partial z}\,\hat{k}$	$\nabla\phi = \dfrac{\partial\phi}{\partial\rho}\,\hat{\boldsymbol{\rho}} + \dfrac{1}{\rho}\dfrac{\partial\phi}{\partial\theta}\,\hat{\boldsymbol{\theta}} + \dfrac{\partial\phi}{\partial z}\,\hat{k}$	$\nabla\phi = \dfrac{\partial\phi}{\partial r}\,\hat{r} + \dfrac{1}{r \sin \varphi}\dfrac{\partial\phi}{\partial\theta}\,\hat{\boldsymbol{\theta}} + \dfrac{1}{r}\dfrac{\partial\phi}{\partial\varphi}\,\hat{\boldsymbol{\varphi}}$
Divergence	$\nabla\cdot\boldsymbol{P} = \dfrac{\partial P_x}{\partial x} + \dfrac{\partial P_y}{\partial y} + \dfrac{\partial P_z}{\partial z}$	$\nabla\cdot\boldsymbol{P} = \dfrac{1}{\rho}\dfrac{\partial}{\partial\rho}(\rho P_\rho) + \dfrac{1}{\rho}\dfrac{\partial P_\theta}{\partial\theta} + \dfrac{\partial P_z}{\partial z}$	$\nabla\cdot\boldsymbol{P} = \dfrac{1}{r^2}\dfrac{\partial}{\partial r}(r^2 P_r) + \dfrac{1}{r \sin \varphi}\dfrac{\partial P_\theta}{\partial\theta}$ $\qquad + \dfrac{1}{r \sin \varphi}\dfrac{\partial}{\partial\varphi}(P_\varphi \sin \varphi)$
Curl	$\nabla\times\boldsymbol{P} = \begin{vmatrix} \hat{\imath} & \hat{\jmath} & \hat{k} \\ \dfrac{\partial}{\partial x} & \dfrac{\partial}{\partial y} & \dfrac{\partial}{\partial z} \\ P_x & P_y & P_z \end{vmatrix}$	$\nabla\times\boldsymbol{P} = \dfrac{1}{\rho}\begin{vmatrix} \hat{\boldsymbol{\rho}} & \rho\hat{\boldsymbol{\theta}} & \hat{k} \\ \dfrac{\partial}{\partial\rho} & \dfrac{\partial}{\partial\theta} & \dfrac{\partial}{\partial z} \\ P_\rho & \rho P_\theta & P_z \end{vmatrix}$	$\nabla\times\boldsymbol{P} = \dfrac{1}{r^2 \sin \varphi}\begin{vmatrix} \hat{r} & r \sin \varphi\,\hat{\boldsymbol{\theta}} & r\hat{\boldsymbol{\varphi}} \\ \dfrac{\partial}{\partial r} & \dfrac{\partial}{\partial\theta} & \dfrac{\partial}{\partial\varphi} \\ P_r & rP_\theta \sin \varphi & rP_\varphi \end{vmatrix}$
Laplacian	$\nabla^2\phi = \dfrac{\partial^2\phi}{\partial x^2} + \dfrac{\partial^2\phi}{\partial y^2} + \dfrac{\partial^2\phi}{\partial z^2}$	$\nabla^2\phi = \dfrac{1}{\rho}\dfrac{\partial}{\partial\rho}\left(\rho\dfrac{\partial\phi}{\partial\rho}\right) + \dfrac{1}{\rho^2}\dfrac{\partial^2\phi}{\partial\theta^2} + \dfrac{\partial^2\phi}{\partial z^2}$	$\nabla^2\phi = \dfrac{1}{r^2}\dfrac{\partial}{\partial r}\left(r^2\dfrac{\partial\phi}{\partial r}\right) + \dfrac{1}{r^2 \sin^2 \varphi}\dfrac{\partial^2\phi}{\partial\theta^2}$ $\qquad + \dfrac{1}{r^2 \sin \varphi}\dfrac{\partial}{\partial\varphi}\left(\sin \varphi\dfrac{\partial\phi}{\partial\varphi}\right)$

EXERCISES

1-5.1. Show that the requirement of orthogonality in a coordinate system implies Eq. (1-139).

1-5.2. Using the transformation relations given in Eq. (1-147), calculate the metric coefficients Eq. (1-148) for the cylindrical coordinate system.

1-5.3. Derive the metric coefficients for the spherical coordinate system.

1-5.4. Force fields having the form

$$F = \frac{A}{r^2}\,\hat{r}$$

are very important both in mechanics and electromagnetics. Show that $F \cdot dr$ is an exact differential and derive the scalar potential ϕ. Using as your contour a unit circle in the $\varphi = \pi/2$ plane, show that Eq. (1-113) is satisfied.

1-5.5. A vector field is given by

$$P = \rho \cos^2 \theta\, \hat{\rho} + \rho \sin \theta\, \hat{\theta}.$$

(a) Evaluate the divergence $\nabla \cdot P$.

(b) Evaluate the contour integral $\oiint_{\mathscr{S}} P \cdot d\sigma$ over the surface of a cylindrical quadrant of unit height and radius a as shown in Fig. 1-15.

(c) Verify the Gauss divergence theorem for this case.

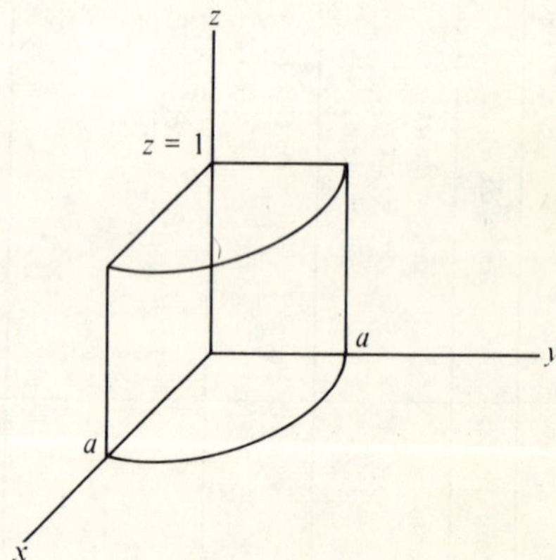

Fig. 1-15

1-5.6. Given the vector field

$$P = \alpha\rho\theta\hat{\rho} + \beta\rho^2\hat{\theta},$$

(a) Evaluate the curl $\nabla \times P$.

(b) Evaluate $\iint_{\mathscr{S}} \nabla \times \boldsymbol{P} \cdot d\boldsymbol{\sigma}$ over a surface in the $\rho\theta$ plane bounded by a circle of radius a centered on the origin.

(c) Evaluate $\oint_{\mathscr{C}} \boldsymbol{P} \cdot d\boldsymbol{r}$ around the circle. Does Stokes' law apply? Explain.

1-5.7. Given the scalar potential function

$$\phi = (ar^2 + br^{-3})\sin 2\varphi \cos\theta,$$

(a) find $\boldsymbol{P} = \nabla\phi$.
(b) find $\nabla \times \boldsymbol{P}$.

1-6 Particle Kinematics

We wish to describe the motion of a *massive particle* that we define to be a body of finite mass $m > 0$ and zero spatial extension. The proper question to ask is: In what way can a particle manifest itself in space? The answer is, only by its *position*, which we can describe in terms of its coordinates (x, y, z) or (ρ, θ, z), or in terms of a radius vector $\boldsymbol{r}(t)$. The motion of a particle must be described in terms of time derivatives of this radius vector. The *velocity* can be defined as

$$\boldsymbol{u} = \frac{d\boldsymbol{r}}{dt} = \dot{\boldsymbol{r}}. \tag{1-151}$$

The term *velocity* will always be taken to imply a vector—that is, a direction as well as a magnitude. The term *speed* usually refers to the magnitude of a velocity. The *acceleration* is defined as

$$\boldsymbol{a} = \frac{d\boldsymbol{u}}{dt} = \frac{d^2\boldsymbol{r}}{dt^2} = \ddot{\boldsymbol{r}}. \tag{1-152}$$

We have introduced a shorthand notation in these two equations, the dot over a quantity to indicate differentiation with respect to time, two dots double differentiation, and so on. Extensive use will be made of this notation.

As we shall shortly see, the description of particle motion does not require that we formally introduce higher derivatives. In order to establish a firm foundation for our kinematics, we must establish only position and time as primary concepts. We can do so most satisfactorily by prescribing an experimental method for defining standards of length and time that may then, in fact as well as in principle, be duplicated anywhere. The first such standard of length was a platinum-iridium bar with two scratches that were defined to be separated by a *meter*. This standard was not a completely satisfactory one for several obvious reasons, and in 1961 an atomic length standard was adopted by international agreement.[6] The *second* is now defined to be $(31,556,925.9747)^{-1}$ of the year 1900.

---

[6] One meter is defined to be equal to 1,650,763.73 wavelengths of the orange spectral line (denoted by $2p_{10}\text{–}5d_5$ in spectroscopic notation) of Kr^{86}.

Having defined the radius vector and its two derivatives, velocity and acceleration, we should see how these vectors are written in our three coordinate systems. In cartesian coordinates the radius vector is

$$\boldsymbol{r} = x\hat{\boldsymbol{i}} + y\hat{\boldsymbol{j}} + z\hat{\boldsymbol{k}} \tag{1-153}$$

as we have seen. The unit vectors $\hat{\boldsymbol{i}}, \hat{\boldsymbol{j}}, \hat{\boldsymbol{k}}$ are constant in direction; hence

$$\boldsymbol{u} = \dot{\boldsymbol{r}} = \dot{x}\hat{\boldsymbol{i}} + \dot{y}\hat{\boldsymbol{j}} + \dot{z}\hat{\boldsymbol{k}} \tag{1-154}$$

and

$$\boldsymbol{a} = \ddot{\boldsymbol{r}} = \ddot{x}\hat{\boldsymbol{i}} + \ddot{y}\hat{\boldsymbol{j}} + \ddot{z}\hat{\boldsymbol{k}}. \tag{1-155}$$

Suppose that we do the same thing in cylindrical coordinates. The basis of this system is a set of unit vectors $\hat{\boldsymbol{\rho}}, \hat{\boldsymbol{\theta}}, \hat{\boldsymbol{k}}$ that point in the direction of increasing ρ, θ, and z, respectively. All these vectors have constant magnitude by definition, but two of them, $\hat{\boldsymbol{\rho}}$ and $\hat{\boldsymbol{\theta}}$, are functions of θ. If we look at the projection of the $\hat{\boldsymbol{\rho}}\hat{\boldsymbol{\theta}}$ plane onto the $\hat{\boldsymbol{i}}\hat{\boldsymbol{j}}$ plane as shown in Fig. 1-16, we find

$$\hat{\boldsymbol{\rho}} = \hat{\boldsymbol{i}} \cos\theta + \hat{\boldsymbol{j}} \sin\theta, \tag{1-156}$$

$$\hat{\boldsymbol{\theta}} = -\hat{\boldsymbol{i}} \sin\theta + \hat{\boldsymbol{j}} \cos\theta, \tag{1-157}$$

from which we can determine

$$\frac{d\hat{\boldsymbol{\rho}}}{d\theta} = \hat{\boldsymbol{\theta}}, \qquad \frac{d\hat{\boldsymbol{\theta}}}{d\theta} = -\hat{\boldsymbol{\rho}}. \tag{1-158}$$

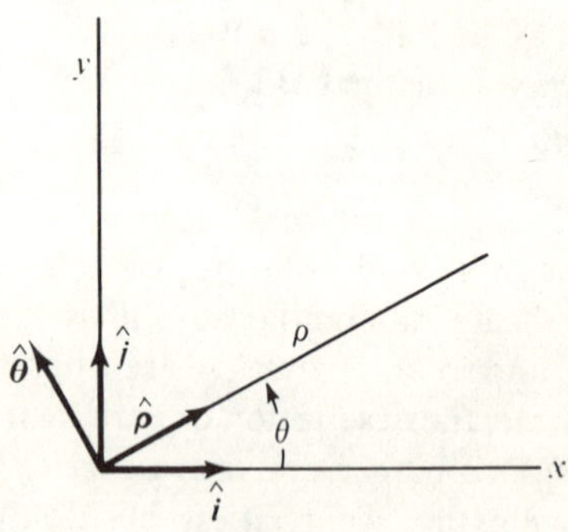

Fig. 1-16

The radius vector in cylindrical coordinates is

$$\boldsymbol{r} = \rho\hat{\boldsymbol{\rho}} + z\hat{\boldsymbol{k}}. \tag{1-159}$$

Differentiating Eq. (1-159) with respect to t

$$\boldsymbol{u} = \dot{\boldsymbol{r}} = \dot{\rho}\hat{\boldsymbol{\rho}} + \rho\dot{\hat{\boldsymbol{\rho}}} + \dot{z}\hat{\boldsymbol{k}}$$

and using Eq. (1-158), we find

$$\boldsymbol{u} = \dot{\rho}\hat{\boldsymbol{\rho}} + \rho\dot{\theta}\hat{\boldsymbol{\theta}} + \dot{z}\hat{\boldsymbol{k}} \tag{1-160}$$

for the velocity vector. Similarly, for the acceleration vector we have

$$a = (\ddot{\rho} - \rho\dot{\theta}^2)\hat{\boldsymbol{\rho}} + (\rho\ddot{\theta} + 2\dot{\rho}\dot{\theta})\hat{\boldsymbol{\theta}} + \ddot{z}\hat{\boldsymbol{k}}. \tag{1-161}$$

Let us do the same sort of calculation for the kinematic vectors expressed in spherical coordinates. Using Fig. 1-17 and Eqs. (1-156) and (1-157), we find

$$\hat{\boldsymbol{r}} = \hat{\boldsymbol{\rho}} \sin \varphi + \hat{\boldsymbol{k}} \cos \varphi = \hat{\boldsymbol{i}} \cos \theta \sin \varphi + \hat{\boldsymbol{j}} \sin \theta \sin \varphi + \hat{\boldsymbol{k}} \cos \varphi, \tag{1-162}$$

$$\hat{\boldsymbol{\theta}} = -\hat{\boldsymbol{i}} \sin \theta + \hat{\boldsymbol{j}} \cos \theta, \tag{1-163}$$

$$\hat{\boldsymbol{\varphi}} = \hat{\boldsymbol{\rho}} \cos \varphi - \hat{\boldsymbol{k}} \sin \varphi = \hat{\boldsymbol{i}} \cos \theta \cos \varphi + \hat{\boldsymbol{j}} \sin \theta \cos \varphi - \hat{\boldsymbol{k}} \sin \varphi. \tag{1-164}$$

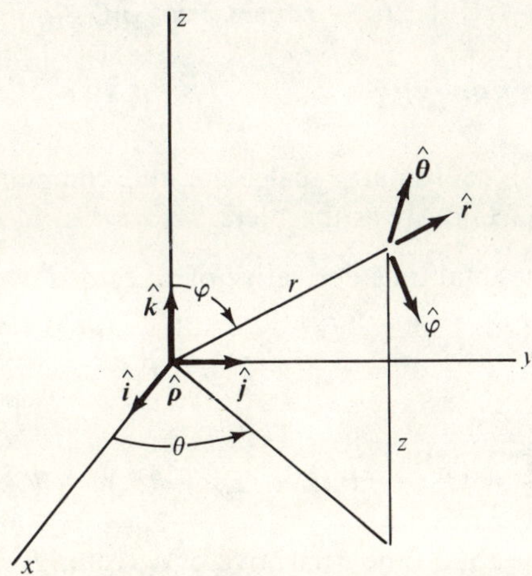

Fig. 1-17

The relations governing the derivatives of the unit vectors are obtained as before by straightforward differentiation of the preceding relations:

$$\frac{\partial \hat{\boldsymbol{r}}}{\partial \theta} = \hat{\boldsymbol{\theta}} \sin \varphi, \qquad\qquad \frac{\partial \hat{\boldsymbol{r}}}{\partial \varphi} = \hat{\boldsymbol{\varphi}},$$

$$\frac{\partial \hat{\boldsymbol{\theta}}}{\partial \theta} = -\hat{\boldsymbol{\rho}} = -\hat{\boldsymbol{r}} \sin \varphi - \hat{\boldsymbol{\varphi}} \cos \varphi, \qquad \frac{\partial \hat{\boldsymbol{\theta}}}{\partial \varphi} = 0, \tag{1-165}$$

$$\frac{\partial \hat{\boldsymbol{\varphi}}}{\partial \theta} = \hat{\boldsymbol{\theta}} \cos \varphi, \qquad\qquad \frac{\partial \hat{\boldsymbol{\varphi}}}{\partial \varphi} = -\hat{\boldsymbol{r}}.$$

The radius vector in spherical coordinates is

$$\boldsymbol{r} = r\hat{\boldsymbol{r}}. \tag{1-166}$$

Differentiating to find the velocity,

$$\dot{\boldsymbol{r}} = \dot{r}\hat{\boldsymbol{r}} + r\dot{\hat{\boldsymbol{r}}},$$

where

$$\dot{\hat{\boldsymbol{r}}} = \frac{d\hat{\boldsymbol{r}}}{dt} = \frac{\partial\hat{\boldsymbol{r}}}{\partial\theta}\,\dot{\theta} + \frac{\partial\hat{\boldsymbol{r}}}{\partial\varphi}\,\dot{\varphi} = \dot{\theta}\sin\varphi\,\hat{\boldsymbol{\theta}} + \dot{\varphi}\hat{\boldsymbol{\varphi}}.$$

Thus

$$\boldsymbol{u} = \dot{\boldsymbol{r}} = \dot{r}\hat{\boldsymbol{r}} + r\dot{\theta}\sin\varphi\,\hat{\boldsymbol{\theta}} + r\dot{\varphi}\hat{\boldsymbol{\varphi}}, \tag{1-167}$$

and similarly

$$\boldsymbol{a} = \ddot{\boldsymbol{r}} = (\ddot{r} - r\dot{\varphi}^2 - r\dot{\theta}^2\sin^2\varphi)\hat{\boldsymbol{r}} + \cdots$$

$$+ (r\ddot{\theta}\sin\varphi + 2\dot{r}\dot{\theta}\sin\varphi + 2r\dot{\theta}\dot{\varphi}\cos\varphi)\hat{\boldsymbol{\theta}} + \cdots$$

$$+ (r\ddot{\varphi} + 2\dot{r}\dot{\varphi} - r\dot{\theta}^2\sin\varphi\cos\varphi)\hat{\boldsymbol{\varphi}}. \tag{1-168}$$

EXERCISES

1-6.1. In cylindrical coordinates, calculate the components of $\dddot{\boldsymbol{r}} = d\boldsymbol{a}/dt$, sometimes known facetiously as the "jerk."

1-6.2. Show that the total time derivative of a vector $\boldsymbol{P}$ written in cylindrical coordinates

$$\boldsymbol{P} = P_\rho\hat{\boldsymbol{\rho}} + P_\theta\hat{\boldsymbol{\theta}} + P_z\hat{\boldsymbol{k}}$$

is

$$\frac{d\boldsymbol{P}}{dt} = (\dot{P}_\rho - \dot{\theta}P_\theta)\hat{\boldsymbol{\rho}} + (\dot{P}_\theta + \dot{\theta}P_\rho)\hat{\boldsymbol{\theta}} + \dot{P}_z\hat{\boldsymbol{k}}. \tag{1-169}$$

1-6.3. Show that the total time derivative of a vector $\boldsymbol{P}$ written in spherical coordinates

$$\boldsymbol{P} = P_r\hat{\boldsymbol{r}} + P_\theta\hat{\boldsymbol{\theta}} + P_\varphi\hat{\boldsymbol{\varphi}}$$

is

$$\frac{d\boldsymbol{P}}{dt} = (\dot{P}_r - \dot{\varphi}A_\varphi - \dot{\theta}\sin\varphi P_\theta)\hat{\boldsymbol{r}} + (\dot{P}_\theta + \dot{\theta}\sin\varphi P_r + \dot{\theta}\cos\varphi P_\varphi)\hat{\boldsymbol{\theta}} + \cdots$$

$$+ (\dot{P}_\varphi + \dot{\varphi}P_r - \dot{\theta}\cos\varphi P_\theta)\hat{\boldsymbol{\varphi}}. \tag{1-170}$$

1-6.4. Use Eq. (1-169) to verify the result of problem 1-6.1, taking as $\boldsymbol{P}$ the acceleration vector given in Eq. (1-161).

1-7 Newton's Laws of Motion

Mechanics was the earliest scientific discipline to be developed, as mentioned earlier. The reason for this early development is found in the fact that everyone in the course of daily life acquires a considerable amount of qualitative knowledge to the motion of bodies. This experience implies that the motions of massive bodies are governed by interaction with their surroundings. For example, it can be found by observing the motion of bodies slid across various surfaces that as the frictional force is reduced to a minimum, the body tends to a constant velocity. This situation suggests that accelerations will result when forces are applied to a body.

Consider, for instance, two ice skaters on a frozen pond. If the two skaters are facing each other and begin to push on opposite ends of a rigid pole, measurements will reveal that the motion of each skater during the application of a constant pushing force will always be such that the ratio of their accelerations will be a constant for any pair of skaters, independent of the force applied:

$$\ddot{\mathbf{r}}_1 = -\alpha \ddot{\mathbf{r}}_2. \tag{1-171}$$

It will also be found that the weights of the two skaters as measured by a scale are associated by the *same* constant

$$W_2 = \alpha W_1. \tag{1-172}$$

The *weight* of a body is, of course, the force exerted on it by the gravitational field in which it finds itself. Galileo is credited with having established that all bodies fall with the same constant acceleration g, neglecting viscous drag. The difference in their weights must then be due to some intrinsic property that we shall call *mass* and define as

$$m_i = \frac{W_i}{g}, \tag{1-173}$$

where the weight and the acceleration of gravity are measured at the same point. Equation (1-172) can now be rewritten in terms of mass as

$$m_2 = \alpha m_1. \tag{1-174}$$

By eliminating the constant α between Eqs. (1-171) and (1-174), we find[7]

$$m_1 \ddot{\mathbf{r}}_1 = -m_2 \ddot{\mathbf{r}}_2. \tag{1-175}$$

We have therefore discovered by the simplest of experiments that the response to a force by a massive body is an acceleration that is equal to the force applied divided by the mass as defined in Eq. (1-174).

---

[7] The mass defined in Eq. (1-173) is clearly based on the response of the body to a gravitational field. The experiment that results in the relation expressed in Eq. (1-175) would give the same result in the absence of a gravitational field. The mass in such a two-body encounter is a measure of the *inertial* rather than the gravitational properties of the bodies. It is not obvious that the inertial and gravitational masses of an object should be equal or even in a constant ratio for all matter. The assumption of the equality of these masses is one of the basic postulates of Einstein's general theory of relativity.

Isaac Newton was the first to formulate these deliberations into a useful theory of mechanics. Newton stated that

> 1. *A massive particle remains at rest or proceeds at a constant speed in a straight line unless it is acted on by an external force.*
> 2. *The rate of change of the momentum of a massive particle is equal to the force applied in both magnitude and direction.*
> 3. *Every action gives rise to an equal and opposed reaction.*

The *momentum* that enters into the second law is defined as the product of the mass and velocity

$$p = mu. \tag{1-176}$$

The first and second laws can then be written

$$F = \frac{dp}{dt}, \tag{1-177}$$

where F is the applied force. Newton's third law is essentially a statement of our experimental finding expressed in Eq. (1-175). Note that the expansion of the right-hand side of Eq. (1-177)

$$\frac{dp}{dt} = \frac{d}{dt}(mu) = u\frac{dm}{dt} + m\frac{du}{dt},$$

is simply equal to the product of the mass and acceleration if we assume the mass to be constant, as is usually the case. With this assumption, Newton's third law can be written

$$F_1 = -F_2. \tag{1-178}$$

The force that enters Newton's second law must be regarded as the total force acting on the mass m. If this force is the vector sum of many forces, these individual forces do not correspond to individual momentum changes. Most of our work will be concerned with the implications of the first and second laws. As noted earlier, the parameter range over which classical mechanics gives answers having acceptable precision is vast. Only when speeds that approach the speed of light and masses or distances that are cosmical or atomic are considered must Newtonian mechanics be replaced by relativity or quantum mechanics.

The third of Newton's laws implies instantaneous action at a distance and must therefore be used selectively, since, as we now know, two bodies cannot interact with each other in less time that it would take a beam of light to pass between them.

When stating a physical law, it is important to specify the reference frame (coordinate system) in which the laws apply. Newton was unable to give satisfactory answers in this area and he was aware of the flaw. The question must be answered by experimentation. When the reference system in which Newton's laws hold is found, then we discover that all other reference systems that are unaccelerated with respect to this system are also suitable for Newtonian mechanics. Such systems in which Newton's laws apply are called *inertial frames of reference*. The transformation laws that apply amongst various systems in steady motion with respect to one another were found by Galileo. These *Galilean transformations* are

$$x' = x - vt$$
$$y' = y$$
$$z' = z \tag{1-179}$$
$$t' = t$$

for the position and time coordinates and

$$u'_x = u_x - v$$
$$u'_y = u_y \tag{1-180}$$
$$u'_z = u_z$$

for the velocity. Figure 1-18 shows the situation described by Eqs. (1-179) and (1-180) wherein the relative velocity of the two systems, v, is directed along the x and x' axes. The unprimed system is stationary for us as observers.

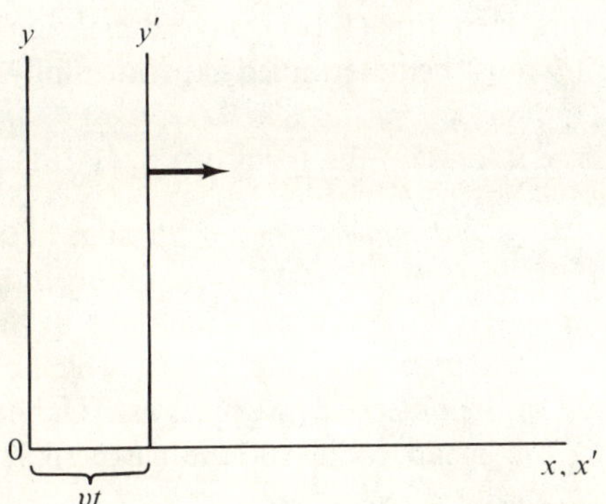

Fig. 1-18 Relative motion of two reference systems.

In the last section of this chapter we describe Newton's law of gravitation. Although it is a theory separate from the theory of motion, the gravitational force is the most common one in our experience and it provides an interesting subject for vector analysis.

1-8 Newtonian Gravitation

In the first section we told how Newton came to adopt a law of universal gravitation wherein the force of attraction varies as one over the square of the distance of separation. He also found that the gravitational force must be proportional to the product of the masses of the particles. This result is apparent from a consideration of the third of Newton's laws.

The law of gravitation has been tested in delicate laboratory experiments, such as the one first carried out by Cavendish in which the gravitational attraction between two lead spheres, one fixed and the other suspended, is measured. Newton's theoretical finding, which we shall verify, that a homogeneous massive sphere behaves gravitationally as though the mass were concentrated at the center has also been tested experimentally. The Newtonian law has been found to be valid over a wide range of masses and separation distances. It is independent of physical properties other than the mass and is also independent of the nature of the medium that separates the masses. The Newtonian law of gravitation is truly a universal law. We must note, however, even though this law tells us *how*, it does not tell us *why*.

The law of gravitation is

$$F = -\frac{\gamma m_1 m_2}{r^2}\,\hat{r}, \tag{1-181}$$

where m_1 and m_2 are the masses of the two particles involved, r is the distance separating the particles, $\hat{r}$ is a unit vector pointing from particle 1 located at the origin of coordinates to particle 2, and γ is called the *gravitational constant*. The minus sign in Eq. (1-181) denotes the attractive nature of the force. Since the quantities entering into Eq. (1-181) have dimensions that are already defined, the constant γ must be determined experimentally. It has been found to equal $6.670 \cdot 10^{-11}$ Nm^2 kg^{-2}.

Of most immediate interest is the form of Eq. (1-181) appropriate to the surface of the earth

$$W = \frac{\gamma M}{R^2}\,m, \tag{1-182}$$

where W is the weight of the object having mass m, M is the mass of the earth ($5.98 \cdot 10^{24}$ kg), R is the mean radius of the earth ($6.37 \cdot 10^6$ m), and the quantity

$$\frac{\gamma M}{R^2} = \frac{6.67 \cdot 10^{-11}\ \text{Nm}^2\ \text{kg}^{-2} \cdot 5.98 \cdot 10^{24}\ \text{kg}}{(6.37 \cdot 10^6)^2\ \text{m}^2}$$

$$= 9.8\ \text{msec}^{-2}, \tag{1-183}$$

a constant acceleration, which we shall abbreviate as[8]

$$g = \frac{\gamma M}{R^2}.$$ (1-184)

Let us define the *gravitational field* vector G at any point in space as the gravitational force per unit mass acting at that point. Thus the field vector at a distance r from a particle of mass m is

$$G = -\frac{\gamma m}{r^2}\hat{r}.$$ (1-185)

If more than one particle is present, then G must be the vector sum of the field produced by each particle

$$G = \gamma \sum_i \frac{m_i}{r_i^2}\hat{r}_i.$$ (1-186)

Consider a particle of mass m located in a volume $\mathscr{V}$ enclosed by a surface $\mathscr{S}$. The integral

$$\oint_{\mathscr{S}} G \cdot d\sigma = -4\pi\gamma m;$$ (1-187)

if more than one particle were present within the volume, the integral would equal $-4\pi\gamma \sum_i m_i$. This finding corresponds to Gauss' law in electrostatics. If we now consider the more general case of a continuous distribution of mass, the density of which may vary from point to point, then Eq. (1-187) becomes

$$\oint_{\mathscr{S}} G \cdot d\sigma = -4\pi \iiint_{\mathscr{V}} \rho \, d\tau,$$ (1-188)

where ρ is the mass density. Using the divergence theorem, Eq. (1-126),

$$\oint_{\mathscr{S}} G \cdot d\sigma = \iiint_{\mathscr{V}} \nabla \cdot G \, d\tau.$$ (1-189)

Comparing the two equations, we find

$$\nabla \cdot G = -4\pi\gamma\rho.$$ (1-190)

In problem 1-5.4 the reader was asked to show that fields having the form of Eq. (1-185) are irrotational. Such being the case,

$$\nabla \times G = 0.$$ (1-191)

Equations (1-190) and (1-191) constitute the field equations for the Newtonian theory of gravitation.

---

[8] The actual value of g varies with latitude from about 9.78 msec^{-2} at the Equator to 9.83 msec^{-2} at the poles.

EXERCISES

1-8.1. Show that the result of integrating G over a closed surface of arbitrary shape enclosing a mass m is given by $-4\pi\gamma m$.

1-8.2. Since the gravitational field is irrotational, we can define a scalar ϕ_G as

$$G = -\nabla\phi_G. \tag{1-192}$$

Show that ϕ_G is given by an equation of the form known as *Poisson's equation*,

$$\nabla^2\phi_G = 4\pi\gamma\rho. \tag{1-193}$$

1-8.3. A homogeneous shell of inner radius r_1 and outer radius r_2 has a mass m. Find the gravitational field G for

(a) $r \leq r_1$,
(b) $r_1 \leq r \leq r_2$,
(c) $r \geq r_2$.

1-8.4. A spherical hollow of radius $R/2$ is made in a lead sphere of radius R as shown in Fig. 1-19. The mass of the hollowed sphere is M. Write an expression for the gravitational field as a function of the distance x along a line passing through the centers of the sphere and the hollow for

(a) $x \leq R$,
(b) $x \geq R$.

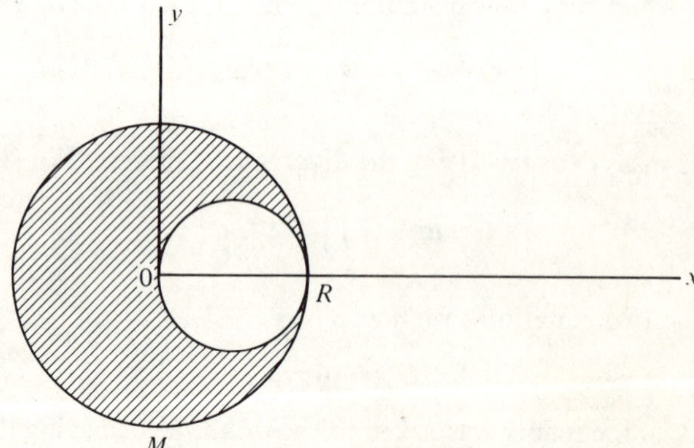

Fig. 1-19

REFERENCES

CAJORI, F., *A History of Physics.* Cambridge, Mass.: MIT Press, 1970.

GALILEO, G., *Dialogue Concerning Two New Sciences*, trans. by H. Crew and A. de Salvio. New York: The Macmillan Co., 1914; New York: Dover Publications, Inc., 1914.

HURD, D. L., and J. J. KIPLING, *The Origins and Growth of Physical Science* (2 vols.). New York: Basic Books, Inc., 1958; Baltimore: Penguin Books, 1964.

MARION, J. B., *Principles of Vector Analysis*. New York: Academic Press, 1965.

MORE, L. T., *Isaac Newton*. New York: Charles Scribner's Sons, 1934; New York: Dover Publications, Inc.

NEWTON, I., *Principia* (2 vols.). Matte's translation revised by F. Cajori. Berkeley and Los Angeles: University of California Press, 1967.

PARTICLE MECHANICS

2-1 General Discussion

Mechanical problems that do not involve externally imposed constraints on the motion of the particle are usually resolved by direct solution of the equation of motion

$$F = \frac{dp}{dt} \qquad (2\text{-}1)$$

or in integral form

$$p_2 - p_1 = \int_{t_1}^{t_2} F\,dt. \qquad (2\text{-}2)$$

The integral on the right-hand side of this equation is known as the *impulse*. Using either form, we can formulate the following conservation theorem.

*If the total force **F** acting on a particle is zero, then the momentum **p** of the particle is conserved.*

Figure 2-1 shows a particle of mass m and momentum p located at the end of a radius vector $r(t)$ drawn from the origin. We define the *angular momentum* L of this particle as

$$L = r \times p. \qquad (2\text{-}3)$$

We further define the *moment* of the force vector about the origin, generally known as the *torque*, as

$$N = r \times F, \qquad (2\text{-}4)$$

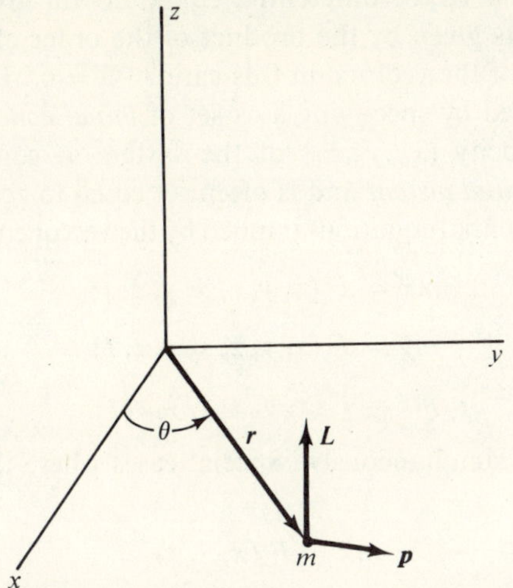

Fig. 2-1 Definition of angular momentum.

where F is the force applied to the particle. An equation for rotational motion analogous to Eq. (2-2) for linear motion can be found by taking the time derivative of Eq. (2-3):

$$\frac{dL}{dt} = N. \tag{2-5}$$

It may also be expressed in integral form

$$L_2 - L_1 = \int_{t_1}^{t_2} N \, dt, \tag{2-6}$$

and a conservation law for angular momentum obtains.

If the total torque N acting on a particle is zero, then the angular momentum L of the particle is conserved.

Another dynamical quantity and a corresponding conservation theorem will be considered in Section 2-4.

The equation with which we shall be concerned in most of our problems is

$$\frac{d^2r}{dt^2} = \frac{1}{m} \cdot F(r, u, t), \tag{2-7}$$

where the mass m is a constant. The mathematicians have been able to furnish an existence theorem that guarantees that solutions to this equation exist, in principle, for any position r, velocity u, and time t. The analytical solution of the equation of motion for all problems of interest is not possible, however.

47

The solution to a vector differential equation will involve a number of integration constants given by the product of the order of the equation and the dimensionality of the vectors, in this case $2 \cdot 3 = 6$. These constants are ordinarily determined by specifying as a set of *initial conditions* the position (x_0, y_0, z_0) and velocity $(\dot{x}_0, \dot{y}_0, \dot{z}_0)$ of the system at some time t_0 that is referred to as the *initial instant* and is often set equal to zero. In general, the three ordinary differential equations implied by the vector equation are coupled

$$m\ddot{x} = F_x(x, y, z, \dot{x}, \dot{y}, \dot{z}, t)$$
$$m\ddot{y} = F_y(x, y, z, \dot{x}, \dot{y}, \dot{z}, t) \qquad (2\text{-}8)$$
$$m\ddot{z} = F_z(x, y, z, \dot{x}, \dot{y}, \dot{z}, t)$$

and must be solved simultaneously. Special cases where these equations are not coupled

$$m\ddot{x} = F_x(x, \dot{x}, t)$$
$$m\ddot{y} = F_y(y, \dot{y}, t) \qquad (2\text{-}9)$$
$$m\ddot{z} = F_z(z, \dot{z}, t)$$

are much more easily handled. Forces that fall into this category fortunately include many of interest, such as the gravitational and electrostatic forces established by point masses and charges that are *central forces*, $\mathbf{F} = F_r(r)\hat{\mathbf{r}}$, viscous forces where $\mathbf{F} = \sum_i F_i(\dot{x}_i)$, and the three-dimensional harmonic oscillator. Most of the problems considered in this book are amenable to solution by simple analytical methods. In order to maintain a balanced perspective, we shall also consider several problems in which a complete solution of the dynamical details cannot be carried out analytically and we must be content with a partial solution. For other cases outside the scope of this book where the nonlinearity of the equation precludes a meaningful analytical analysis, the reader is referred to Kamke's encyclopedic work, which lists more than 1500 solved differential equations in such a way that a particular equation of interest can readily be found.[1] A good treatment of numerical methods can be found in the book by Levy and Baggott.[2]

EXERCISES

2-1.1. Prove Eq. (2-5).

2-1.2. Show that central force motion occurs in a fixed plane.

---

[1] E. Kamke, *Differentialgleichungen* (3rd ed.). New York: Chelsea Publishing Co., 1948.

[2] H. Levy and E. A. Baggott, *Numerical Solutions of Differential Equations.* New York: Dover Publications, Inc.

2-1.3. A yacht having a displacement (weight) of 15,000 lb is drifting toward a collision with a pier at a speed of 2 ft/sec. Over what minimum distance can a 150-lb crewman hope to stop the yacht by exerting an average force equal to his weight?

2-2 Particle Motion under a Constant Force

The best practical example of particle motion under a constant applied force is offered by a study of projectile motion in the earth's gravitational field. In order to approximate this force field as uniform, we must restrict our attention to flights near the earth's surface. Moreover, in order to ascertain the extent to which the nonuniformity of the gravitational field manifests itself as a function of altitude h above the earth's surface, we shall expand Eq. (1-185) for the gravitational field in a Taylor series, retaining only the first two terms:

$$G(r) = G(R) + \frac{dG}{dr}\bigg|_{r=R}(r - R) + \cdots$$

$$= \frac{\gamma m}{R^2} - \frac{2\gamma mh}{R^3}. \tag{2-10}$$

If we then require that the field at an altitude h differ from the field at the earth's surface ($r = R$) by less than a fraction ϵ, we find for the maximum altitude

$$h \le \frac{R\epsilon}{2}. \tag{2-11}$$

For $\epsilon = 1\%$, the altitude is approximately 32 km.

Let us examine the simplest case of one-dimensional motion in the direction of the gravitational field—that is, *falling bodies*. The equation of motion is

$$\ddot{z} = \frac{1}{m}\left(-\frac{\gamma Mm}{R^2}\right) = -g, \tag{2-12}$$

where g is the acceleration of gravity ≈ 9.8 msec^{-2}. Integrating successively with respect to time, we find for the speed

$$u_z = \dot{z} = u_0 - gt \tag{2-13}$$

where u_0 is the speed at $t = 0$ and

$$z = z_0 + u_0 t - \tfrac{1}{2}gt^2 \tag{2-14}$$

gives the height z as a function of the initial height z_0, the initial speed u_0, and the time t.

Let us turn to a more general discussion of the motion of projectiles in a uniform field. In this discussion, as in the one above, we shall ignore the effect of air resistance. Since the only force is that of gravity in the $-z$ direction, the motion will be two dimensional and will remain in the plane defined by the initial velocity $u_0(u_{0x}, u_{0y}, u_{0z})$ and the z axis. In practical instances, many additional factors may need to be considered. Some of them, such as the effect of the nonuniformity of the gravitational field, effects due to rotation of the earth, and air resistance, will be discussed in subsequent sections. Other effects, such as those due to the spin of a projectile fired from a rifled gun, are treated in texts on exterior ballistics.

A particle of mass m is projected from ground level, $z = 0$, with an initial velocity u_0 lying in the xz plane. This velocity vector is directed at an angle α above the horizontal. The scalar equations of motion are

$$\ddot{z} = -g, \tag{2-15}$$

$$\ddot{x} = 0. \tag{2-16}$$

Integrating once with respect to time, we find

$$\dot{z} = -gt + u_0 \sin \alpha, \tag{2-17}$$

$$\dot{x} = u_0 \cos \alpha. \tag{2-18}$$

Integrating again, we obtain

$$z = u_0 t \sin \alpha - \tfrac{1}{2}gt^2, \tag{2-19}$$

$$x = u_0 t \cos \alpha. \tag{2-20}$$

The altitude z will be zero at the end of the flight. Using Eq. (2-19), we find for the flight time

$$\tau = \frac{2u_0}{g} \sin \alpha. \tag{2-21}$$

By eliminating t between Eqs. (2-19) and (2-20), the equation of the projectile path is found to be

$$z = -\frac{gx^2}{2u_0^2 \cos^2 \alpha} + x \tan \alpha. \tag{2-22}$$

We can find the horizontal range R by setting $z = 0$ in Eq. (2-22) and solving equation for $x = R$. The result is

$$R = \frac{2u_0^2}{g} \sin \alpha \cos \alpha = \frac{u_0^2}{g} \sin 2\alpha \tag{2-23}$$

from which we see that the range is a maximum at $\alpha = \pi/4$ and that this maximum range is u_0^2/g.

The path described by Eq. (2-22) is a *parabola*. This fact can be seen more easily if we shift our origin to the highest point of the path—that is, to the point where $z = 0$. If we set $x = z = 0$ and $t = 0$ at this point, then Eq. (2-19) must become

$$z = -\tfrac{1}{2}gt^2. \tag{2-24}$$

Eliminating t between Eqs. (2-24) and (2-20), the path equation for the new coordinate system is

$$x^2 = -\left(\frac{2u_0^2}{g}\cos^2\alpha\right)z, \tag{2-25}$$

which is clearly a parabola.

EXERCISES

2-2.1. A man observes a ball to rise above and then fall back down past a window having a height l. If the ball is in view through the window for a total time span τ, find the height above the top of the window that the ball rose.

2-2.2. What is the time of flight of a projectile fired with a muzzle velocity u_0 at an angle α above the horizontal if the launching site is located at a height h above the target?

2-2.3. Derive an equation for the maximum range of a gun mounted at a height h above the target plane.

2-2.4. A gun fires two shots, each with the same muzzle velocity u_0. The first shot is at an angle of elevation α, the second at an elevation angle $\beta < \alpha$ at a time τ later. What must τ be such that the two shots collide in midair?

2-2.5. What angle of elevation α must be used by a gun firing from the foot of a hill of slope angle θ on a target T located straight up the hill at maximum range as shown in Fig. 2-2?

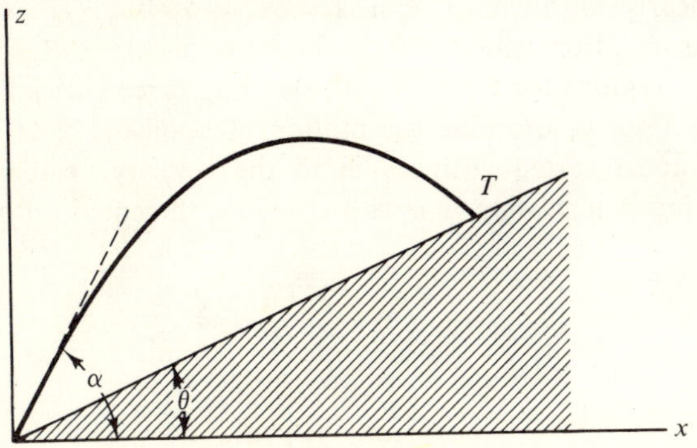

Fig. 2-2

2-2.6. A gun fires a projectile in such a way that it attains a range R in the horizontal plane of the gun site and a maximum height h above this plane. Show that the *maximum* horizontal range for the same muzzle velocity is

$$R_m = 2h + \frac{R^2}{8h}. \tag{2-26}$$

2-2.7. A projectile is fired from a gun at an elevation angle of $45°$ with a muzzle velocity u_0. As the projectile reaches the highest point of its trajectory, it explodes into two fragments of equal mass. One of the fragments falls vertically from rest. How far from the gun does the other fragment land?

2-3 Particle Motion under a Velocity-Dependent Force

The drag force that is generated when a body moves through a viscous medium was first shown by Newton to have the form

$$\boldsymbol{F}_D = -\tfrac{1}{2}\rho u^2 A C \hat{u}, \tag{2-27}$$

where ρ is the mass density of the medium, u is the speed, A is the area of the body projected onto the plane normal to $\boldsymbol{u}$, $\hat{u}$ is a unit vector in the direction of motion, and C is the drag coefficient. This drag coefficient is a function both of the shape and smoothness of the body and of a dimensionless function of the speed of the body u, its typical dimension L, the density of the medium ρ, and its viscosity η:

$$R_e = \frac{\rho u L}{\eta}. \tag{2-28}$$

This number, known as *Reynolds number*, is a measure of the relative importance of inertial and viscous effects. For $R_e \ll 1$ viscous effects dominate, and the drag coefficient is found to be proportional to R_e^{-1}; thus the drag increases linearly with speed. For $1 < R_e < 5 \cdot 10^5$, $C \propto R_e^{-1/2}$ and the drag varies as $u^{3/2}$. For values of R_e in the range $5 \cdot 10^5 < R_e < 5 \cdot 10^6$, there is a transition region; for $R_e > 5 \cdot 10^6$, the drag force varies as $u^{9/5} \approx u^2$.

In this section we examine the motion of bodies subject to a velocity-dependent force directed antiparallel to the velocity vector. For a body initially moving in a straight line with speed u_0, the equation of motion is

$$m\frac{du}{dt} = F(u), \tag{2-29}$$

which we can integrate

$$\int_{u_0}^{u} \frac{du}{F(u)} = \int_{t_0}^{t} \frac{dt}{m} \tag{2-30}$$

to find the speed u as a function of u_0 and $(t - t_0)/m$. A further integration gives us $x(t)$

$$x = x_0 + \int_{t_0}^{t} f\left(u_0, \frac{t - t_0}{m}\right) dt. \qquad (2\text{-}31)$$

We shall examine first the case of pure viscous drag at low speeds, $R_e \ll 1$. Then

$$m \frac{du}{dt} = -ku. \qquad (2\text{-}32)$$

Integrating as indicated above, we find

$$u = u_0 e^{-kt/m}, \qquad (2\text{-}33)$$

where $t_0 = 0$. Note that even for long times the body never comes to rest but only approaches doing so asymptotically. Integrating again, we have

$$x = \int_0^t u_0 e^{-kt/m} dt = \frac{mu_0}{k}(1 - e^{-kt/m}). \qquad (2\text{-}34)$$

For long times, x approaches a definite value

$$\lim_{t \to \infty} x = \frac{mu_0}{k}. \qquad (2\text{-}35)$$

Let us examine a slightly more complicated example of viscous drag by considering a *slowly* falling body. The equation of motion is now

$$m \frac{du}{dt} = mg - ku, \qquad (2\text{-}36)$$

where we have taken the coordinate to be positive downward. Integrating

$$\int_0^u \frac{du}{(mg/k) - u} = \int_0^t \frac{k}{m} dt, \qquad (2\text{-}37)$$

we find

$$\ln\left(\frac{mg}{k} - u\right) - \ln\left(\frac{mg}{k}\right) = \frac{kt}{m}.$$

Taking the antilog, we obtain

$$u = \frac{mg}{k}(1 - e^{-kt/m}), \qquad (2\text{-}38)$$

which has the same form as Eq. (2-34). For long times, the acceleration $du/dt \to 0$ and the velocity $u \to mg/k$, which is called the *terminal velocity*. This situation can be seen directly by examining the equation of motion,

Eq. (2-36). Integrating a second time and using the initial condition $y(0) = 0$, we find for y

$$y = \frac{mg}{k}\left[t - \frac{m}{k}\left(1 - e^{-kt/m}\right)\right]. \tag{2-39}$$

One important means of checking on ourselves is to examine the behavior of complicated results in limiting cases. We would expect, for example, that for a very short time following the drop the first term on the right-hand side of Eq. (2-36) should dominate. We can analyze Eq. (2-39) for short times by expanding the exponential term in a power series in t

$$e^x = \sum_{n=0}^{\infty} \frac{x^n}{n!} \tag{2-40}$$

and looking at the first couple of terms:

$$y = \frac{mgt}{k} - \left(\frac{m}{k}\right)^2 g\left[1 - \left(1 - \frac{kt}{m} + \frac{1}{2!}\left(\frac{kt}{m}\right)^2 - \frac{1}{3!}\left(\frac{kt}{m}\right)^3 + \cdots\right)\right]$$

$$\approx \frac{1}{2}gt^2 - \frac{1}{6}\frac{gk}{m}t^3 \tag{2-41}$$

for $t \ll m/k$. The first term in this expansion corresponds to free fall; the second term provides the lowest-order correction.

As we saw earlier, the drag is more nearly a quadratic function of velocity for higher Reynolds numbers where viscous effects are small. In this region skin friction and the energy expended in transverse acceleration of the medium provide most of the drag. For this case,

$$m\frac{du}{dt} = mg - \lambda u^2. \tag{2-42}$$

Integrating, we obtain

$$u = \sqrt{\frac{mg}{\lambda}}\tanh\left(\sqrt{\frac{g\lambda}{m}}\,t\right). \tag{2-43}$$

Since the hyperbolic tangent approaches unity for large arguments, the terminal velocity is $\sqrt{mg/\lambda}$, a result that could also have been gleaned by inspection from Eq. (2-42). Integrating again, we find[3]

$$y = \frac{m}{\lambda}\ln\left[\cosh\left(\sqrt{\frac{g\lambda}{m}}\,t\right)\right]. \tag{2-44}$$

---

[3] The reader can find a comprehensive set of hyperbolic function identities in M. Abramowitz and I. A. Stegun (Eds.), *Handbook of Mathematical Functions*. New York: Dover Publications, Inc., 1965, pp. 83–89.

Let us reexamine the case of a projectile moving near the surface of the earth under the influence of the uniform gravitational field and air resistance. As a result of the large Reynolds numbers experienced by a projectile, the drag force must be approximately proportional to u^2 throughout the flight, excepting only the last foot or so of altitude attained. We will assume that the motion is two dimensional. The horizontal component of the motion is governed by

$$\frac{du_x}{dt} = -\frac{\lambda}{m} u_x^2.$$

(2-45)

If the initial speed in the x direction is u_{x0} and the initial instant is given by $t = 0$, then we find upon integration

$$u_x = u_{x0}\left(\frac{\lambda u_{x0}}{m} t + 1\right)^{-1},$$

(2-46)

and
$$x = \frac{m}{\lambda} \ln \left(\frac{\lambda u_{x0}}{m} t + 1\right).$$

(2-47)

Consider next the vertical component of the motion. As we know, the drag force is always opposed to the motion. If the drag force were proportional to u_y or any odd power of u_y, then things would take care of themselves automatically as u_y changes signs at the zenith of the trajectory. With a quadratic drag force, the sign of the drag force is independent of the sign of u_y. The problem of the vertical motion must therefore be split into two problems. For the upward motion

$$\frac{du_y}{dt} = -g - \frac{\lambda}{m} u_y^2,$$

(2-48)

where the y axis is taken to be increasing upward. For the downward motion we will use

$$\frac{du_y}{dt} = g - \frac{\lambda}{m} u_y^2,$$

(2-49)

where the time scale is restarted at $t = 0$ at the instant that u_y passes through zero and is considered to increase downward. Integration Eq. (2-48)

$$\int_{u_{y0}}^{u_y} \frac{du_y}{(mg/\lambda) + u_y^2} = -\frac{\lambda}{m} \int_0^t dt,$$

we find

$$u_y = \sqrt{\frac{mg}{\lambda}} \tan \left[\arctan \left(\frac{\lambda u_{y0}^2}{mg}\right)^{1/2} - \sqrt{\frac{\lambda g}{m}}\, t\right]$$

(2-50)

for the vertical speed. The time taken to reach maximum altitude ($u_y = 0$) can be found by setting the argument of the tangent function equal to zero.

$$\tau = \sqrt{\frac{m}{\lambda g}}\ \arctan\left(\frac{\lambda u_{y0}^2}{mg}\right)^{1/2}. \tag{2-51}$$

Integrating again, we find

$$y = \frac{m}{\lambda}\ln\left\{\left(\frac{\lambda u_{y0}^2}{mg} + 1\right)\cos\left[\arctan\left(\frac{\lambda u_{y0}^2}{mg}\right)^{1/2} - \sqrt{\frac{\lambda g}{m}}\,t\right]\right\}. \tag{2-52}$$

The maximum altitude attained by the projectile can easily be determined.

$$y_m = \frac{m}{\lambda}\ln\left(\frac{\lambda u_{y0}^2}{mg} + 1\right). \tag{2-53}$$

Since Eq. (2-49) is identical in form with Eq. (2-42), the integrals are given by Eqs. (2-43) and (2-44). By using the substitution

$$\frac{du_y}{dt} = \frac{du_y}{dy}\frac{dy}{dt} = u_y\frac{du_y}{dy}$$

to change the independent variable in Eq. (2-49) to y

$$u_y\frac{du_y}{dy} = g - \frac{\lambda}{m}u_y^2, \tag{2-54}$$

we find upon integration

$$u_y^2 = \frac{mg}{\lambda}(1 - e^{-2\lambda y/m}). \tag{2-55}$$

By substituting from Eq. (2-53), we find for the vertical component of the impact speed U

$$U = u_T\left\{1 - \left[1 + \left(\frac{u_{y0}}{u_T}\right)^2\right]^{-2}\right\}^{1/2}, \tag{2-56}$$

where u_T is the terminal speed $u_T^2 = mg/\lambda$.

EXERCISES

2-3.1. Examine the case of a high-speed torpedo that loses its propulsive power at $t = 0$ and begins to slow under the action of a drag force

$$\mathbf{F} = -\lambda u^2\hat{\mathbf{u}}.$$

Integrate the equation of motion to find $u(t)$ and $x(t)$. Note that the distance traveled $x \to \infty$ for long times, which is contrary to experience. Explain.

2-3.2. Assume that the drag on a homogeneous sphere of density ρ_s falling near the earth is proportional to the surface area of the sphere and to the nth power of the speed. Show that the terminal velocity of the sphere is proportional to the nth root of the radius of the sphere.

2-3.3. Eliminate u_y between Eqs. (2-43) and (2-55) in order to verify Eq. (2-44).

2-3.4. Analyze the motion of a projectile under the assumption that the drag varies linearly with the speed. Derive the equation for the trajectory $y(x)$. How would you go about determining the maximum range as a function of the angle of elevation?

2-3.5. A ball is thrown horizontally from a moving truck with a speed u_{x0} at right angles to the motion of the truck at speed u_{y0}. If the initial height of the ball is h, calculate the point of impact of the ball, assuming

    (a) no air resistance,
    (b) a drag force proportional to u,
    (c) a drag force proportional to u^2.

2-4 Particle Motion under a Position-Dependent Force

The basic classical forces of nature, gravitational and electrical, are both functions of position. This type of force is therefore of major interest to us. The equation of motion can be written

$$m \frac{du}{dt} = F(r) \tag{2-57}$$

if we maintain our assumption that the mass of the particle is constant. Our approach to the integration of this equation is to define a new dynamical quantity that can be measured, the *energy*.

We can define the *work* done by the external force $F(r)$ acting on the particle as it is moved from a position r_1 to a position r_2 in the force field as

$$W = \int_{r_1}^{r_2} F \cdot dr. \tag{2-58}$$

Replacing the force by the left-hand side of Eq. (2-57) and writing $dr = u\,dt$, we find

$$\int_{r_1}^{r_2} F \cdot dr = m \int_{t_1}^{t_2} \frac{du}{dt} \cdot u\,dt = \frac{m}{2} \int_{t_1}^{t_2} \frac{d}{dt} u^2\,dt,$$

or

$$W = \frac{m}{2}(u_2^2 - u_1^2). \tag{2-59}$$

The scalar quantity $mu^2/2$ is called the *kinetic energy* of the particle and is denoted by T. The work done is therefore equal to the change in the kinetic energy.

If the force field is irrotational, the work done in moving a particle around a closed path is zero

$$\oint \mathbf{F} \cdot d\mathbf{r} = 0 \qquad (2\text{-}60)$$

and a scalar potential function can be defined

$$\mathbf{F} = -\nabla V. \qquad (2\text{-}61)$$

The quantity V is referred to as the *potential energy*. This potential energy only has significance when we consider *changes* in it. For this reason, the location of the zero level of V is arbitrary. If the force is *conservative* (to use the physically descriptive synonym for *irrotational*), then

$$\int_{r_1}^{r_2} \mathbf{F} \cdot d\mathbf{r} = -\int_{r_1}^{r_2} \nabla V \cdot d\mathbf{r} = -\int_{v_1}^{v_2} dV,$$

or
$$W = V_1 - V_2. \qquad (2\text{-}62)$$

We can combine this relation with that expressed in Eq. (2-59) to find

$$T_1 + V_1 = T_2 + V_2, \qquad (2\text{-}63)$$

which we may state as an energy conservation theorem.

If *the force acting on a particle is irrotational, then the total energy of the particle*

$$E = T + V \qquad (2\text{-}64)$$

is conserved.

In one dimension, the equation of motion is

$$m \frac{du_x}{dt} = F(x). \qquad (2\text{-}65)$$

This equation can be integrated by making use of Eq. (2-59). We find

$$u^2 = u_0^2 + \frac{2}{m} \int_0^x F(x)\, dx. \qquad (2\text{-}66)$$

We saw that a conservative force field is one in which the work done in moving a particle around a closed path is zero. In one dimension, a closed path amounts to shuffling back and forth on the same track:

$$\oint F(x)\, dx = \int_{x_1}^{x_2} F(x)\, dx + \int_{x_2}^{x_1} F(x)\, dx$$

$$= \int_{x_1}^{x_2} F(x)\, dx - \int_{x_1}^{x_2} F(x)\, dx$$

$$= 0. \qquad (2\text{-}67)$$

Hence *all one-dimensional position-dependent forces are conservative*. A potential $V(x)$ can therefore be defined such that

$$V(x) - V(0) = -\int_0^x F(x)\, dx \tag{2-68}$$

and

$$E = \tfrac{1}{2}mu^2 + V(x) = \text{const.} \tag{2-69}$$

Solving Eq. (2-69) for the speed u,

$$u = \frac{dx}{dt} = \sqrt{\frac{2}{m}}\,[E - V(x)]^{1/2}, \tag{2-70}$$

we can now integrate for x, which must be disentangled algebraically from

$$\sqrt{\frac{m}{2}}\int_0^x \frac{dx}{\sqrt{E - V(x)}} = t. \tag{2-71}$$

In this formulation, the initial instant and position have been taken to be zero. The initial speed is arbitrary and is contained in the constant E.

EXERCISES

2-4.1. Show that the rate of change of the kinetic energy of a particle is given by

$$\boldsymbol{F} \cdot \boldsymbol{u} = \frac{dT}{dt}, \tag{2-72}$$

where $\boldsymbol{F}$ is the external applied force and $\boldsymbol{u}$ is the instantaneous particle velocity.

2-4.2. A particle of mass m is acted on by a force

$$F = ax^2 - bx.$$

Find the potential function $V(x)$. Describe the motion for any and all values of u_0, the initial speed.

2-4.3. Solve the equation for one-dimensional motion of a body of mass m subject to a repulsive force

$$F = \frac{A}{x^3}.$$

2-4.4. Show that the gravitational potential derived by using Eq. (1-164) reduces to the potential for motion in a uniform field in the limit as $r \to R$

$$\lim_{r \to R} \phi_G(r) = mgh,$$

where $h = r - R$, R is the radius of the earth, and g is the acceleration of gravity at $r = R$.

2-4.5. A body falling from a great height can no longer be considered to move in a uniform field. The force is

$$F = -\frac{\gamma M m}{r^2}.$$

(a) Derive the gravitational potential directly by integration of the force equation. Fix the value of the potential at $r \to \infty$ as zero.

(b) Neglecting air resistance, find the speed $u(t)$.

2-4.6. A body is projected upward from the earth's surface with an initial speed that causes the total energy E to be positive. Using the results of problem 2-4.5, show that the projectile will continue to move outward forever with a speed that decreases to approach

$$\lim_{r \to \infty} u(r) = \sqrt{\frac{2E}{m}}. \tag{2-73}$$

2-4.7. If $E < 0$, show that a particle projected upward in the earth's gravitational field will attain a height above the surface

$$h = r - R = \frac{\gamma M m}{|E|} - R$$

and will then fall back toward the earth.

2-4.8. Discuss the motion of a body projected upward from the earth's surface with $E = 0$.

2-4.9. Find the gravitational potential at any point on the axis of a disk of radius a, thickness τ, and mass density ρ.

2-4.10. Imagine a smooth tunnel through the earth connecting *any* two points on its surface. How long does it take for an object of mass m dropped into one end of the tunnel to arrive at the other end? Describe the motion.

2-5 Static and Kinetic Friction

If a body experiences a force having components that are both normal and parallel to a *rough* surface on which the body is at rest, it is possible for the reaction of the surface on the body to have a component parallel to the surface as well as a normal component. This situation is shown in Fig. 2-3.

Suppose that $F_\perp$ is just the weight of the body. Then the normal reaction $R_\perp$ is equal to the weight in magnitude and is opposite in direction. Experience confirms that a block placed on a rough surface can tolerate a small horizontal force $F_\parallel$ without moving. So there must be a reaction force $R_\parallel$ equal and opposite to $F_\parallel$ exerted by the surface on the block. It is the *frictional reaction.*

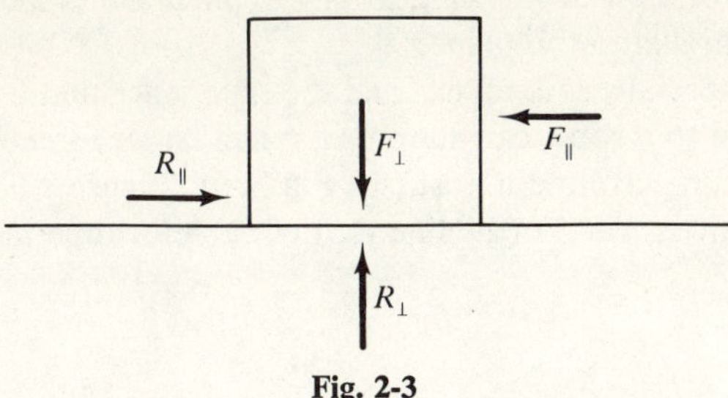

Fig. 2-3

If the applied horizontal force is increased, a value will be found beyond which the block will begin to move. Experimentally, this limiting horizontal force is found to be nearly independent of the contact area between the block and the surface and to depend only on the nature of the contacting surfaces. In addition, it is found that the force necessary to produce motion is directly proportional to the force normal to the surface—in this case, the weight. A *static friction coefficient* μ_s can be defined that is approximately constant for a given pair of contact surfaces and approximately independent of the normal force,

$$F_{\max} = \mu_s R_\perp. \qquad (2\text{-}74)$$

It must be stressed that μ_s is defined only when slipping is about to occur. The *angle of friction*, ϵ, is defined as

$$\tan \epsilon = \mu_s. \qquad (2\text{-}75)$$

The quantities ϵ and μ_s can be determined for a given pair of surfaces by the action of an inclined plane as shown in Fig. 2-4. In this figure α is the angle of inclination of the plane and W is the weight of the block acting vertically downward. The force that tends to cause the block to slide down the plane is $W \sin \alpha$. If the angle α is increased, it is found that sliding will begin at $\alpha = \epsilon$. Even though the friction force in this case is directed up the inclined plane, such is not always true. The force of friction is always such as to

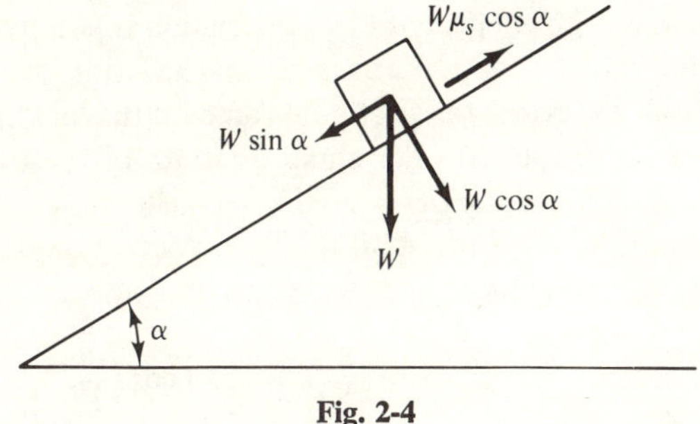

Fig. 2-4

oppose the vector sum of the tangentially applied forces and may point in any direction parallel to the plane.

Everyone (especially sailors) has had the experience that a relatively small force F_1, applied to a rope, can support a much larger force F_2 by wrapping the rope a few turns around a post (see Fig. 2-5). Figure 2-6 shows schematically a small increment Δs (the line AB) of cable or rope in contact with a

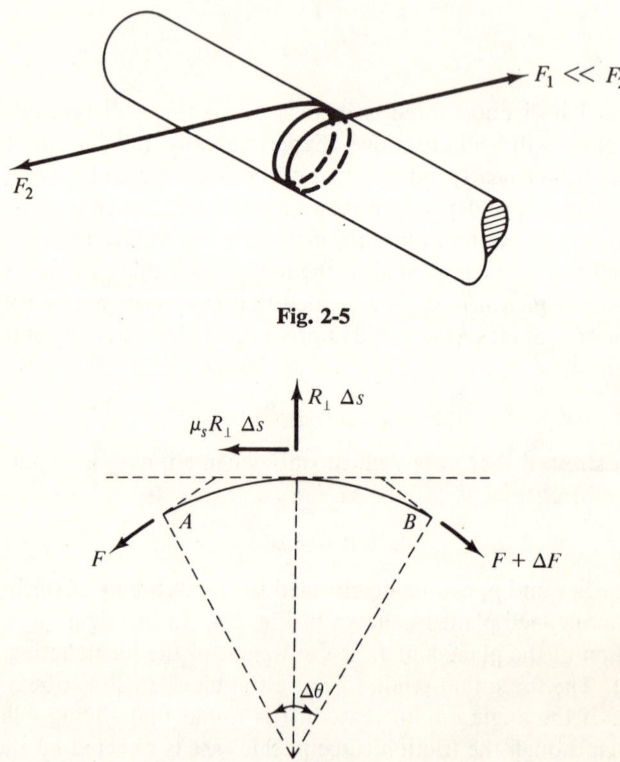

Fig. 2-5

Fig. 2-6

rough curved surface. The coefficient of static friction is μ_s and the rope is on the verge of slipping to the right. Let the normal reaction of the surface be $R_\perp$ per unit length of the rope; the frictional force on the length Δs of rope is then $\mu_s R_\perp \Delta s$ directed to the left. For small Δs, both of these forces can be considered to act at the midpoint C of the segment. Since the rope is in static equilibrium, the vector sum of all the forces acting on it must be zero. Considering force components tangent at C, we find

$$F \cos \frac{\Delta\theta}{2} + \mu_s R_\perp \Delta s = (F + \Delta F) \cos \frac{\Delta\theta}{2}. \tag{2-76}$$

Neglecting products of infinitesimals, it reduces to

$$\Delta F = \mu_s R_\perp \Delta s. \tag{2-77}$$

Similarly, taking the sum of force components normal at C,

$$R_\perp \Delta s = F \sin \frac{\Delta\theta}{2} + (F + \Delta F) \sin \frac{\Delta\theta}{2}, \tag{2-78}$$

which reduces to

$$R_\perp \Delta s = 2F \sin \frac{\Delta\theta}{2}$$

$$\approx 2F\left[\frac{\Delta\theta}{2} - \frac{(\Delta\theta)^3}{3! \cdot 2^3} + \cdots\right]$$

$$\approx F\Delta\theta \tag{2-79}$$

to first order in infinitesimals. Taking the limit as $\Delta\theta \to d\theta$, Eqs. (2-77) and (2-79) can be combined to give

$$dF = \mu_s F \, d\theta. \tag{2-80}$$

Integrating, we find

$$\int_{F_0}^{F} \frac{dF}{F} = \int_0^\theta \mu_s \, d\theta = \ln \frac{F}{F_0} = \mu_s\theta,$$

or

$$F = F_0 e^{\mu\theta}. \tag{2-81}$$

In this formula F_0 is the force that must be applied to support the larger force F; θ is the angle between the normals to the curved surface at the points A and B where the rope enters and leaves the surface. If the contact surface is smooth, then $\mu_s \to 0$ and, consequently, $(dF/d\theta) \to 0$.

Occasionally it is useful to know the normal reaction of the surface as a function of its radius of curvature. Rewriting Eq. (2-79) in the limit $\Delta\theta \to d\theta$, we find

$$R_\perp = \frac{F}{ds/d\theta}, \tag{2-82}$$

or

$$R_\perp = \frac{F}{r}, \tag{2-83}$$

where $r = ds/d\theta$ is the radius of curvature of the surface.

The concept of *kinetic friction* can be introduced in much the same way as for static friction. The frictional force that acts when one surface slides over another is almost independent of the contact area and is nearly independent of the speed of the sliding for moderate normal forces. As before, the frictional force is found to oppose the motion and to be nearly proportional to the normal force

$$f = \mu N, \tag{2-84}$$

where N is the normal force at the contact surface and μ is the *coefficient of kinetic friction*. This coefficient can be measured by an inclined plate arrangement much as we did for the static coefficient. Referring again to Fig. 2-4, we regard the block to be in motion under the action of the downward force $W \sin \alpha$ and the upward force $W\mu \cos \alpha$. The numerical value of μ is found by adjusting the angle of inclination α until the block moves down the plane with a constant speed. If this angle is $\alpha = \epsilon$, then $\tan \epsilon = \mu$ as before and we have defined an *angle of kinetic friction*.

For a given pair of contact surfaces, it is always found that

$$\mu_s > \mu, \tag{2-85}$$

although they are of the same order of magnitude. The coefficient varies from about 0.03 for very smooth surfaces to approximately unity for sandpaper on sandpaper.

Suppose that we apply a force F to accelerate a block of mass m sliding on a rough horizontal surface. The equation of motion is

$$m\ddot{x} = F - \mu mg \tag{2-86}$$

in which μ is assumed constant. Multiplying this equation by $\dot{x}\,dt = dx$ and integrating

$$\int_{u_1}^{u_2} mu\,du = \int_{x_1}^{x_2} F\,dx - \int_{x_1}^{x_2} \mu mg\,dx,$$

where we have written

$$m\ddot{x}\dot{x}\,dt = m\dot{x}\,\frac{d\dot{x}}{dt}\,dt = mu\,du,$$

we find for constant F

$$F \cdot (x_2 - x_1) = T_2 - T_1 + \mu mg(x_2 - x_1). \tag{2-87}$$

The left-hand side of this equation is simply the total work done by the force F. This work is divided between an increase in the kinetic energy and the energy dissipated by friction appearing as nonrecoverable heat in the block and the surface.

EXERCISES

2-5.1. A cubical block of mass m is at rest on a rough inclined plane as shown in Fig. 2-7. The angle of inclination is α and the coefficient of static friction

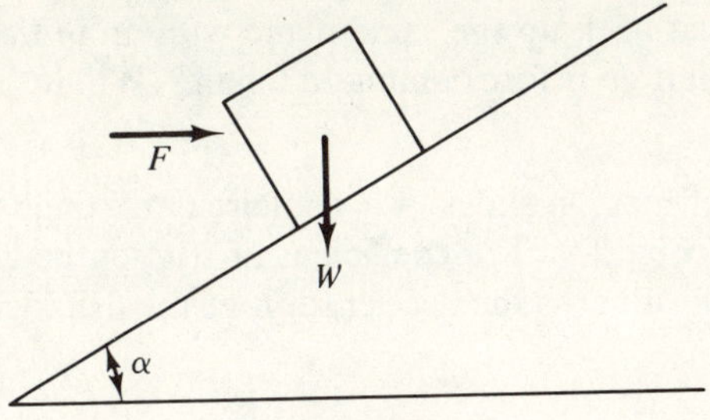

Fig. 2-7

is μ_s. The externally applied horizontal force F passes through the geometrical center of the block. Find the minimum and maximum values of F consistent with static equilibrium.

2-5.2. A block is placed on a plank and the end of the plank is raised until the block begins to slide, whereupon the angle of inclination is fixed at $\alpha = \epsilon$. The block is observed to slide a distance s in time t. Determine the coefficients of static and kinetic friction between the plank and the block.

2-5.3. A block of mass m_1 is stacked on top of a block of mass m_2. It has been determined that a horizontal force $F \geq F_c$ applied to the top block will cause it to slip on the bottom one. Let the block stack be placed on a frictionless table. Find the maximum force that can be applied to the lower block such that the blocks will move together.

2-5.4. Two masses m_1 and m_2 are linked by a massless cord drawn over a frictionless pulley as shown in Fig. 2-8. If the masses are on the verge of slipping to the left, find the coefficient of friction μ_s (the two surface contacts are assumed to have identical friction properties).

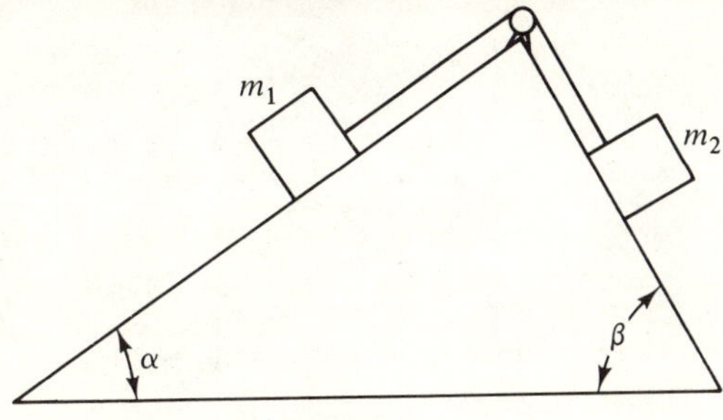

Fig. 2-8

2-5.5. Reconsider problem 2-5.4, assuming that the two contact surfaces have different coefficients of static friction μ_1 and μ_2. Write the appropriate equations in terms of known or measured quantities and solve them for μ_1 and μ_2.

65

2-5.6. A block slides down a plane having an angle of inclination α at constant speed. It is launched back up the same plane with an initial speed u_0. How far up the plane will it go before coming to a halt? Will it then begin to slide down again?

2-5.7. A cylinder of mass m slides in a semicircular trough having the same radius as shown in Fig. 2-9. The coefficient of friction between the cylinder and the trough is μ. What is the acceleration of the cylinder?

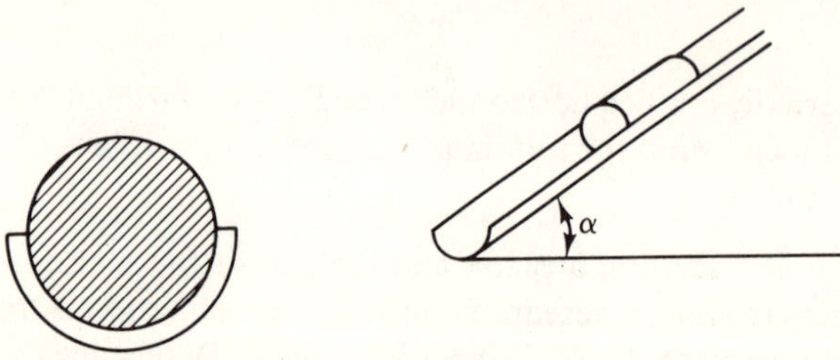

Fig. 2-9

2-5.8. A massless string with weights W_2 and W_1 ($W_2 > W_1$) at its ends is draped across a rough log in a plane normal to the axis. The coefficients of friction between the string and the log are μ and μ_s. What is the criterion for the system to be inequilibrium? If this criterion is not satisfied, what is the acceleration of the system? What are the tensions in various parts of the cord?

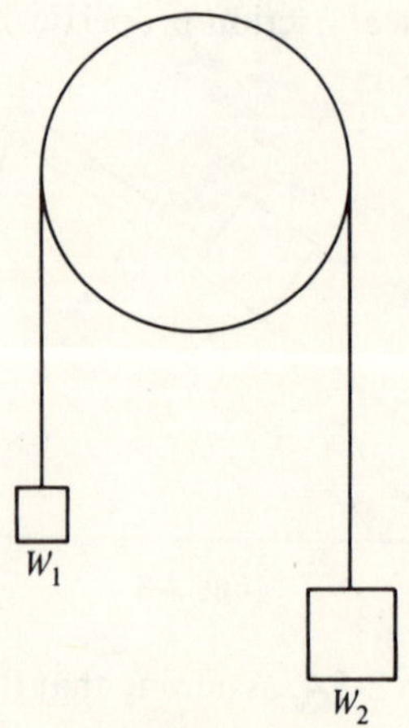

Fig. 2-10

2-5.9. A block of mass m_1 is connected by a string to a mass m_2 hanging over the edge of the rough table on which block m_1 slides. Assume that the string contacts the table only at the edge. The coefficient of friction between the block and the table is μ_1, that between the string and the table, μ_2. Determine the acceleration of the system.

2-6 Motion of Charged Particles in Electric and Magnetic Fields

Here we consider the motion of a charged particle under the action of electric and magnetic forces. In doing so, we are concerned only with the mechanics involved and do not delve into the physics of the electromagnetic field. Knowledge of the trajectories followed by charged particles in various electric and magnetic field configurations is of considerable interest in the fields of astrophysics and plasma physics.

The basic statement of Newton's second law for a particle in an electromagnetic field is

$$\frac{d^2 \boldsymbol{r}}{dt^2} = \frac{q}{m}\left[\boldsymbol{E}(\boldsymbol{r}, t) + \boldsymbol{u} \times \boldsymbol{B}(\boldsymbol{r}, t)\right], \tag{2-88}$$

where q is the charge of the particle, m is its mass, $\boldsymbol{E}$ is the electric field strength, $\boldsymbol{u}$ is the particle velocity, and $\boldsymbol{B}$ is the magnetic flux density. In electromagnetics two different systems of units are in common usage. The *MKSQ* system, in terms of which we shall write our equations, measures lengths in meters, mass in kilograms, time in seconds, and charge in coulombs. The coulomb force law analogous to the law of universal gravitation is defined in this system by defining charge as a basic unit. There must then be a constant of proportionality analogous to γ in the gravitation law to make the units work out. The MKSQ system is the practical one, since electrical quantities are measured in volts, amperes, and similar units. The Gaussian CGS system is somewhat more physical and is popular with physicists.

Let us look at some examples of motion in uniform static fields. If the magnetic field is zero and there is only a uniform electric field in the x direction, then in analogy with falling bodies

$$x = \frac{qEt^2}{m} + u_{x0}t + x_0, \tag{2-89}$$

$$y = u_{y0}t + y_0, \tag{2-90}$$

$$z = u_{z0}t + z_0. \tag{2-91}$$

The magnetic force is new to our experience. If

$$\boldsymbol{B}(\boldsymbol{r}, t) = B\hat{\boldsymbol{k}} \tag{2-92}$$

is a constant, then the components of the vector equation of motion (assuming $E = 0$) are

$$\ddot{x} = \frac{qB}{m}\,\dot{y}, \qquad \ddot{y} = -\frac{qB}{m}\,\dot{x}, \qquad \ddot{z} = 0. \tag{2-93}$$

This is our first example of scalar component equations that are coupled owing to the vector product $u \times B$ that defines the magnetic force. Differentiating the first two of Eqs. (2-93)

$$\dddot{x} = \omega_c \ddot{y}, \qquad \dddot{y} = -\omega_c \ddot{x},$$

where ω_c, known as the *cyclotron frequency*, is

$$\omega_c = \frac{qB}{m}, \tag{2-94}$$

x and y can be eliminated, and the third-order uncoupled equations are

$$\dddot{x} + \omega_c^2 \dot{x} = 0, \qquad \dddot{y} + \omega_c^2 \dot{y} = 0. \tag{2-95}$$

Since the dependent variables x and y do not appear in these equations, they can be written in terms of the velocity components as

$$\ddot{u}_x + \omega_c^2 u_x = 0, \qquad \ddot{u}_y + \omega_c^2 u_y = 0. \tag{2-96}$$

The general solution to these equations is given by

$$u_x = C_1 \cos \omega_c t + C_2 \sin \omega_c t, \tag{2-97}$$

$$u_y = C_3 \cos \omega_c t + C_4 \sin \omega_c t. \tag{2-98}$$

In carrying out this integration, we have, in effect, specified the initial acceleration components to be zero, thereby eliminating two of the eight integration constants. The other initial conditions are

$$\begin{aligned} x(0) &= x_0, & y(0) &= 0, & z(0) &= 0 \\ \dot{x}(0) &= 0, & \dot{y}(0) &= u_{y0}, & \dot{z}(0) &= u_{z0} \end{aligned} \tag{2-99}$$

which are quite general. We find for the solution to Eqs. (2-93)

$$x = x_0 \cos \omega_c t,$$

$$y = \left(\frac{u_{y0}}{\omega_c}\right) \sin \omega_c t, \tag{2-100}$$

$$z = u_{z0} t.$$

By adjusting the origin of coordinates so that $x_0 = u_{0y}/\omega_c$, the trajectory of the particle can be shown to consist of circular motion centered on the origin in the xy plane

$$x^2 + y^2 = \left(\frac{u_{y0}}{\omega_c}\right)^2 \tag{2-101}$$

and a constant speed along the z axis—that is, a constant pitch spiral.

This same information can also be deduced in another way with less work. Since the magnetic force F is a vector that is always at right angles to the velocity u, then $F \cdot u = 0$; and as a result of Eq. (2-72), the kinetic energy $T = \frac{1}{2}mu^2$ and the magnitude of the velocity u are constant. Taking the dot product of the equation of motion with u, we see that

$$u \cdot \frac{du}{dt} = u \cdot a = u \cdot \frac{q}{m} u \times B = 0. \tag{2-102}$$

Thus a particle moving at constant speed and subject to a constant acceleration always at right angles to the trajectory must move in a circle (ignoring the steady motion parallel to the magnetic field). The radius of this circle is found by equating the radial acceleration with $\ddot{\rho} = 0$ as given in Eq. (1-131) to F/m:

$$\rho\dot{\theta}^2 = \frac{u_\perp^2}{\rho} = \frac{qu_\perp B}{m}. \tag{2-103}$$

Solving for the radius ρ, we find

$$\rho_c = \frac{mu_\perp}{qB} = \frac{u_\perp}{\omega_c}, \tag{2-104}$$

where ρ_c is known as the *cyclotron radius* and $u_\perp$ is the magnitude of the velocity normal to B:

$$u_\perp = u - u_z\hat{k}. \tag{2-105}$$

Let us complicate matters by superposing a uniform electric field

$$E = E\hat{i} \tag{2-106}$$

on the uniform magnetic field specified in Eq. (2-92). The equations of motion for a charged particle in this combination of fields are

$$\ddot{x} = \omega_c\dot{y} + a, \qquad \ddot{y} = -\omega_c\dot{x}, \tag{2-107}$$

where $a = qE/m$ and we are ignoring motion in the z direction. Taking as our initial conditions

$$\begin{aligned} x(0) &= 0, & y(0) &= 0, \\ \dot{x}(0) &= 0, & \dot{y}(0) &= 0, \\ \ddot{x}(0) &= a, & \ddot{y}(0) &= 0, \end{aligned} \tag{2-108}$$

the solutions to Eqs. (2-107) can be found by proceeding as with Eqs. (2-93). They are

$$x = \frac{a}{\omega_c^2} (1 - \cos \omega_c t), \tag{2-109}$$

$$y = \frac{a}{\omega_c^2} (\sin \omega_c t - \omega_c t). \tag{2-110}$$

These are the equations of an ordinary cycloid in parametric form (see Fig. 2-11). This is the path of a point located on the rim of a circle having a radius

$$\frac{a}{\omega_c^2} = \frac{m}{q}\frac{E}{B^2},\tag{2-111}$$

as it rolls along the negative y axis. Comparing this relation with Eq. (2-104) for the cyclotron radius, we see that the orbital speed must be

$$u_\perp = \frac{E}{B}.\tag{2-112}$$

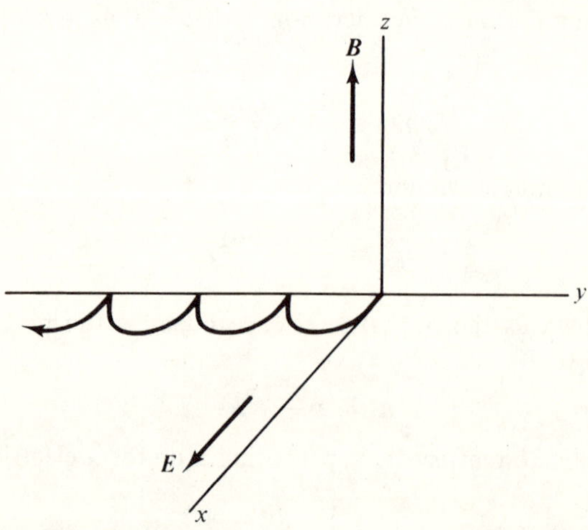

Fig. 2-11 Particle motion in crossed electric and magnetic fields.

The nature of this motion can be understood in the following terms. A charged particle initially at rest at the origin is accelerated in the x direction (if its charge $q > 0$) by the electric field. As it acquires velocity, it begins to "see" the magnetic field, which starts to deflect it in the $E \times B$ direction. When the deflection has proceeded to the point that $u \perp E$, the power input to the particle from the electric field

$$P = \mathbf{F} \cdot \mathbf{u} = q\mathbf{E} \cdot \mathbf{u}\tag{2-113}$$

approaches zero. The particle continues to deflect and returns to the y axis with zero speed to begin the next cycle.

The important point is that the net effect when we average over the motion in the x direction is a steady *drift* in the $E \times B$ direction at a velocity given by

$$\mathbf{w}_E = \frac{\mathbf{E} \times \mathbf{B}}{B^2}.\tag{2-114}$$

Let us show that this is so. We write the general vector equation of motion for a particle in the presence of a magnetic and electric field, both assumed to be uniform but not necessarily orthogonal:

$$\frac{du}{dt} = \frac{q}{m}\left[(u \times B) + E\right]. \tag{2-115}$$

Next, decompose the velocity vector into components parallel and perpendicular to the magnetic field:

$$u = u_{\parallel} + u_{\perp}. \tag{2-116}$$

Equation (2-115) can be decomposed in the same way and we find

$$\frac{du_{\parallel}}{dt} = \frac{q}{m} E_{\parallel} \tag{2-117}$$

and

$$\frac{du_{\perp}}{dt} = \frac{q}{m}\left[u_{\perp} \times B + E_{\perp}\right]. \tag{2-118}$$

Equation (2-117) describes uniform acceleration along the magnetic field, and we can write the solution by inspection. In order to dissect Eq. (2-118), we shall introduce a further decomposition on $u_{\perp}$:

$$u_{\perp} = w_E + u_0, \tag{2-119}$$

where w_E is the drift velocity defined in Eq. (2-114). Inserting Eq. (2-119) into Eq. (2-118), we have

$$\frac{du_0}{dt} = \frac{q}{m}\left[\frac{(E_{\perp} \times B) \times B}{B^2} + u_0 \times B + E_{\perp}\right] \tag{2-120}$$

The expansion of the vector triple product yields

$$(E_{\perp} \times B) \times B = B(E_{\perp} \cdot B) - E_{\perp}(B \cdot B) = -B^2 E_{\perp}; \tag{2-121}$$

hence

$$\frac{du_0}{dt} = \frac{q}{m} u_0 \times B, \tag{2-122}$$

which is simply gyromotion around the magnetic field. In other words, if the particle motion is viewed from a coordinate system moving in the negative y direction at a speed $w_E = E_{\perp}/B$, the electric field is transformed away.

In a more general way, we might have replaced the electrical force qE by any external force F. We would have found that the general formula for the drift velocity is

$$w = \frac{F_{\perp} \times B}{qB^2}. \tag{2-123}$$

This force may, for example, be the gravitational force. The drift velocity

$$w_g = \frac{m}{q} \frac{g_\perp \times B}{B^2} \tag{2-124}$$

in this case depends on the ratio of m/q, unlike the electric drift w_E, which is the same for all charged particles.

Before leaving the subject of particle motion under electromagnetic forces, consider the situation in which the electric field is zero and the magnetic field is inhomogeneous. Suppose that the inhomogeneity is transversal as shown in Fig. 2-12,

$$B = B(x)\hat{k}. \tag{2-125}$$

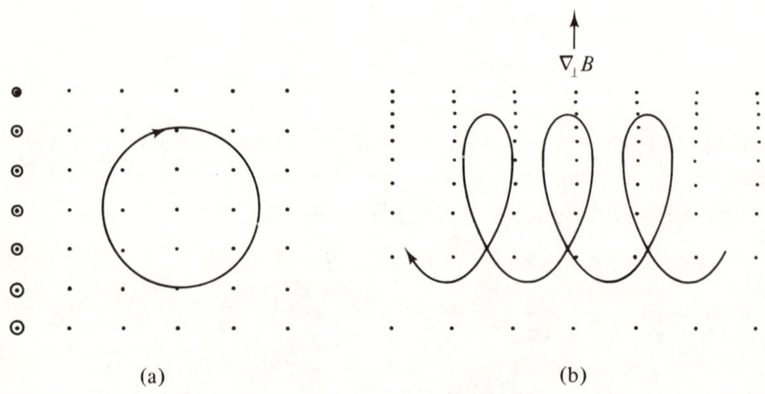

(a)   (b)

Fig. 2-12 Particle motion in uniform (a) and transversally inhomogeneous (b) magnetic fields.

Here we are looking down on the xy plane. On the left-hand side of the figure we have depicted a uniform magnetic field as a uniform array of points, and the circular gyration of a positive particle is shown. On the right we see an inhomogeneous field that decreases in the positive x direction. The transverse motion of a positively charged particle in this field features a drift induced by the difference in the magnetic deflecting force over the distance of a cyclotron orbit. The force that gives rise to this drift is

$$F_\perp = -\tfrac{1}{2}mu_\perp^2 \frac{\nabla_\perp B}{B}. \tag{2-126}$$

A transversal inhomogeneity in the magnetic field is equivalent to a curvature in the field lines. The radius of this curvature is given by

$$R = \left| \frac{\nabla_\perp B}{B} \right|^{-1}. \tag{2-127}$$

A particle moving along a curved field will experience a *centripetal* acceleration given by $u_\parallel^2/R$. The corresponding *centrifugal* force that contributes to the drift is

$$F_\perp = -mu_\parallel^2 \frac{\nabla_\perp B}{B}.$$

(2-128)

The net drift velocity for particles in a magnetic field having a transversal inhomogeneity is therefore

$$w = \frac{1}{\omega_c B^2}\left(u_\parallel^2 + \frac{u_\perp^2}{2}\right)(B \times \nabla_\perp B).$$

(2-129)

If the variation of the magnetic field in space is sufficiently gradual such that the cyclotron radius is much less than the scale length of the variation, say

$$\rho_c^{-1} \gg \left|\frac{\nabla B}{B}\right|,$$

(2-130)

then a quantity known as the *magnetic moment* μ_m of the charged particle will be *nearly* a constant of the motion.[4] The magnetic moment of an electrical current I that encloses an area A is IA. Since the current generated by an orbiting charge is

$$I = \frac{q\omega_c}{2\pi},$$

(2-131)

where ω_c is the cyclotron frequency (an angular frequency, hence the factor of 2π), then

$$\mu_m = I\pi\rho_c^2 = \frac{q\omega_c}{2\pi} \cdot \pi \cdot \frac{u_\perp^2}{\omega_c^2} = \frac{\frac{1}{2}mu_\perp^2}{B}.$$

(2-132)

Suppose that we examine the nature of particle motion in a magnetic field that also has a longitudinal gradient

$$B = B_\rho(\rho, z)\hat{\rho} + B_z(\rho, z)\hat{k}.$$

(2-133)

One of the laws of electromagnetism states that the magnetic field is solenoidal. So

$$\nabla \cdot B = 0.$$

(2-134)

Expanding Eq. (2-134) in cylindrical coordinates, we find

$$\frac{1}{\rho}\frac{\partial}{\partial\rho}(\rho B_\rho) + \frac{\partial B_z}{\partial z} = 0.$$

(2-135)

---

[4] A general proof of this statement and a discussion of this class of "almost" constants of the motion known as *adiabatic invariants* will be forthcoming in Chapter 3.

In order to integrate this equation and satisfy the condition given in Eq. (2-130) for the invariance of μ_m, we will assume that $\partial B_z/\partial z$ is constant with respect to ρ over the cross section of the particle orbit and that

$$\frac{\partial B_z}{\partial z} \approx \frac{\partial B}{\partial z} = \nabla_\parallel B. \tag{2-136}$$

Equation (2-135) can then be integrated to yield

$$B_\rho = -\frac{\rho}{2}\nabla_\parallel B. \tag{2-137}$$

In this approximation we take the parallel direction to refer to the average direction of $\boldsymbol{B}$—namely, the z direction—rather than the local direction of $\boldsymbol{B}$. Thus the parallel component of the equation of motion is

$$\frac{d\boldsymbol{u}_\parallel}{dt} = \frac{q}{m}(\boldsymbol{u}\times\boldsymbol{B})_\parallel = \frac{q}{m}u_\perp B_\rho\hat{\boldsymbol{k}}.$$

Letting $\rho = \rho_c = u_\perp/\omega_c$ in Eq. (2-137) and substituting from this equation for B_ρ into the equation of motion, we find

$$m\frac{d\boldsymbol{u}_\parallel}{dt} = -\mu_m\,\boldsymbol{\nabla}_\parallel B. \tag{2-138}$$

So a charged particle that moves along the magnetic field into a region of higher B will find itself repelled. Since $\mu_m = \frac{1}{2}mu_\perp^2/B$ stays constant, energy must be transferred from longitudinal motion to transverse motion as B increases. The rate of this increase is simply

$$\boldsymbol{F}_\parallel\cdot\boldsymbol{u}_\parallel = -\mu_m\boldsymbol{u}_\parallel\cdot\boldsymbol{\nabla}_\parallel B = -\frac{d}{dt}(\tfrac{1}{2}mu_\parallel^2). \tag{2-139}$$

Since the total energy is limited to

$$T_0 = \tfrac{1}{2}m(u_\parallel^2 + u_\perp^2), \tag{2-140}$$

the particle must come to a halt in its longitudinal motion when the magnetic field reaches the value

$$B \to \frac{T_0}{\mu_m}. \tag{2-141}$$

It is then reflected and begins to gain longitudinal motion in the opposite direction. This process is known as the *magnetic mirror* effect. A system using two electrical coils to generate a magnetic mirror configuration, as shown in Fig. 2-13, was one of the earliest schemes envisioned to trap a superheated gas of charged particles (known as a *plasma*) in order to isolate it from contact with material walls.

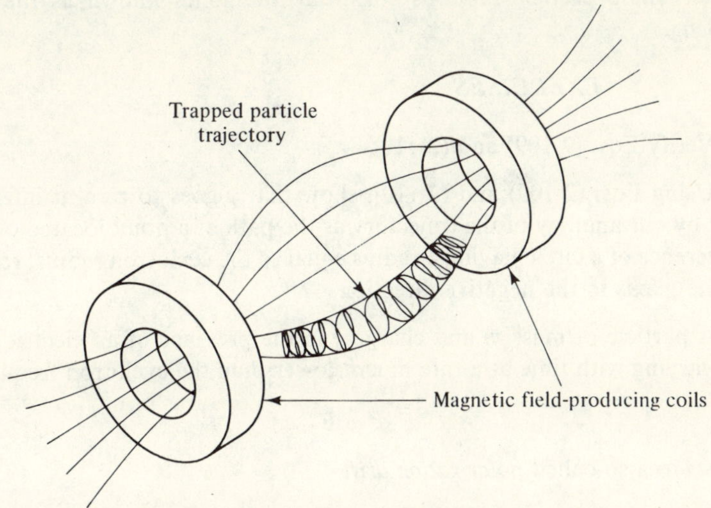

Fig. 2-13 A magnetic mirror.

Particles from the sun impinge on the magnetic field of the earth, which has the approximate configuration shown in Fig. 2-14. The particles become trapped in an oscillation back and forth between mirror points in the high latitudes. Those particles that enter the mirror system with a high enough value of $\frac{1}{2}mu_\parallel^2$ such that the field cannot reflect them will be dumped into the top of the atmosphere at these latitudes. They collide with atmospheric molecules and excite them, thereby producing the auroral displays. The curvature of the field lines leads to a transversal inhomogeneity drift that causes positively charged particles to move around the earth in a westerly

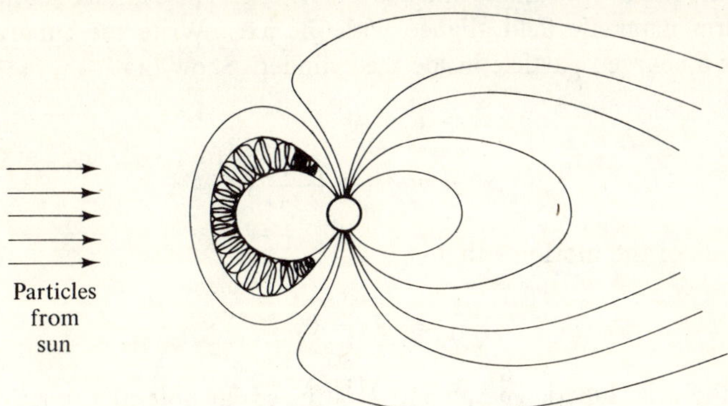

Fig. 2-14 Trapping of high-energy particles from the sun in the geomagnetic field.

75

direction. These particle motions constitute the zones known as the *Van Allen Belts*.

EXERCISES

2-6.1. Verify Eqs. (2-109) and (2-110).

2-6.2. Using Eqs. (2-109) and (2-110), show that y goes to zero at intervals implied by our analogy of the trajectory as the path of a point located on the circumference of a circle having a radius equal to the cyclotron radius, rolling along the y axis in the negative direction.

2-6.3. A particle of mass m and charge q in the presence of an electric field that is varying with time at a rate much slower than the cyclotron frequency

$$\dot{E} \ll \omega_c E$$

is subject to a so-called *polarization drift*

$$w_p = \frac{m\dot{E}_\perp}{qB^2} \tag{2-142}$$

in addition to the electric drift. If the particle is at rest in a uniform magnetic field and an electric field is switched on and rises to its final value E at a sufficiently slow rate, show that the energy acquired by the particle due to polarization drift simply accounts for the electric drift energy $\frac{1}{2}m(E_\perp/B)^2$.

2-6.4. An electric field is established between a cylindrical metal shell and a fine wire drawn along its axis:

$$E = \frac{\alpha}{\rho}\,\hat{\rho},$$

where ρ is the radial coordinate and α is a constant. The cylinder is immersed in a uniform magnetic field aligned with the axis. Write the equations of motion for a charged particle inside the cylinder. Show that

$$\mathscr{E} = T + qV$$

and
$$\mathscr{L} = m\rho^2\dot{\theta} + \frac{qB\rho^2}{2}$$

are constants of the motion—that is,

$$\frac{d\mathscr{E}}{dt} = \frac{d\mathscr{L}}{dt} = 0.$$

Here T is the total kinetic energy and V is the scalar potential function that described the electric field

$$E = -\nabla V.$$

2-6.5. A charged particle acquires an energy qV in falling through a potential V. Referring to Fig. 2-11, show that the total kinetic energy acquired by a particle of charge q in a crossed field configuration starting from rest at the origin is

$$|qV| = qxE = m\left(\frac{E_\perp}{B}\right)^2. \tag{2-143}$$

Reconcile this result with that obtained in problem 2-6.3.

2-6.6. A long straight wire carries current along the z axis of a cylindrical coordinate system. The magnetic field established by this current is

$$\mathbf{B} = \frac{\beta}{\rho}\,\hat{\mathbf{\theta}},$$

where β is a constant, ρ is the radial coordinate, and $\hat{\mathbf{\theta}}$ is the unit vector in the azimuthal direction. A positively charged particle is injected into the system with a kinetic energy T along a line parallel to the z axis and at a distance s from the wire. Calculate the drift motion of the particle.

2-6.7. Show that

$$\frac{d}{dt}\left(\tfrac{1}{2}mu_\parallel^2\right) = -\mu_m\frac{dB}{dt} \tag{2-144}$$

for a particle in a mirror field.

2-6.8. Using Eq. (2-144) and the law of conservation of energy, show that the magnetic moment is an invariant—that is,

$$\frac{d\mu_m}{dt} = 0. \tag{2-145}$$

2-6.9. A charged particle is injected into a magnetic mirror, such as that shown in Fig. 2-13, along a line that makes an angle with the axis of symmetry. If the magnetic field at the point midway between the coils that we take to be the point of injection is denoted by B_0 and that under the throat of the coils is B_m, show that, in order to be confined within the system, the injection angle must obey the condition

$$\theta < \arcsin\sqrt{\frac{B_0}{B_m}}. \tag{2-146}$$

2-7 The Principle of Superposition

Mechanical problems involving time-dependent forces are of such importance that a significant portion of the following chapter will be devoted to a specific example, the driven harmonic oscillator. In this section we consider the treatment of problems involving arbitrary time-dependent forces in a most general way.

A most useful theorem called the *principle of superposition* applies to all differential equations that are linear in their dependent variable and its derivatives. This theorem can be stated in the following way. Let there be a set of functions $x_k(t)$, where $k = 1, 2, 3, \ldots, n$ and n may be finite or infinite. These functions are all solutions of the linear differential equation

$$\mathscr{L}x_k(t) = F_k(t), \tag{2-147}$$

where $\mathscr{L}$ is a general linear differential operator.

Let

$$F(t) = \sum_{k=1}^{n} F_k(t). \tag{2-148}$$

The principle of superposition then states that the function

$$x(t) = \sum_{k=1}^{n} x_k(t) \tag{2-149}$$

will satisfy

$$\mathscr{L}x(t) = F(t). \tag{2-150}$$

In other words, the sum of solutions to any linear differential equation is itself a solution. This statement is easily proved. Operating on Eq. (2-149) with $\mathscr{L}$, we find

$$\mathscr{L}x(t) = \mathscr{L} \sum_{k} x_k(t)$$

$$= \sum_{k} \mathscr{L}x_k(t)$$

$$= \sum_{k} F_k(t)$$

$$= F(t),$$

where the linear differential operator $\mathscr{L}$ and the summation operator commute and Eqs. (2-147) and (2-148) have been used. This is a very important theorem that enables us to solve problems involving time-dependent forces that can be expressed as a sum of forces $F_k(t)$, for each of which solutions to the problem can be found.

An important special case arises when the force is periodic—that is,

$$F(t + \tau) = F(t), \tag{2-151}$$

where τ is the period of the force. In general, this force is not sinusoidal; it can, however, be represented as the sum of sine and cosine terms with periods of τ, $\tau/2$, $\tau/3$, etc.—that is, all possible sine and cosine terms having periods that are a whole fraction of τ, for example,

$$F(t) = A_0 + \sum_{k=1}^{\infty} \left[A_k \cos\left(\frac{2\pi kt}{\tau}\right) + B_k \sin\left(\frac{2\pi kt}{\tau}\right) \right], \qquad (2\text{-}152)$$

where A_0, A_k, and B_k are constants to be determined. This is known as *Fourier series* representation. We define

$$\omega_k = \frac{2\pi k}{\tau}. \qquad (2\text{-}153)$$

Using the Euler identity

$$e^{\pm i\phi} = \cos\phi \pm i\sin\phi, \qquad (2\text{-}154)$$

where $i = \sqrt{-1}$, we can rewrite Eq. (2-152) as

$$F(t) = A_0 + \sum_{k=1}^{\infty} \left[\frac{A_k}{2}(e^{i\omega_k t} + e^{-i\omega_k t}) - \frac{iB_k}{2}(e^{i\omega_k t} - e^{-i\omega_k t}) \right]$$

$$= A_0 + \sum_{k=1}^{\infty} \tfrac{1}{2}[(A_k - iB_k)e^{i\omega_k t} + (A_k + iB_k)e^{-i\omega_k t}]$$

$$= \sum_{k=-\infty}^{\infty} C_k e^{i\omega_k t}. \qquad (2\text{-}155)$$

Here, for $k > 0$, we have redefined the real constants A_0, A_k, B_k in terms of a complex constant C_k such that

$$C_{\pm k} = \tfrac{1}{2}(A_k \mp iB_k),$$
$$C_0 = A_0. \qquad (2\text{-}156)$$

The problem now consists of determining C_k for a given function $F(t)$. Fortunately, doing so is a simple matter. Multiply Eq. (2-155) by $e^{-2\pi ijt/\tau}$, where j is an index analogous to k, and integrate with respect to t over the limits $-\tau/2$ to $+\tau/2$:

$$\int_{-\tau/2}^{\tau/2} F(t)e^{-2\pi ijt/\tau}\, dt = \int_{-\tau/2}^{\tau/2} \sum_{k=-\infty}^{\infty} e^{-2\pi ijt/\tau} C_k e^{2\pi ikt/\tau}\, dt. \qquad (2\text{-}157)$$

Assuming the ability of the summation and integration operations to commute, we have a sum of integrals of a simple sinusoid over its period or some integral multiple thereof. Such integrals are equal to zero for any $j \neq k$. The only nonzero integral in the sum occurs when $j = k$ for which the right-hand side of Eq. (2-157) reduces to

$$\int_{-\tau/2}^{\tau/2} C_j\, dt = \tau C_j. \qquad (2\text{-}158)$$

Thus

$$C_j = \frac{1}{\tau} \int_{-\tau/2}^{\tau/2} F(t)e^{-2\pi i j t/\tau}\, dt. \tag{2-159}$$

If this integral exists, as it does for any periodic function of practical interest, the problem is solved, since the solution of linear equations with purely sinusoidal right-hand sides is a simple matter.

Let us extend this method to *nonperiodic* functions. Returning to our definition of ω_k given in Eq. (2-153) and denoting the difference between successive values of ω_k as $\delta\omega = 2\pi/\tau$, we can write Eq. (2-159) as

$$C_k = \frac{\delta\omega}{2\pi} \int_{-\tau/2}^{\tau/2} F(t')e^{-i\omega_k t'}\, dt', \tag{2-160}$$

where t' is a dummy variable introduced to avoid conflict in the following equations. Substituting Eq. (2-160) for C_k into Eq. (2-155), we have

$$F(t) = \frac{1}{2\pi} \sum_{k=-\infty}^{\infty} \left[e^{i\omega_k t}\, \delta\omega \int_{-\tau/2}^{\tau/2} F(t')e^{-i\omega_k t'}\, dt' \right]. \tag{2-161}$$

In order to generalize for nonperiodic functions, we let the period $\tau \to \infty$. Then

$$\delta\omega = \frac{2\pi}{\tau} \to 0. \tag{2-162}$$

The summation thus becomes an integral as the difference in successive frequencies approaches zero, and Eq. (2-161) goes over into

$$F(t) = \frac{1}{2\pi} \int_{-\infty}^{\infty} e^{i\omega t}\, d\omega \int_{-\infty}^{\infty} F(t')e^{-i\omega t'}\, dt'. \tag{2-163}$$

Next, we can define

$$\mathscr{F}(\omega) = \frac{1}{\sqrt{2\pi}} \int_{-\infty}^{\infty} F(t)e^{-i\omega t}\, dt \tag{2-164}$$

such that Eq. (2-163) becomes

$$F(t) = \frac{1}{\sqrt{2\pi}} \int_{-\infty}^{\infty} \mathscr{F}(\omega)e^{i\omega t}\, d\omega. \tag{2-165}$$

The function $\mathscr{F}(\omega)$, known as the *Fourier transform* of $F(t)$, is an amplitude function describing the distribution of frequencies in the continuous range of sinusoids that make up $F(t)$.

Another method of solving linear equations having right-hand sides that are arbitrary functions of t is the *Green's function* method. In Fig. 2-15 we have decomposed the force $F(\tau)$ represented by the curve drawn as a function of τ in terms of a large number of impulses of duration $\Delta\tau$:

$$F(\tau)\Delta\tau \equiv \text{impulse}. \tag{2-166}$$

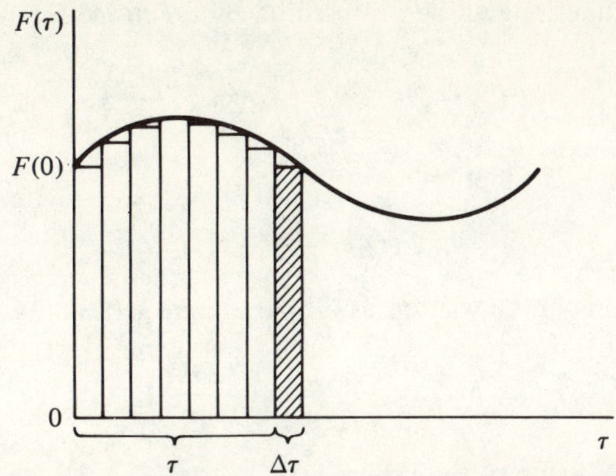

Fig. 2-15 Decomposition of an impulse of arbitrary shape into a sum of rectangular impulses.

We assume that $F(\tau)$ vanishes for $\tau < 0$. The problem described by the equation

$$\mathscr{L}x(t) = F(t) \tag{2-167}$$

is now recast by defining a Green's function $G(t, \tau)$ as the response of the system at time t to a *unit* impulse delivered at time τ, where

$$0 < \tau < t. \tag{2-168}$$

The response $x(t)$ to all the impulses applied from time $\tau = 0$ to $\tau = t$ is

$$x(t) = \sum_{\tau=0}^{t} F(\tau)\,\Delta\tau\, G(t, \tau). \tag{2-169}$$

If the excitation is a continuous function, we can pass to the limit $\Delta\tau \to 0$, and

$$x(t) = \int_{0}^{t} F(\tau)G(t, \tau)\, d\tau. \tag{2-170}$$

The problem now becomes the determination of the Green's function G for the specified differential operator $\mathscr{L}$, plus the solution of the integral of Eq. (2-170). The reader is referred to a text on integral equations for the details of such solutions.

EXERCISES

2-7.1. If $F(t)$ is an *even* function

$$F(t) = F(-t), \tag{2-171}$$

show that the function can be represented by a *Fourier cosine series*

$$F(t) = A_0 + \sum_{k=1}^{\infty} A_k \cos\left(\frac{2\pi kt}{\tau}\right),\tag{2-172}$$

and if $F(t)$ is an *odd* function

$$F(t) = -F(-t),\tag{2-173}$$

then the function can be written as a *Fourier sine series*

$$F(t) = \sum_{k=1}^{\infty} B_k \sin\left(\frac{2\pi kt}{\tau}\right).\tag{2-174}$$

2-7.2. Verify the value of the integral

$$\int_{-\pi}^{\pi} e^{-int} e^{imt}\, dt = 2\pi\delta_{nm},\tag{2-175}$$

where n and m are integers and

$$\delta_{nm} = \begin{cases} 0, & n \neq m \\ 1, & n = m \end{cases}\tag{2-176}$$

is the *Kronecker delta.* This relation is the *orthogonality condition* for circular functions.

2-7.3. Expand the function $F(t) = t$ in a Fourier series over the interval $-\tau/2 \le t \le \tau/2$ to show that

$$F(t) = 2\sum_{n=1}^{\infty} \frac{(-1)^{n+1}}{n}\sin nt.$$

Plot the sum of the first three terms and compare with the original function.

2-7.4. If $\mathscr{F}(\omega)$ is the Fourier transform of $F(t)$, show that $-i\omega\mathscr{F}(\omega)$ is the Fourier transform of dF/dt.

2-7.5. Show that

$$G(t, \tau) = \begin{cases} t, & 0 \le t < \tau \\ \tau, & \tau < t \le 1 \end{cases}$$

is the appropriate Green's function for the differential operator $\mathscr{L} = d^2/dt^2$ and the boundary conditions

$$x(0) = 0$$

$$\dot{x}(0) = 0.$$

REFERENCES

ARFKEN, GEORGE, *Mathematical Methods for Physicists* (2nd ed.). New York: Academic Press, 1970. Fourier series and integral transformations are covered in Chapters 14 and 15, respectively. Integral equations and Green's functions are discussed in Chapter 16.

PRANDTL, L., and O. G. TIETJENS, *Applied Hydro- and Aeromechanics*. New York: Dover Publications, Inc., 1957. Chapter 5 presents a comprehensive discussion of viscous drag.

SCHMIDT, GEORGE, *Physics of High Temperature Plasmas*. New York: Academic Press, 1966. Chapter 2 presents a detailed development of the motion of a charged particle in an electromagnetic field.

OSCILLATIONS AND WAVES

One of the most common types of motion of physical systems is *harmonic oscillation.* Examples of this kind of motion include the behavior of a mass driven by a spring that is alternately compressed and extended, small oscillations of a pendulum, and the motions of charges in electrical circuits containing inductive and capacitive elements. As we shall see when we come to study the stability of equilibria, this type of oscillatory motion is approximated by *all* systems in stable equilibrium, static or dynamic, when given a *small* displacement from equilibrium. The common denominator in all such cases is the presence of a *restoring force* that is a linear function of the displacement.

Linear wave motion is an extension of the notion of oscillations in time to include a *propagation* of the oscillation. In this chapter we examine linear oscillator motion and wave propagation in strings under tension and also discuss large-amplitude nonlinear oscillations and waves from a general point of view.

3-1 The Undamped Simple Harmonic Oscillator

Let us begin with a simple pendulum moving in a vertical plane under the action of a uniform gravitational field. This system consists of a point mass m suspended from a fixed point by a massless string of length l, free to move in a plane about its equilibrium position as shown in Fig. 3-1. The forces acting on the mass are the tension in the string T and the force of gravity mg. Because our only concern is with motion along the arc of a circle,

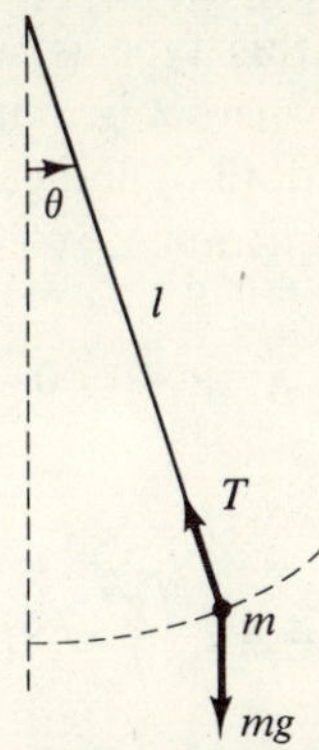

Fig. 3-1 A simple pendulum.

it is most convenient to use polar coordinates to describe the angular motion. And since $\dot{\rho} = 0$, we have

$$ml\ddot{\theta} = -mg \sin \theta. \qquad (3\text{-}1)$$

This equation is not an easy one to solve and the motion it describes is not simple harmonic, for the right-hand term is not a linear function of θ. So we must make a small-angle approximation. Expanding the sine function in a Maclaurin series, we find[1]

$$\sin \theta = \theta - \frac{\theta^3}{3!} + \frac{\theta^5}{5!} - \cdots. \qquad (3\text{-}2)$$

Retaining only the lowest-order term, Eq. (3-1) becomes in the small displacement limit

$$\ddot{\theta} + \frac{g}{l}\,\theta = 0. \qquad (3\text{-}3)$$

---

[1] The reader who is not already familiar with power series expansions of such functions as $(1 + x)^n$, $\sin x$, $\cos x$ should acquaint himself with the more commonly encountered ones. The Taylor expansion of the function $f(x)$ about the point x_0 is

$$f(x) = f(x_0) + \frac{df}{dx}\bigg|_{x=x_0} (x - x_0) + \frac{d^2f}{dx^2}\bigg|_{x=x_0} \frac{(x - x_0)^2}{2!} + \frac{d^3f}{dx^3}\bigg|_{x=x_0} \frac{(x - x_0)^3}{3!} + \cdots$$

$$= \sum_{n=0}^{\infty} \frac{f^{(n)}(x_0)}{n!} (x - x_0)^n.$$

If $x_0 = 0$, the series is known as a *Maclaurin series* and

$$f(x) = \sum_{n=0}^{\infty} \frac{f^{(n)}(0)}{n!} x^n.$$

This is a second-order linear homogeneous equation with constant coefficients. The solution to equations of this type is usually found by assuming an exponential form, say $\theta = e^{\lambda t}$, where λ is a quantity to be determined. Substituting this assumption into Eq. (3-3) and carrying out the indicated operations, we obtain the *auxiliary equation*

$$\lambda^2 + \frac{g}{l} = 0 \tag{3-4}$$

from which we find for λ

$$\lambda_{1,2} = \pm i\sqrt{\frac{g}{l}} = \pm i\omega_0, \tag{3-5}$$

where $i = \sqrt{-1}$ and we have introduced the notation $\omega_0^2 = g/l$. Since the equation is a linear one, the most general solution is given by the sum

$$\theta = Ae^{\lambda_1 t} + Be^{\lambda_2 t} = Ae^{i\omega_0 t} + Be^{-i\omega_0 t}, \tag{3-6}$$

where A and B are arbitrary constants. Using the Euler relation given in Eq. (2-154), we find

$$\theta = (A + B)\cos\theta + i(A - B)\sin\theta$$

$$= C\cos\theta + D\sin\theta. \tag{3-7}$$

Although the exponential solution is convenient for some purposes, it has the disadvantage that A and B are complex even though, as we shall show for this case, C and D are not. A more straightforward, if less general, method for solving Eq. (3-3) is to multiply by an integration factor $2\dot\theta$. Then

$$2\dot\theta\ddot\theta = -2\omega_0^2\dot\theta\theta,$$

which integrates to

$$\dot\theta^2 = -\omega_0^2\theta^2 + \alpha = \omega_0^2(\beta^2 - \theta^2),$$

where the integration constant α has been redefined: $\alpha = \omega_0^2\beta^2$. Taking the square root and separating variables, we have

$$\frac{d\theta}{\sqrt{\beta^2 - \theta^2}} = \omega_0\,dt, \tag{3-8}$$

which we integrate directly to find

$$\theta = \beta\sin(\omega_0 t + \phi). \tag{3-9}$$

In this equation β and ϕ are our integration constants. For many purposes, Eq. (3-9) is a convenient form for the solution. The answer can also be expressed in the form of Eq. (3-7). Using the familiar trigonometric identity for the sine of the sum of two angles,

$$\theta = \beta \sin (\omega_0 t + \phi)$$

$$= \beta \sin \omega_0 t \cos \phi + \beta \cos \omega_0 t \sin \phi$$

$$= C \cos \omega_0 t + D \sin \omega_0 t, \tag{3-10}$$

where

$$C = \beta \sin \phi \tag{3-11}$$

and

$$D = \beta \cos \phi \tag{3-12}$$

are real.

Since the gravitational force is conservative, the motion of the pendulum can also be analyzed from an energy point of view with the knowledge that the total energy E is a constant of the motion. We write the total energy of the pendulum as the sum of the kinetic and potential energies

$$T + V = \tfrac{1}{2}ml^2\dot{\theta}^2 + mgl(1 - \cos \theta) = E. \tag{3-13}$$

It is left for the reader to show that this equation is equivalent to Eq. (3-1). In terms of our solution, Eq. (3-9), the kinetic energy is

$$T = \tfrac{1}{2}ml^2\dot{\theta}^2 = \tfrac{1}{2}ml^2\omega_0^2\beta^2 \cos^2 (\omega_0 t + \phi). \tag{3-14}$$

The potential energy, defined to be zero at $\theta = 0$, is

$$V = -\int_0^\theta Fl d\theta, \tag{3-15}$$

where the restoring force acting in the minus θ direction is $-ml\omega_0^2\theta$ in our linearized approximation. Thus

$$V = \int_0^\theta ml\omega_0^2\theta l \, d\theta = \tfrac{1}{2}ml^2\omega_0^2\theta^2$$

$$= \tfrac{1}{2}ml^2\omega_0^2\beta^2 \sin^2 (\omega_0 t + \phi). \tag{3-16}$$

Accordingly, the total energy E is

$$E = T + V = \tfrac{1}{2}ml^2\omega_0^2\beta^2[\cos^2 (\omega_0 t + \phi) + \sin^2 (\omega_0 t + \phi)]$$

$$= \tfrac{1}{2}m(l\omega_0\beta)^2, \tag{3-17}$$

a constant, proportional both to the square of the amplitude and the square of the frequency. Clearly, the time average value of the total energy $\langle E \rangle$ is simply equal to E. If we take τ_0 to be the *period* of the oscillation, then the time average value of the kinetic energy is

$$\langle T \rangle = \frac{\frac{1}{2}m(l\omega_0\beta)^2}{\tau_0} \int_0^{\tau_0} \cos^2(\omega_0 t + \phi)\, dt$$

$$= \tfrac{1}{4}m(l\omega_0\beta)^2 \tag{3-18}$$

and

$$\langle V \rangle = \frac{\frac{1}{2}m(l\omega_0\beta)^2}{\tau_0} \int_0^{\tau_0} \sin^2(\omega_0 t + \phi)\, dt$$

$$= \tfrac{1}{4}m(l\omega_0\beta)^2. \tag{3-19}$$

From Eqs. (3-14) and (3-16) we see that when T is a maximum and equal to E, $V = 0$ and vice versa.

EXERCISES

3-1.1. The error x entailed in truncating the expansion of $\sin\theta$ at the first-order term is approximately given by the difference in the first- and third-order approximations divided by the third-order approximation. Using this prescription, calculate the maximum angles in degrees such that $x = 0.1\%, 0.5\%, 1\%$, and 5%.

3-1.2. Using Eq. (3-13) as the starting point, derive Eq. (3-1).

3-1.3. Show that the second-order approximation to the potential energy term in Eq. (3-13)

$$V = mgl(1 - \cos\theta)$$

agrees with the potential energy calculated in Eq. (3-16)

$$V = \tfrac{1}{2}m(l\omega_0\theta)^2.$$

3-1.4. Assuming that the pendulum string of length l is replaced by a rigid massless rod of the same length, calculate the energy that must be imparted to the pendulum bob in order for it to be able just to swing in a complete circle. If we go back to the string again, how much energy is required to keep the bob in a circle without "dropping out"?

3-1.5. A pendulum that normally has a period τ_0 is found to have a period shorter by 1% when operated in the boxcar of a train moving at speed u on a circular track. What is the radius of the track?

3-1.6. A horizontal surface oscillates horizontally along the x axis, moving back and forth through a distance of 30 cm with a period of 5 sec. What is the least coefficient of static friction between the oscillating surface and a massive block placed on it such that the block does not slip?

3-1.7. A slingshot consists of a strip of rubber having an unstretched length L and spring constant k, the two ends of which are tied to the prongs for a forked stick at a distance l apart, $l < L$. A rock of mass m is placed in the middle of the rubber strip and drawn back until the rubber is stretched to a length λL. The sling is then released. Find the speed with which the rock leaves the slingshot.

3-1.8. One end of a perfectly elastic, massless hacksaw blade is embedded in a wall and the other end is free to oscillate in the vertical plane as shown in Fig. 3-2. It is found that a 500-g mass placed on the blade tip leads to a 2-cm deflection. We drop a lump of putty having a mass of 750 g from a height of 20 cm onto the blade tip, the hacksaw blade being initially at rest in the horizontal plane. The putty sticks to the blade tip with no energy loss. Derive an equation describing the motion of the blade tip.

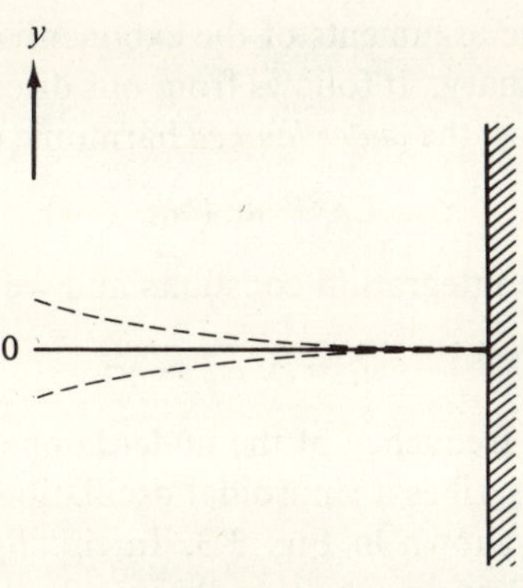

Fig. 3-2

3-2 The Damped Harmonic Oscillator

Macroscopic mechanical processes are never entirely friction free; all pendulums eventually run down. It is therefore of practical interest to consider some cases of harmonic motion in which a damping force proportional to the velocity is present. Our equation of motion is given by

$$m\ddot{x} + b\dot{x} + kx = 0, \tag{3-20}$$

where x may refer to the linear coordinate describing the motion of a mass on a spring or the angular motion of a pendulum; b and k are constant coefficients that describe the strength of the damping and restoring forces, respectively.

The exponential method of solution is the most convenient one here. We substitute $x = e^{\lambda t}$ and find for the auxiliary equation

$$m\lambda^2 + b\lambda + k = 0, \tag{3-21}$$

which has as its solutions

$$\lambda_{1,2} = -\frac{b}{2m} \pm \left[\left(\frac{b}{2m}\right)^2 - \frac{k}{m}\right]^{1/2} = -\gamma \pm \sqrt{\gamma^2 - \omega_0^2}. \tag{3-22}$$

The quantity γ is known as the *damping coefficient*, and ω_0 will be recognized as the angular frequency of the corresponding undamped oscillator. In these terms, the general solution to Eq. (3-20) is

$$x = e^{-\gamma t}[Ae^{\sqrt{\gamma^2 - \omega_0^2}\,t} + Be^{-\sqrt{\gamma^2 - \omega_0^2}\,t}]. \tag{3-23}$$

The quantities γ^2 and ω_0^2 are positive definite; thus we can distinguish three cases of interest, depending on whether $(\gamma^2 - \omega_0^2)$ is less than, greater than, or equal to zero.

If $\gamma^2 < \omega_0^2$, then the arguments of the exponentials in the bracketed term of Eq. (3-23) are imaginary. It follows from our discussion in the first section that the solution for this, the *underdamped* harmonic oscillator, can be written

$$x = Ce^{-\gamma t}\sin(\omega_1 t + \phi), \tag{3-24}$$

where C and ϕ are the integration constants and we have defined

$$\omega_1 = \sqrt{\omega_0^2 - \gamma^2} \tag{3-25}$$

as the natural angular frequency of the underdamped system.

Equation (3-24) describes a sinusoidal oscillation bounded by the curves $Ce^{-\gamma t}$ and $-Ce^{-\gamma t}$ as shown in Fig. 3-3. In this figure we have taken the *phase angle ϕ* to be $\pi/2$ as an arbitrary choice. So the solid curve represents

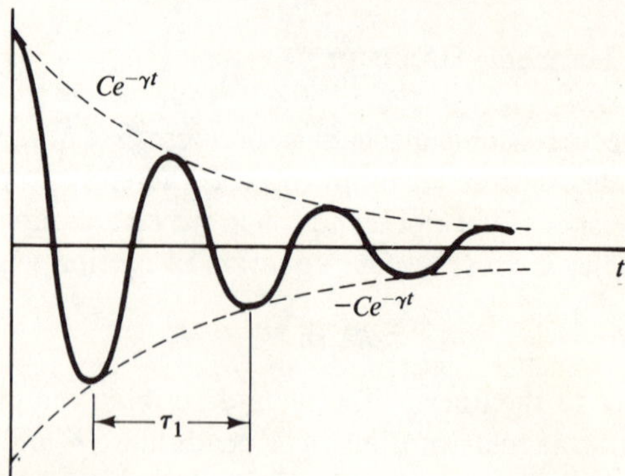

Fig. 3-3 The underdamped harmonic oscillator.

the function $Ce^{-\gamma t} \cos \omega_1 t$. Clearly, this curve will only contact the exponential envelopes at points where $\cos \omega_1 t = \pm 1$. In terms of t, these points are given by

$$t_n = \frac{n\pi}{\omega_1}, \qquad n = 0, 1, 2, \ldots, \tag{3-26}$$

and are separated by half a period $\tau_1/2$. The maxima and minima of the function

$$x = Ce^{-\gamma t} \cos \omega_1 t \tag{3-27}$$

do not occur exactly at these points. It is left as an exercise to show that the maxima and minima occur at times given by

$$t = \frac{1}{\omega_1} \arctan\left(-\frac{\gamma}{\omega_1}\right), \tag{3-28}$$

which are also separated by $\pi/\omega_1 = \tau_1/2$.

The ratio between any two successive maxima is given by

$$\frac{Ce^{-\gamma t}}{Ce^{-\gamma(t+\tau_1)}} = e^{\gamma\tau_1}. \tag{3-29}$$

The natural logarithm of this ratio is called the *logarithmic decrement* of the motion and is given by

$$\delta = \gamma\tau_1 = \frac{b\pi}{m\omega_1}. \tag{3-30}$$

The energy of a damped harmonic oscillator is, of course, not constant. The damping term is dissipative in nature. If the damping coefficient $\gamma \ll \omega_0$ such that the undamped natural frequency ω_0 is nearly equal to ω_1, we can then use the findings of Eq. (3-17) to express the energy of the lightly damped system as

$$E \approx \tfrac{1}{2}kC^2 e^{-2\gamma t} \tag{3-31}$$

and we see that the energy decays exponentially at twice the rate at which the amplitude decays.

Consider next the *overdamped* case where $\gamma^2 > \omega_0^2$ and the arguments of the exponential functions are real. In this case, the solution may be written

$$x = Ae^{-\gamma_1 t} + Be^{-\gamma_2 t}, \tag{3-32}$$

where A and B are the integration constants and

$$\gamma_{1,2} = -\gamma \pm \sqrt{\gamma^2 - \omega_0^2}. \tag{3-33}$$

Since the sum of monotonically decaying exponentials of different argument and amplitude is not necessarily monotonic, it is often more convenient to express such a solution in terms of hyperbolic functions. The relationship between these functions and exponentials with real arguments is quite analogous to the relationship between the circular functions and exponentials of imaginary argument. So we can write for the overdamped oscillator (which does not oscillate)

$$
\begin{aligned}
x &= e^{-\gamma t}(Ae^{\alpha t} + Be^{-\alpha t}) \\
&= e^{-\gamma t}(A \cosh \alpha t + A \sinh \alpha t + B \cosh \alpha t - B \sinh \alpha t) \\
&= e^{-\gamma t}[(A + B) \cosh \alpha t + (A - B) \sinh \alpha t] \\
&= e^{-\gamma t}(C \cosh \alpha t + D \sinh \alpha t),
\end{aligned}
\tag{3-34}
$$

where we define $\alpha = \sqrt{\gamma^2 - \omega_0^2}$. It is left as an exercise for the reader to show that this solution can also be written

$$
x = \beta e^{-\gamma t} \sinh (\alpha t + \phi),
\tag{3-35}
$$

where β and ϕ are integration constants playing the role of amplitude and phase angle, respectively.

Finally, we examine the singular case of *critical damping* where $\gamma^2 = \omega_0^2$. As we can see from Eq. (3-23), only one solution results:

$$
x = Ae^{-\gamma t}.
\tag{3-36}
$$

It is left as an exercise to show that another solution is

$$
x = Bte^{-\gamma t}.
\tag{3-37}
$$

Such being the case, the general solution for the critically damped oscillator is

$$
x = (A + Bt)e^{-\gamma t}.
\tag{3-38}
$$

This equation contains three possible situations of interest, depending on the magnitudes of the constants A and B. These situations are revealed by differentiating Eq. (3-38) with respect to time and setting the resulting equation equal to zero:

$$
\dot{x} = -\gamma(A + Bt)e^{-\gamma t} + Be^{-\gamma t} = 0
$$

from which we find for t

$$
t = \frac{1}{\gamma} - \frac{A}{B}.
\tag{3-39}
$$

If $A/B = 1/\gamma$, the curve described by Eq. (3-38) has a maximum at $t = 0$. If $A/B < 1/\gamma$, the maximum appears at some positive value of t given by Eq. (3-39). If $A/B > 1/\gamma$, then Eq. (3-38) possesses no maxima for any positive value of t. The case of critical damping is of practical interest to designers of meter movements, pneumatic returns for doors, capacitive electrical discharge circuits, and so on, since it is a feature of this condition that the mechanical or electrical system in question is returned to equilibrium in the shortest time without overshoot.

3-2.1. Show that the times corresponding to maxima and minima of the curve of an underdamped harmonic oscillator described by

$$x = Ce^{-\gamma t} \cos \omega_1 t$$

are given by Eq. (3-28). If $\gamma = x\omega_0$, where $x < 1$, calculate the location of the first maximum in terms of the angle $\omega_1 t$ for $x = 0.1$ and 0.5.

3-2.2. The frequency of an underdamped harmonic oscillator is adjusted to be equal to half the frequency experienced by the oscillator without damping. What is the logarithmic decrement of this system?

3-2.3. A mass m is subject to a linear restoring force $-kx$ and a damping force $-b\dot{x}$. This mass is given an initial displacement x_0 and an initial velocity $-u_0$ directed back toward the equilibrium point. Find the maximum value of $|u_0|$ such that the mass does not overshoot the equilibrium point in the critically damped and overdamped cases.

3-2.4. Show that the solution to the overdamped harmonic oscillator can be written in the form given in Eq. (3-35). Using this equation, evaluate the constants β and ϕ for the following sets of initial conditions:

(a) $x(0) = 0$,

   $\dot{x}(0) = u_0$.

(b) $x(0) = x_0$,

   $\dot{x}(0) = 0$.

3-2.5. Show that Eq. (3-37) satisfies the equation

$$\ddot{x} + 2\gamma\dot{x} + \omega_0^2 x = 0$$

for the case of critical damping.

3-2.6. Soaring insurance rates have forced automobile manufacturers to redesign bumpers to offer greater impact protection. Suppose that it is desired to design a spring and hydraulic damper for a bumper such that an auto weighing 4000 lb and hitting a solid wall at a speed of 5 mph will experience a maximum average deceleration of $\frac{1}{2}g$. What is the maximum travel of the bumper? What values must be given to the spring constant k and damping coefficient b in order that the bumper return to its equilibrium position in as short a time as possible after the auto rebounds from the wall?

3-2.7. A mass m, subject to a linear restoring force $-kx$ and a damping force μmg due to sliding friction, is given an initial displacement A and released. Find the motion. Remember that the friction force changes sign at each half

93

period; therefore each half period must be regarded as a new problem using the final x and $\dot{x}$ of the preceding half period as initial conditions.

3-2.8. A mass m is suspended from light elastic string having spring constant k. The system is immersed in a viscous liquid that imposes a damping constant b on the motion. Assuming the motion to be underdamped, calculate the maximum initial displacement from equilibrium such that the tension of the string does not become zero during the upward swing (second half cycle). What effect would a larger initial displacement have on the motion?

3-3 The Driven Harmonic Oscillator

We now wish to expand the results of the first two sections to consider a damped harmonic oscillator subject to a driving force that we shall initially consider to be sinusoidal. The equation for such a system is

$$\ddot{x} + \frac{b}{m}\dot{x} + \omega_0^2 x = \frac{F}{m}\cos\omega t, \tag{3-40}$$

where F is the amplitude of the driving force and ω is its frequency. This equation is still linear with constant coefficients but is no longer homogeneous. The general solution to such an equation is the sum of the homogeneous solution that we found in the last section and the *particular integral* that satisfies the inhomogeneous equation (3-40). Experience in these matters leads us to guess that this particular integral will be composed of harmonic terms having frequency ω. If there were no damping—that is, no first derivative term—then simple inspection reveals that a single cosine term would do the job. The presence of damping necessitates the addition of a sine term that amounts to a phase shift between the driver and the displacement. We can therefore assume a particular solution of the form

$$x_p = \alpha\cos\omega t + \beta\sin\omega t, \tag{3-41}$$

where the constants α and β will be determined by substituting this trial function into Eq. (3-40). Doing so, we obtain

$$-\alpha\omega^2\cos\omega t - \beta\omega^2\sin\omega t - \alpha\omega\frac{b}{m}\sin\omega t + \beta\omega\frac{b}{m}\cos\omega t$$

$$+ \alpha\omega_0^2\cos\omega t + \beta\omega_0^2\sin\omega t = \frac{F}{m}\cos\omega t$$

$$\text{or} \quad \left(-\alpha\omega^2 + \beta\omega\frac{b}{m} + \alpha\omega_0^2 - \frac{F}{m}\right)\cos\omega t$$

$$+ \left(-\beta\omega^2 - \alpha\omega\frac{b}{m} + \beta\omega_0^2\right)\sin\omega t = 0. \tag{3-42}$$

In order for this equation to hold true for all values of t, the coefficients of the sine and cosine terms must vanish separately. This step provides us with a pair of simultaneous equations from which α and β can be determined. Carrying out this little algebra, we find

$$\alpha = \frac{F\omega b}{\omega^2 b^2 + m^2(\omega_0^2 - \omega^2)^2},$$

$$\beta = \frac{Fm(\omega_0^2 - \omega^2)}{\omega^2 b^2 + m^2(\omega_0^2 - \omega^2)^2}. \tag{3-43}$$

This information is best made use of by recasting our trial function in a more convenient form. It can easily be shown by trigonometric identity that

$$x_p = \sqrt{\alpha^2 + \beta^2} \cos(\omega t - \varphi), \tag{3-44}$$

where

$$\varphi = \arctan \frac{\beta}{\alpha} \tag{3-45}$$

is completely equivalent to the function expressed in Eq. (3-41). Using the form of Eqs. (3-44) and (3-45), together with Eqs. (3-43), we can write our particular integral as

$$x_p = \frac{F}{[m^2(\omega_0^2 - \omega^2)^2 + \omega^2 b^2]^{1/2}} \cos(\omega t - \varphi), \tag{3-46}$$

where the phase angle φ is given by

$$\varphi = \arctan \frac{\omega b}{m(\omega_0^2 - \omega^2)}. \tag{3-47}$$

In order to find the general solution of Eq. (3-40), we add Eq. (3-46) to the general solution of the homogeneous equation. For the present purpose, we shall only consider the underdamped (that is, the oscillatory) solution of the homogeneous part. Our general solution for the harmonically driven oscillator is then

$$x = Ce^{-bt/2m} \cos(\omega_1 t + \phi) + \frac{F}{[m^2(\omega_0^2 - \omega^2)^2 + \omega^2 b^2]^{1/2}} \cos(\omega t - \varphi), \tag{3-48}$$

where C and ϕ are the constants of integration. The first term of this equation describes oscillation of the system at its natural frequency, and, because of the exponential damping term, its contribution is negligible for large values of t. For this reason, we refer to this term as the *transient* solution. The second term describes the *steady-state* solution.

It is evident that the amplitude of the steady-state solution can be varied over a wide range by fixing the natural frequency of the oscillator ω_0 and allowing the frequency ω of the driver to vary. The system is said to be tuned to a condition of *resonance* when ω is made to approach a value such that the amplitude is a maximum. Since $F \propto x$ and $u = \dot{x} \propto \omega x$ for a linear oscillator, the energy input rate given by Fu is proportional to the *square* of the amplitude of the oscillator. Introducing the quantity

$$Z = m^2(\omega_0^2 - \omega^2)^2 + \omega^2 b^2 \tag{3-49}$$

in terms of which the square of the displacement may be written

$$x^2 = \frac{F^2}{Z} \cos^2(\omega t - \varphi). \tag{3-50}$$

The amplitude F^2/Z will be a maximum when Z is minimized. Differentiating Eq. (3-49) with respect to ω^2 and setting the result equal to zero,

$$\frac{dZ}{d\omega^2} = -2m^2(\omega_0^2 - \omega^2) + b^2 = 0$$

from which we find for the value of $\omega = \omega_2$ corresponding to minimum Z

$$\omega_2^2 = \omega_0^2 - \frac{b^2}{2m^2}. \tag{3-51}$$

This frequency is known as the *resonant frequency*. It differs from the natural frequency of both the damped and undamped oscillator, $\omega_1^2 = \omega_0^2 - b^2/4m^2$ and ω_0. In order to study the approach of the system to resonance, let us introduce yet another quantity having the dimensions of a frequency

$$\Omega^2 = \omega^2 - \omega_2^2, \tag{3-52}$$

in terms of which the frequency of the driver is

$$\omega^2 = \Omega^2 + \omega_2^2 = \Omega^2 + \omega_0^2 - \frac{b^2}{2m^2}. \tag{3-53}$$

Using this expression for ω, the amplitude of Eq. (3-50) becomes

$$\frac{F^2}{Z} = \frac{F^2}{m^2(\omega_0^2 - \omega^2)^2 + \omega^2 b^2}$$

$$= \frac{F^2}{m^2[(b^2/2m^2) - \Omega^2]^2 + b^2[\Omega^2 + \omega_0^2 - (b^2/2m^2)]}$$

$$= \frac{F^2}{m^2\Omega^2 + b^2[\omega_0^2 - (b/2m)^2]}$$

$$= \frac{F^2}{m^2\Omega^2 + b^2\omega_1^2}. \tag{3-54}$$

Since we are interested only in frequencies near the resonant value, we can approximate Eq. (3-52) as

$$\Omega^2 = (\omega - \omega_2)(\omega + \omega_2) \approx 2\omega(\omega - \omega_2) \approx 2\omega_1(\omega - \omega_2), \tag{3-55}$$

assuming that the damping is small enough to allow the approximation $\omega \approx \omega_1$. Substituting Eq. (3-55) into Eq. (3-54) yields

$$\frac{F^2}{Z} = \frac{F^2}{4m^2\omega_1^2(\omega - \omega_2)^2 + b^2\omega_1^2} = \frac{F^2/\omega_1^2}{4m^2(\omega - \omega_2)^2 + b^2}. \tag{3-56}$$

The maximum value of F^2/Z occurs for $\omega = \omega_2$ for which $\Omega = 0$. From Eq. (3-54) we see that this condition corresponds to

$$\left(\frac{F^2}{Z}\right)_{\max} = \frac{F^2}{b^2\omega_1^2}. \tag{3-57}$$

If we denote $(F^2/Z)_{\max}$ by A_0^2 and the value of F^2/Z for the lightly damped system near resonance as given by Eq. (3-56) by A^2, then the ratio A^2/A_0^2 is

$$\frac{A^2}{A_0^2} = \frac{b^2}{4m^2(\omega - \omega_2)^2 + b^2}. \tag{3-58}$$

This expression is plotted as a function of $\omega - \omega_2$ in Fig. 3-4. The half width of the resonant peak at half maximum, referred to as the *resonance half width*, is also shown on the figure. We see from Eq. (3-58) that its value is $b/2m$. Thus the sharpness of the resonance is inversely proportional to the dissipation constant b.

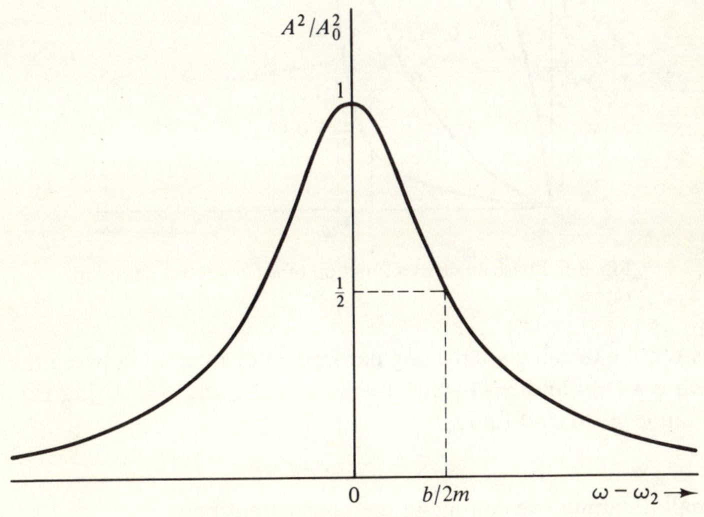

Fig. 3-4

We now wish to examine the phase of the displacement relative to that of the driving force. From Eq. (3-47) we see that $\omega = 0$ corresponds to $\varphi = 0$, that $\omega = \omega_0$ gives $\varphi = \pi/2$, and that as $\omega \rightarrow \infty$, $\varphi \rightarrow \pi$. The behavior of the phase angle for an oscillator with small damping is plotted in Fig. 3-5.

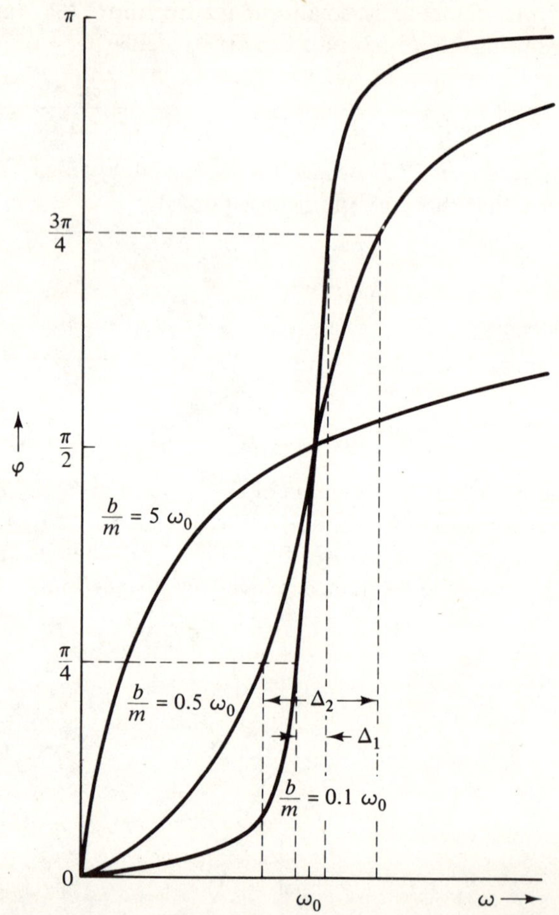

Fig. 3-5 Phase angle as a function of ω for several damping rates.

It is instructive to calculate the angular frequency interval between the points at which $\varphi = \pi/4$ and $3\pi/4$—that is, $\tan \varphi = +1$ and -1. Using Eq. (3-47), we set $\tan \varphi = \pm 1$ and find

$$m(\omega_0^2 - \omega^2) = \pm \omega b. \tag{3-59}$$

For small damping, we can make the approximation

$$\omega_0^2 - \omega^2 = (\omega_0 + \omega)(\omega_0 - \omega) \approx 2\omega(\omega_0 - \omega). \tag{3-60}$$

So we find in the limit that $\omega \approx \omega_0$ at $\tan \varphi = \pm 1$

$$\omega - \omega_0 = \pm \frac{b}{2m}, \tag{3-61}$$

which is just equal to the half width of the resonant peak. The frequency bandwidths marked Δ_1 and Δ_2 are equal to $b_1/m_1\omega_0$ and $b_2/m_2\omega_0$, respectively, however large the band may be. The band is only centered with respect to ω_0 to a first-order approximation. For $b = 0$ the phase angle is given by

$$\varphi = \begin{cases} 0, & 0 \le \omega < w_0, \\ \pi, & \omega_0 < \omega \le \infty. \end{cases} \tag{3-62}$$

Consider next the energy properties of the sinusoidally driven oscillator. We have shown that the square of the amplitude reaches a maximum at a driving frequency ω_2 given by Eq. (3-51). At this frequency a maximum average potential energy will be stored in the system. We now wish to know at which frequency of the driving force will the average kinetic energy of the oscillator be a maximum. The kinetic energy is

$$T = \tfrac{1}{2}m\dot{x}^2 = \frac{\tfrac{1}{2}mF^2\omega^2}{[m^2(\omega_0^2 - \omega^2)^2 + \omega^2 b^2]} \sin^2(\omega t - \varphi)$$

$$= \frac{m}{2}\frac{F^2}{Z}\omega^2 \sin^2(\omega t - \varphi). \tag{3-63}$$

Since the average value of $\sin^2(\omega t - \varphi)$ is $\tfrac{1}{2}$, the average kinetic energy will be

$$\langle T \rangle = \frac{m}{4}\frac{F^2}{Z}\omega^2. \tag{3-64}$$

In order to find the maximum value of this quantity, we take the derivative of $F^2\omega^2/Z$ with respect to ω^2 and set the result equal to zero. It is left as an exercise to show that this procedure yields $\omega = \omega_0$ as the driving frequency corresponding to maximum average kinetic energy. The fact that the average kinetic and potential energies are maximized at different driving frequencies is not surprising, for the system, because of the damping, is not conservative. If the damping factor b is allowed to approach zero, then the driving frequency for which the average potential is a maximum will also be ω_0 as can be seen from Eq. (3-51).

Let us turn next to the question of power inputs into work and dissipative loss. The power equation is found by multiplying Eq. (3-40) by $m\dot{x}$:

$$m\dot{x}\ddot{x} + b\dot{x}^2 + m\omega_0^2 x\dot{x} = F\dot{x} \cos \omega t. \tag{3-65}$$

This equation may be rewritten

$$\frac{d}{dt}(T + V) + b\dot{x}^2 = \dot{x}F \cos \omega t, \tag{3-66}$$

where
$$T = \tfrac{1}{2}m\dot{x}^2 \tag{3-67}$$

and
$$V = \tfrac{1}{2}m\omega_0^2 x^2 \tag{3-68}$$

are the instantaneous kinetic and potential energies, respectively. The first term on the left of Eq. (3-66) is the time rate of change of the total energy of the oscillator. The second term gives the rate at which energy is being dissipated by the damping force. The term on the right-hand side describes the rate at which energy is being supplied to (or withdrawn from) the system by the driving force. If we write $\dot{x}$ explicitly, the right-hand side of Eq. (3-66) may be rewritten

$$F\dot{x}\cos\omega t = -\left(\frac{F^2\omega}{Z^{1/2}}\right)\cos\omega t\,\sin(\omega t - \varphi)$$

$$= \frac{F^2\omega}{Z^{1/2}}(\cos^2\omega t\,\sin\varphi - \cos\omega t\,\sin\omega t\,\cos\varphi). \tag{3-69}$$

Focusing our attention on the parenthetical expression containing the sinusoidal terms, it can be seen that for times when

$$\tan\varphi < \tan\omega t \tag{3-70}$$

the sign of Eq. (3-69) changes from positive to negative and the driving force is extracting energy from the system. If we average Eq. (3-69) with respect to time, we see that the second term in parentheses averages to zero; thus

$$\langle F\dot{x}\cos\omega t\rangle = \frac{F^2\omega}{2Z^{1/2}}\sin\varphi \tag{3-71}$$

and the driving term is, on the average, supplying energy to the system rather than withdrawing it. It can easily be shown that the average rate at which power is being supplied to the system as given by Eq. (3-71) is simply equal to the average rate at which energy is being dissipated $\langle b\dot{x}^2\rangle$ as one might expect.

In electrical circuit theory, where driven oscillator systems with damping are also encountered, it is often useful to define a quantity Q as 2π times the maximum energy stored in the system divided by the energy lost in one period. In mechanical terms, this is

$$Q = \frac{2\pi(\tfrac{1}{2}m\dot{x}^2)_{\max}}{\langle b\dot{x}^2\rangle\tau_1} = \frac{\pi m F^2\omega^2/Z}{(b/2)(F^2\omega^2/Z)(2\pi/\omega_1)} = \frac{m\omega_1}{b}, \tag{3-72}$$

which is inversely proportional to the resonance width.

Since the equations that govern driven oscillators with damping are linear, the principle of superposition applies. A periodic driving force of arbitrary form can be Fourier analyzed by using Eq. (2-152), where the coefficients A_0, A_k, B_k are given by

$$A_0 = \frac{1}{\tau} \int_0^\tau F(t)\, dt \tag{3-73}$$

$$A_k = \frac{2}{\tau} \int_0^\tau F(t) \cos\left(\frac{2\pi k t}{\tau}\right) dt \tag{3-74}$$

$$B_k = \frac{2}{\tau} \int_0^\tau F(t) \sin\left(\frac{2\pi k t}{\tau}\right) dt \tag{3-75}$$

and τ is the period of the force. Nonperiodic forces applied to harmonic oscillators can be handled by attacking the problem as a whole, using the Fourier integral transform or Green's function techniques.

EXERCISES

3-3.1. Prove that Eqs. (3-44) and (3-45) constitute a particular solution to Eq. (3-40) that is equivalent to the one given in Eq. (3-41).

3-3.2. Show that the driver frequency that results in a maximum value of the average kinetic energy in a sinusoidally driven damped harmonic oscillator is just equal to the natural frequency ω_0 of the undamped system.

3-3.3. Show explicitly that the average power input to a sinusoidally driven damped harmonic oscillator is merely equal to the rate at which energy is being dissipated.

3-3.4. A critically damped oscillator with mass m and damping constant b is given an initial displacement A. Determine the time taken for half the initial energy of the oscillator to be dissipated.

3-3.5. An electron bound in an atom is subject to a force

$$F = eE \cos \omega t,$$

where e is the charge of the electron and the atom is irradiated by a plane electromagnetic wave of frequency ω and associated electric field strength E The damping coefficient due to radiation by the accelerated electron can be calculated from electromagnetic theory to be

$$b = \frac{2}{3}\frac{e^2\omega^2}{c^3}$$

in Gaussian CGS units. Using $e = 4.8 \cdot 10^{-10}$ esu and $\omega = 10^{16}$, calculate the full width of the resonance at half maximum. This is the natural width of a spectral line.

3-3.6. Determine the general solution for the motion of an underdamped harmonic oscillator having mass m, damping constant b, and spring constant k when the driving force is a constant F_0. Solve the inhomogeneous equation by assuming a form for the particular solution and then determining the constant coefficients.

3-3.7. Show that the Fourier expansion coefficients given in Eqs. (3-73) to (3-75) are equivalent to Eq. (2-159).

3-3.8. An underdamped harmonic oscillator is subject to a force given by

$$F(t) = \begin{cases} 0, & -\dfrac{\tau}{2} \le t < 0, \\[2ex] F_0, & 0 < t \le \dfrac{\tau}{2}. \end{cases}$$

Solve for the motion, using the method of Fourier series.

3-3.9. An underdamped oscillator is subject to an impulse—that is, a large force of very short duration. The oscillator is initially at rest at $x = 0$ and can be assumed not to move appreciably during the application of the impulse. If the impulse is delivered at $t = 0$, show that the motion is given by

$$x = \frac{p_0}{m\omega_1}\, e^{-\gamma t} \sin \omega_1 t,$$

where $p_0 = mu_0 = \int F\, dt$.

3-4 Adiabatic Invariance

In Section 2-6 and problem 2-6.8 we showed that the magnetic moment

$$\mu_m = \frac{\frac{1}{2}mu_\perp^2}{B} = \frac{q}{m}\frac{T_\perp}{\omega_c} \tag{3-76}$$

of a charged particle gyrating in a magnetic field is a constant of the motion if the magnetic field, as seen from the particle reference system, varies sufficiently slowly with respect to the period of gyration. The magnetic moment is, under these conditions, an example of an *adiabatic invariant*.

The classical example of an adiabatic invariant, first discussed by Einstein and Lorentz, is a pendulum, the suspending string of which is slowly being shortened or lengthened. Here, as in the case of the gyrating particle, it can be shown that the ratio of the energy of the pendulum to its frequency remains a constant. This finding has an important corollary in quantum theory, where the energy of a quantum oscillator is given by

$$E = nh\nu \tag{3-77}$$

and ν is the circular frequency of the oscillation, h is Planck's constant, and n is the number of quanta associated with the oscillator.

We wish to formulate the notion of adiabatic invariance in a somewhat more general way. Envision a particle that moves in a one-dimensional potential well with energy E between the turning points x_1 and x_2 as shown in Fig. 3-6. If the potential is unchanging, the equation of motion can be written

$$m\ddot{x} + \frac{dV}{dx} = 0 \tag{3-78}$$

from which we find by multiplying by $\dot{x}$ and integrating

$$\tfrac{1}{2}m\left(\frac{dx}{dt}\right)^2 + V(x) = E \tag{3-79}$$

the energy integral. Solving Eq. (3-79) for dt and carrying out a contour integration over a complete period, the oscillation period is found to be

$$\tau = \oint dt = \oint \frac{dx}{[(2/m)(E - V)]^{1/2}}. \tag{3-80}$$

Let us now define the *action integral*

$$J = \oint p_x \, dx = \oint m \frac{dx}{dt} \, dx = \oint [2m(E - V)]^{1/2} \, dx. \tag{3-81}$$

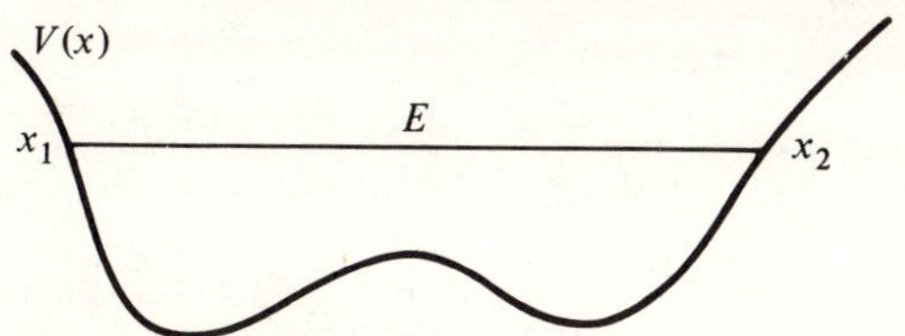

Fig. 3-6 Energy diagram for a particle in a potential well.

With these preliminary definitions in hand, we would like to examine the particle motion in a potential field that slowly (with respect to τ) varies. We will express the potential in terms of x and a parameter a, which is a slowly changing function of time

$$V = V[x, a(t)]. \tag{3-82}$$

Our purpose is to show that the action integral J is an adiabatic invariant. The time derivative of J is

$$\frac{dJ}{dt} = \frac{d}{dt} 2 \int_{x_1}^{x_2} [2m(E - V)]^{1/2}\, dx, \tag{3-83}$$

where x_1 and x_2 are the turning points. Our assumption of the "slowness" of the variation amounts to the assumption that the particle energy E remains an integral of the motion during a period of oscillation and varies only over many periods. This means that $a(t)$ is fixed when evaluating the integral in Eq. (3-83). The turning points and the integrand are time dependent for $t \gg \tau$, however. Bearing this point in mind, we carry out the indicated differentiation in Eq. (3-83):

$$\frac{1}{2}\frac{dJ}{dt} = [2m(E - V)]^{1/2}|^{x_2} \frac{dx_2}{dt} - [2m(E - V)]^{1/2}|^{x_1} \frac{dx_1}{dt}$$

$$+ \int_{x_1}^{x_2} \frac{d}{dt} [2m(E - V)]^{1/2}\, dx. \tag{3-84}$$

Since $E - V$ vanishes at the turning points, the first two terms on the right-hand side of this equation are zero. The integrand in the third term is

$$\frac{d}{dt} [2m(E - V)]^{1/2} = \left[\frac{2}{m}(E - V)\right]^{-1/2}\left(\frac{dE}{dt} - \frac{dV}{dt}\right). \tag{3-85}$$

The parenthetical term can be evaluated as follows.

$$\frac{dV}{dt} = \frac{\partial V}{\partial x}\frac{dx}{dt} + \frac{\partial V}{\partial a}\frac{da}{dt}, \tag{3-86}$$

and using Eq. (3-79),

$$\frac{dE}{dt} = \left(m\frac{d^2x}{dt^2} + \frac{\partial V}{\partial x}\right)\frac{dx}{dt} + \frac{\partial V}{\partial a}\frac{da}{dt}, \tag{3-87}$$

where, by virtue of Eq. (3-78), we see that the term in parentheses vanishes. Inserting the results of Eqs. (3-86) and (3-87) into Eq. (3-85), we find

$$\frac{d}{dt}[2m(E - V)]^{1/2} = \left[\frac{2}{m}(E - V)\right]^{-1/2}\left(-\frac{\partial V}{\partial x}\frac{dx}{dt}\right)$$

$$= \left\{\frac{2}{m}\left[\frac{m}{2}\left(\frac{dx}{dt}\right)^2\right]\right\}^{-1/2}\left(-\frac{\partial V}{\partial x}\frac{dx}{dt}\right)$$

$$= -\frac{\partial V}{\partial x}. \tag{3-88}$$

Substituting this result back into Eq. (3-84), we find the result

$$\frac{dJ}{dt} = -\oint \frac{\partial V}{\partial x}\, dx = 0. \tag{3-89}$$

Note that, in addition to postulating negligible time changes over a period of oscillation, we have also made the tacit assumption that the turning points vary continuously. The emergence of a potential peak that confines the particle to one side or the other of the general well as shown in Fig. 3-7 must be excluded, however slow the time variation of V may be. In like manner, the inverse process that leads to the abrupt escape of a particle from a region where it was previously confined will similarly cause it to "forget" its formerly invariant action integral.

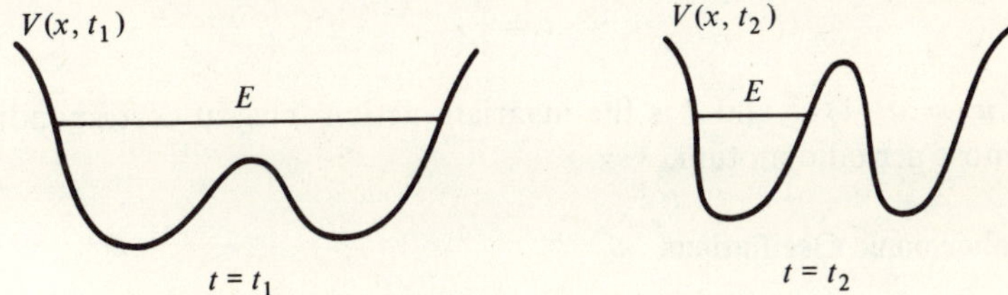

Fig. 3-7 Growth of a potential peak that destroys the adiabatic invariance of the action integral.

The derivation given above may be refined in order to answer the question: How much does an adiabatic invariant vary? Landau and Lifshitz show in their book on mechanics that the change in the action integral ΔJ is proportional to $\dot{a}^2 \Delta t$, and so ΔJ can be very small for a long time if $\Delta a = \dot{a}\,\Delta t$ is finite.[2]

EXERCISES

3-4.1. Plot the kinetic energy of a particle that moves with a total energy E in the potential well shown in Fig. 3-6 between the turning points x_1 and x_2.

3-4.2. For the explicit case of a one-dimensional undamped harmonic oscillator with slowly varying parameters, show that the constancy of J is equivalent to the constancy of E/ω_0.

3-4.3. A ball is bounced into a slowly converging channel as shown in Fig. 3-8. The motion in the y direction is almost periodic. If we consider the x coordinate as $a(t)$, the parameter describing the slow change of the potential

---

[2] L. D. Landau and E. M. Lifshitz, *Mechanics* (2nd ed.). Oxford: Pergamon Press, 1969, pp. 154–158.

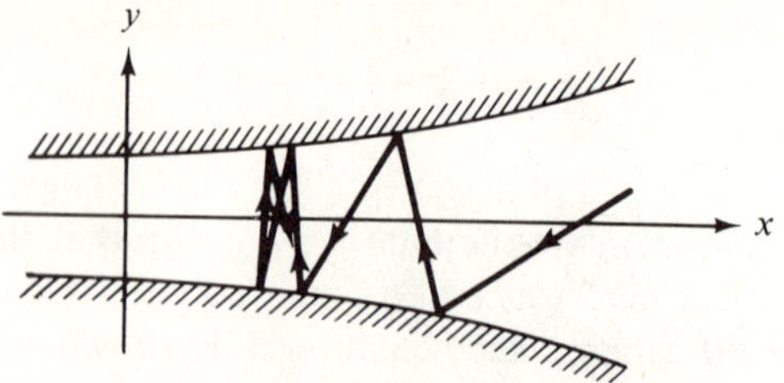

Fig. 3-8 Motion of a ball in a converging channel.

such that $V = V(y, a)$, show that the channel height $l(x)$ corresponding to the maximum penetration of the ball into the channel is given by

$$l = \frac{J}{u},$$

where $u^2 = u_x^2 + u_y^2$ and J is the invariant action integral corresponding to the almost periodic motion.

3-5 Anharmonic Oscillations

Oscillator systems are only simple harmonic if the restoring force is a linear function of the displacement and the damping force is a linear function of the velocity. The general equation for an undriven anharmonic oscillator to which these restrictions do not apply is

$$m\ddot{x} + g(\dot{x}) + f(x) = 0, \qquad (3\text{-}90)$$

where $g(\dot{x})$ and $f(x)$ are nonlinear functions of their respective arguments.

Since mathematicians have not yet seen fit to provide physicists with a nonlinear calculus, general solutions to Eq. (3-90) are not possible, and approximation methods must be employed to obtain particular solutions in most cases. Such solutions are dealt with in textbooks on nonlinear mechanics. We can, however, show that, in general, the oscillation period for nonlinear systems is a function of the amplitude of the oscillation.

In the absence of damping and driving terms, Eq. (3-90) becomes

$$m\ddot{x} = -f(x). \qquad (3\text{-}91)$$

Since $\ddot{x} = \dot{x}\, d\dot{x}/dx$, Eq. (3-91) can be rewritten

$$m\dot{x}\, d\dot{x} = -f(x)\, dx, \qquad (3\text{-}92)$$

which can be integrated to yield

$$\tfrac{1}{2}m\dot{x}^2 = -\int_{x_m}^{x} f(x)\, dx, \qquad (3\text{-}93)$$

where $\dot{x} = 0$ at $x = x_m$, the maximum displacement. If the integration over x on the right-hand side of this equation is taken from x_m to a still positive value of x nearer the origin, the sign of $\dot{x}$ will be negative. Reversing the limits of integration and taking the square root, we find

$$\frac{dx}{dt} = -\left[\frac{2}{m} \int_x^{x_m} f(x)\, dx \right]^{1/2} = -\sqrt{\frac{2}{m}}\, [F(x_m) - F(x)]^{1/2}, \qquad (3\text{-}94)$$

from which the period can be found on integration to be

$$\tau = 4\sqrt{\frac{m}{2}} \int_0^{x_m} \frac{dx}{[F(x_m) - F(x)]^{1/2}}. \qquad (3\text{-}95)$$

Clearly, the period of such an oscillation depends on the amplitude x_m. For the familiar case of a linear restoring force, $f(x) = kx$, where k is a constant, this dependence on x_m drops out.

Let us examine the case of a pendulum that is not limited to small amplitude. The equation of motion prior to our previous linearizing assumption of small angular amplitude was given in Eq. (3-1) as

$$ml\ddot{\theta} = -mg \sin\theta. \qquad (3\text{-}96)$$

Multiplying this equation by $l\dot{\theta}$ and integrating, the corresponding energy integral is found:

$$E = T + V = \tfrac{1}{2}ml^2\dot{\theta}^2 - mgl\cos\theta. \qquad (3\text{-}97)$$

The linearizing assumption amounts to limiting the total energy E to a value only slightly larger than $-mgl$. In the present case, we shall extend our analysis to cover energies up to mgl, which amounts to allowing the angular amplitude to approach $\theta_m = \pi$. Solving Eq. (3-97) for $\dot{\theta} = d\theta/dt$ and integrating, we have

$$\sqrt{\frac{2g}{l}}\, t = \int_0^\theta \frac{d\theta}{(E/mgl + \cos\theta)^{1/2}}, \qquad (3\text{-}98)$$

where the initial angle has been taken to be zero for convenience. If θ_m corresponds to the maximum angular amplitude, then

$$E = -mgl\cos\theta_m \qquad (3\text{-}99)$$

and Eq. (3-96) can be rewritten

$$\sqrt{\frac{2g}{l}}\, t = \int_0^\theta \frac{d\theta}{(\cos\theta - \cos\theta_m)^{1/2}}. \qquad (3\text{-}100)$$

Fortunately, this equation can be solved in terms of a tabulated function. To this end, we first recast the problem by making use of the identity

$$\cos\theta = 1 - 2\sin^2\frac{\theta}{2} \qquad (3\text{-}101)$$

to rewrite Eq. (3-100) as

$$\sqrt{\frac{g}{l}}\, t = \int_0^\theta \frac{d\theta}{[\sin^2(\theta_m/2) - \sin^2(\theta/2)]^{1/2}}. \tag{3-102}$$

We then make the following change in the angular variable. Let

$$\sin \varphi = \frac{\sin \theta/2}{\sin \theta_m/2} = \frac{\sin \theta/2}{a}. \tag{3-103}$$

Since θ varies between the limits $-\theta_m \le \theta \le \theta_m$, we see that φ will vary from 0 to 2π for one complete cycle. So our equation can now be written

$$\sqrt{\frac{g}{l}}\, t = \int_0^\varphi \frac{d\varphi}{(1 - a^2 \sin^2 \varphi)^{1/2}}, \tag{3-104}$$

which is the standard form for an elliptic integral of the first kind.[3] It is left as an exercise to show that the period of the large-amplitude pendulum is, to second order in $a = \sin(\theta_m/2)$,

$$\tau = 2\pi \sqrt{\frac{l}{g}} \left(1 + \frac{a^2}{4} + \cdots\right). \tag{3-105}$$

Next, we wish to consider an example of a driven anharmonic oscillator. We shall choose a mass attached to a spring, the restoring force of which varies nonlinearly with the displacement, specifically

$$k(x)x = \kappa x \pm \epsilon x^3, \tag{3-106}$$

where κ and ϵ are positive constants. The choice of the positive sign on the quartic term corresponds to a "hard" spring for which the stiffness increases with increasing displacement. The negative sign refers to a "soft" spring for which the stiffness decreases with x. If we take the driving term to have a cosine dependence on time, the equation of motion becomes

$$\ddot{x} = -\frac{\kappa x}{m} \mp \frac{\epsilon}{m} x^3 + \frac{F_0}{m} \cos \omega t. \tag{3-107}$$

We shall assume ϵ to be sufficiently small in order to use the method of successive approximations to find a particular integral.

As a starting point, we will assume as a first approximation that

$$x_1 = A \cos \omega t. \tag{3-108}$$

---

[3] See M. Abramowitz and Irene A. Stegun, *Handbook of Mathematical Functions*. New York: Dover Publications, Inc., 1965, pp. 587–616.

Substituting this guess into the right-hand side of Eq. (3-107), we obtain an equation for the second approximation:

$$\ddot{x}_2 = \frac{1}{m}(F_0 - \kappa A)\cos \omega t \mp \frac{\epsilon A^3}{m}\cos^3 \omega t. \tag{3-109}$$

Using the trigonometric identity

$$\cos^3 \theta = \tfrac{3}{4}\cos \theta + \tfrac{1}{4}\cos 3\theta, \tag{3-110}$$

Eq. (3-109) becomes

$$\ddot{x}_2 = \frac{1}{m}\left[\left(F_0 - \kappa A \mp \frac{3\epsilon A^3}{4}\right)\cos \omega t \mp \frac{\epsilon A^3}{4}\cos 3\omega t\right]. \tag{3-111}$$

Integrating twice, we find as a second approximation to the solution of Eq. (3-107)

$$x_2 = \frac{1}{m\omega^2}\left[\left(\kappa A - F_0 \pm \frac{3\epsilon A^3}{4}\right)\cos \omega t \pm \frac{\epsilon A^3}{36}\cos 3\omega t\right], \tag{3-112}$$

which should suffice if ϵ is not too large.

The nature of the resonance that arises when the driving frequency is varied differs from that of a driven linear oscillator. In the present case, even in the absence of damping, the amplitude cannot increase without limit for a given driving frequency. This situation occurs because if the driving frequency is tuned to the natural frequency of the system for a given amplitude, it will no longer be equal to the natural frequency at some larger amplitude. If we plot the amplitude (or the square of the amplitude) as a function of the driving frequency ω, the resonant peak will be found to tilt due to the presence of the nonlinear term ϵx^3. If the positive sign is taken as corresponding to a "hard" spring, the natural frequency will increase with increasing amplitude and the resonant peak will lean toward higher values of ω. For the "soft" spring case, ω_0 decreases with increasing amplitude and the tilt is toward lower values of ω. This tilting of the resonant peak leads to the interesting consequence that x and x^2 are *not* single valued with respect to ω for a small range of ω values near the resonant peak. Thus the variation of ω over this range can lead to abrupt changes in the amplitude of the motion.

EXERCISES

3-5.1. Show by substituting $f(x) = kx$, where k is a constant, into Eq. (3-92) and carrying out the integration that the resulting period is

$$\tau = 2\pi\sqrt{\frac{m}{k}},$$

which is independent of x_m.

3-5.2. Given the homogeneous linear differential equation with variable coefficients

$$\ddot{x} + f(t)\dot{x} + g(t)x = 0$$

for which $x_1 = \alpha(t)$ and $x_2 = \beta(t)$ are both known to be solutions, show that the linear combination

$$x_3 = c_1\alpha(t) + c_2\beta(t)$$

is also a solution.

3-5.3. Show that two known solutions to a nonlinear differential equation such as

$$\ddot{x} + ax^2 = 0$$

cannot be superposed to give a third solution.

3-5.4. Expand the integrand of Eq. (3-104) in a power series in a^2 and integrate term by term to show that the period of the large-amplitude pendulum is given by Eq. (3-105).

3-5.5. Consider a mass m fixed, as shown in Fig. 3-9, to the center of a light elastic string of force constant k. When the string is in the configuration AOB, the tension on each side of the mass is T_0, corresponding to a length $2l$. The mass is driven by a force applied in the x direction, $F_0 \cos \omega t$. Gravity is neglected. Show that the equation of motion for the mass is given by

$$m\ddot{x} = -\frac{2T_0x}{l} - \frac{kl - T_0x^3}{l^3} + F_0 \cos \omega t,$$

an equation of the same form as Eq. (3-105).

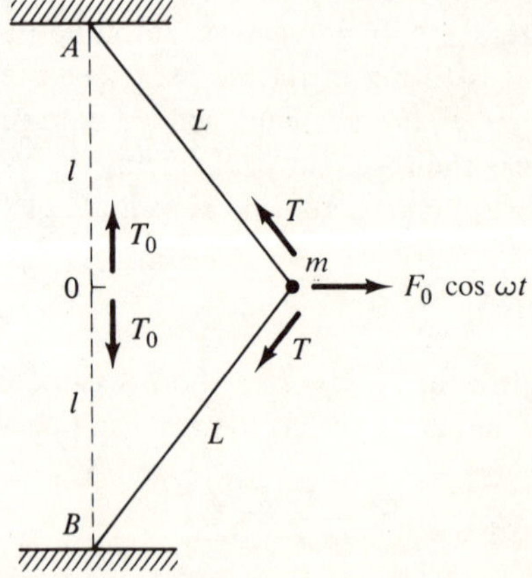

Fig. 3-9

3-6 The Vibrating String

In this section and the following one we deviate from our discussion of the oscillatory motions of a particle to consider a system with many degrees of freedom—namely, the vibratory and wave motion of a homogeneous massive string or cable of length L, fixed at its endpoints and under tension. Points on the string are designated by the x coordinate measured from the left-hand end. The position $x = L$ corresponds to the right-hand termination. Assuming that vibratory motion takes place in two dimensions only, we will label displacements of the string by $y(x, t)$. In addition, we assume that such displacements are quite small compared with the length such that the tension τ of the string can be assumed to be unchanged by the lateral displacements.

In order to derive the equation of motion of the string, we will consider an infinitesimal segment of string between the points x and $x + dx$ as shown in Fig. 3-10. If we take the mass per unit length of the string to be σ, then the mass of this segment is $\sigma\, dx$. The y component of the tension acting on any point of the displaced string is

$$\tau_y = \tau \sin \theta, \tag{3-113}$$

where θ is the angle between the tangent to the string at the point considered and the x axis. Our assumption of small displacement is embodied in the assumption that θ is sufficiently small such that

$$\sin \theta \approx \tan \theta = \frac{\partial y}{\partial x}, \tag{3-114}$$

the local slope of the string. The net force on the segment in the positive y direction due to the difference in the vertical components of the tension at either end can thus be written

$$dF = \tau_y\big|_{x+dx} - \tau_y\big|_x$$
$$= \tau \frac{\partial y}{\partial x} + \frac{\partial}{\partial x}\left(\tau \frac{\partial y}{\partial x}\right) dx - \tau \frac{\partial y}{\partial x} = \frac{\partial}{\partial x}\left(\tau \frac{\partial y}{\partial x}\right) dx. \tag{3-115}$$

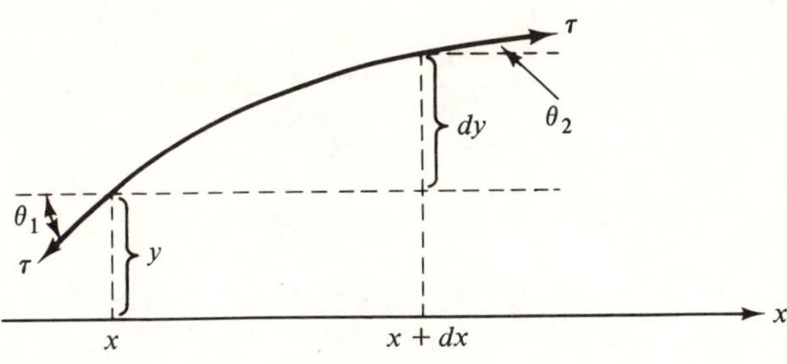

Fig. 3-10

If we had not limited ourselves to small θ and hence small slopes $\partial y/\partial x$, a net horizontal component of force on the segment due to tension would have arisen to cause horizontal motion. If there is also, in addition to tension forces, an external force f per unit length acting on the string (for example, the gravitational force $f = \sigma g$), the equation of motion will be

$$\sigma \, dx \frac{\partial^2 y}{\partial t^2} = \frac{\partial}{\partial x}\left(\tau \frac{\partial y}{\partial x}\right) dx + f \, dx. \tag{3-116}$$

Unless τ is very small or f very large, such external forces can usually be neglected. Since the assumption of small displacement implies constant tension τ, Eq. (3-116) with $f = 0$ becomes

$$\frac{\partial^2 y}{\partial t^2} - \frac{\tau}{\sigma} \frac{\partial^2 y}{\partial x^2} = 0. \tag{3-117}$$

This equation is a linear homogeneous partial differential equation for the displacement function $y(x, t)$. We are interested in solving it for a given displacement function $y(x, 0) = y_0(x)$ and initial displacement velocity $\partial y/\partial t\,|_{x,0} = u_{0y}(x)$. In addition, because of the fixed endpoints, the solution must obey the boundary conditions

$$y(0, t) = y(L, t) = 0. \tag{3-118}$$

We shall first try to find solutions that satisfy the boundary conditions without regard to the initial conditions. In solving Eq. (3-117), we shall use the method of *separation of variables*, which is one of the few general methods of attacking linear partial differential equations. Fortunately, it works for many of the equations of interest in physics (but not all). We will assume that solutions exist that can be written as the product of a function of x and a function of t:

$$y(x, t) = X(x)T(t). \tag{3-119}$$

Inserting this assumption into Eq. (3-115) and carrying out the indicated differentiations give

$$X\frac{d^2 T}{dt^2} = \frac{\tau}{\sigma} T \frac{d^2 X}{dx^2}$$

or
$$\frac{1}{T}\frac{d^2 T}{dt^2} = \frac{\tau}{\sigma}\frac{1}{X}\frac{d^2 X}{dx^2}. \tag{3-120}$$

We note that the left-hand side of this equation is a function only of t and that the right-hand side is a function only of x. Since each side is independent of the independent variable of the other side, then both sides can only be equal to a constant. We will take this constant to be negative definite, this choice to be justified a posteriori. Let

$$\frac{d^2T}{dt^2} = -\omega^2 T,$$
(3-121)

and

$$\frac{d^2X}{dx^2} = -\frac{\sigma\omega^2}{\tau}X.$$
(3-122)

We have therefore replaced our second-order partial differential equation by two second-order ordinary differential equations and are back on familiar ground. The two ordinary equations are immediately recognizable as harmonic oscillator equations. The solution to the first in the most convenient form is

$$T(t) = A\cos\omega t + B\sin\omega t,$$
(3-123)

where A and B are the integration constants. Similarly, we find for Eq. (3-122)

$$X(x) = C\cos kx + D\sin kx,$$
(3-124)

where we have introduced

$$k = \sqrt{\frac{\sigma}{\tau}}\,\omega.$$
(3-125)

In order for the boundary conditions given in Eq. (3-118) to hold for all times, $X(x)$ must satisfy

$$X(0) = C = 0,$$
$$X(L) = D\sin kx = 0.$$
(3-126)

The second of these equations requires that either $D = 0$, which leads to a trivial solution, or

$$\sin kx = 0,$$
(3-127)

which can only be true if $kL = n\pi$, where n is a positive nonvanishing integer. Hence

$$\omega_n = \frac{n\pi}{L}\sqrt{\frac{\tau}{\sigma}}, \qquad n = 1, 2, 3, \ldots.$$
(3-128)

It is left as an exercise to prove that if the separation constant had been chosen to be positive definite (say ω^2) or zero, it would not have been possible to satisfy the boundary conditions specified in Eq. (3-118).

The circular frequencies $\nu_n = \omega_n/2\pi$ are the *normal* frequencies of vibration of the string. For any given value of n, a solution to Eq. (3-117) is found by substituting Eqs. (3-123) and (3-124) into Eq. (3-119) and making use of Eqs. (3-126) and (3-128):

$$y_n(x, t) = A_n \sin k_n x \cos \omega_n t + B_n \sin k_n x \sin \omega_n t, \qquad (3\text{-}129)$$

where ω_n is given by Eq. (3-128) and

$$k_n = \frac{n\pi}{L}. \qquad (3\text{-}130)$$

The constant D has been set equal to one, which is equivalent to lumping it into A and B. For any given integer n, this solution describes a *normal mode* of vibration of the string.

Now let us see how the initial conditions are satisfied. Using our solution, Eq. (3-127), the initial position and velocity are

$$y_n(x, 0) = y_0(x) = A_n \sin k_n x \qquad (3\text{-}131)$$

and
$$\left. \frac{\partial y_n}{\partial t} \right|_{x,0} = u_{0y}(x) = B_n \omega_n \sin k_n x. \qquad (3\text{-}132)$$

These results imply that only for these particular initial conditions will the string vibrate in one of its normal modes. Here the principle of superposition for solutions of linear equations comes to our rescue. Using Eq. (3-129), we can construct solutions of a more general type by adding solutions corresponding to different values of n:

$$y(x, t) = \sum_{n=1}^{\infty} (A_n \sin k_n x \cos \omega_n t + B_n \sin k_n x \sin \omega_n t). \qquad (3\text{-}133)$$

The initial conditions for this general solution are

$$y_0(x) = \sum_{n=1}^{\infty} A_n \sin k_n x \qquad (3\text{-}134)$$

and
$$u_{0y}(x) = \sum_{n=1}^{\infty} B_n \omega_n \sin k_n x. \qquad (3\text{-}135)$$

The success of the calculation now depends on whether we can express the desired initial conditions for the string by Eqs. (3-134) and (3-135), using Fourier analysis. Using the orthogonality condition

$$\int_0^L \sin \frac{n\pi x}{L} \sin \frac{m\pi x}{L} = \frac{L}{2} \delta_{nm}, \qquad (3\text{-}136)$$

it can easily be shown that the coefficients are

$$A_m = \frac{2}{L} \int_0^L y_0(x) \sin k_m x \, dx \qquad (3\text{-}137)$$

$$\omega_m B_m = \frac{2}{L} \int_0^L u_{0y}(x) \sin k_m x \, dx, \qquad (3\text{-}138)$$

where $y_0(x)$ and $u_{0y}(x)$ can now be specified to be any continuous functions of x over the interval $0 < x < L$.

As an example, consider the case shown in Fig. 3-11 where the string is initially displaced at the midpoint a distance s ($s \ll L$) and then released with no initial velocity. The initial conditions are

$$
y_0(x) = \begin{cases} \dfrac{2s}{L} x, & 0 < x < \dfrac{L}{2}, \\[2mm] \dfrac{2s}{L} (L - x), & \dfrac{L}{2} < x < L. \end{cases} \tag{3-139}
$$

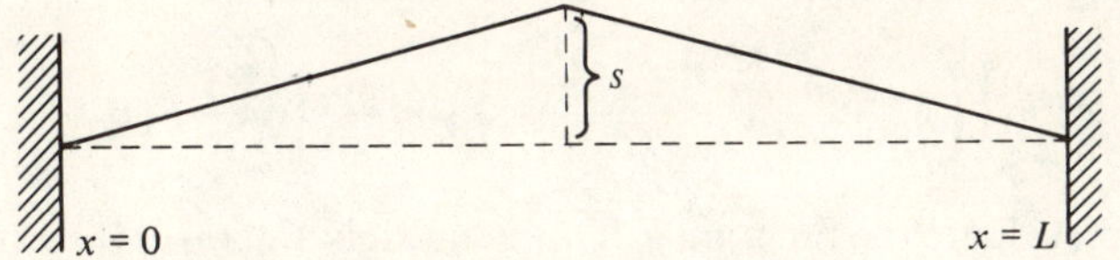

Fig. 3-11

Obviously all the B_m are zero. The A_m are given by

$$
A_m = \frac{2}{L} \left[\frac{2s}{L} \int_0^{L/2} x \sin \frac{m\pi x}{L} \, dx + \frac{2s}{L} \int_{L/2}^{L} (L - x) \sin \frac{m\pi x}{L} \, dx \right]
$$

$$
= \begin{cases} \dfrac{8s}{m^2 \pi^2} \sin \dfrac{m\pi}{2}, & m \text{ odd} \\[2mm] 0, & m \text{ even} \end{cases} \tag{3-140}
$$

and the general solution must therefore be

$$
y(x, t) = \frac{8s}{\pi^2} \left(\cos \frac{\pi}{L} \sqrt{\frac{\tau}{\sigma}} \, t \sin \frac{\pi x}{L} - \frac{1}{9} \cos \frac{3\pi}{L} \sqrt{\frac{\tau}{\sigma}} \, t \sin \frac{3\pi x}{L} \right.
$$

$$
\left. + \frac{1}{25} \cos \frac{5\pi}{L} \sqrt{\frac{\tau}{\sigma}} \, t \sin \frac{5\pi x}{L} - \cdots \right). \tag{3-141}
$$

Only the odd harmonics have been excited in order to avoid a node at the midpoint.

Let us expand our horizons to consider the motion of a string subject to damping and driving forces. We will consider the damping force to be proportional to the velocity. This force, acting on a segment dx of the string, is

$$
-\beta \frac{\partial y}{\partial t} \, dx,
$$

where β is the damping coefficient (per unit length). The negative sign guarantees that the damping force is directed oppositely to $\partial y / \partial t$. The driving force per unit length will be taken as $F(x, t)$ acting in the y direction. With these refinements, the equation of motion becomes

$$\sigma \frac{\partial^2 y}{\partial t^2} + \beta \frac{\partial y}{\partial t} - \tau \frac{\partial^2 y}{\partial x^2} = F(x, t). \tag{3-142}$$

Making use of the experience gained with the free undamped string, we assume that the solution can be written

$$y(x, t) = \sum_{n=1}^{\infty} g_n(t) \sin \frac{n\pi x}{L}. \tag{3-143}$$

Substituting this assumption into Eq. (3-142), we obtain

$$\sum_{n=1}^{\infty} \left[\left(\ddot{g}_n + \frac{\beta}{\sigma} \dot{g}_n + \frac{n^2 \pi^2}{L^2} \frac{\tau}{\sigma} g_n \right) \sin \frac{n\pi x}{L} \right] = \frac{1}{\sigma} F(x, t). \tag{3-144}$$

In order to evaluate the function $g_n(t)$, we multiply Eq. (3-144) by $\sin (m\pi x/L)$ and integrate from 0 to L:

$$\sum_{n=1}^{\infty} \left[\left(\ddot{g}_n + \frac{\beta}{\sigma} \dot{g}_n + \frac{n^2 \pi^2}{L^2} \frac{\tau}{\sigma} g_n \right) \int_0^L \sin \frac{n\pi x}{L} \sin \frac{m\pi x}{L} \, dx \right]$$

$$= \frac{1}{\sigma} \int_0^L F(x, t) \sin \frac{m\pi x}{L} \, dx, \tag{3-145}$$

which, by virtue of Eq. (3-136), yields

$$\ddot{g}_m + \frac{\beta}{\sigma} \dot{g}_m + \frac{m^2 \pi^2}{L^2} \frac{\tau}{\sigma} g_m = \frac{2}{\sigma L} f_m(t), \tag{3-146}$$

where

$$f_m(t) = \int_0^L F(x, t) \sin \frac{m\pi x}{L} \, dx. \tag{3-147}$$

We note that $f_m(t)$ is simply the mth Fourier coefficient of the Fourier expansion of $F(x, t)$.

Equation (3-146) is quite similar in form to Eq. (3-42) except for the fact that the driving force in the present case is more general in nature. We can therefore write the transient or homogeneous solution for the underdamped case as

$$g_m = \exp \left(-\frac{\beta t}{2\sigma} \right) (A_m \sin \omega_m t + B_m \cos \omega_m t), \tag{3-148}$$

where A_m and B_m are the integration constants and

$$\omega_m = \sqrt{\omega_{0m}^2 - \frac{\beta^2}{4\sigma^2}} = \sqrt{\frac{m^2 \pi^2}{L^2} \frac{\tau}{\sigma} - \frac{\beta^2}{4\sigma^2}}. \tag{3-149}$$

If the driving function is of the form $F(x, t) = \varphi(x) \cos \omega t$, then, after evaluating the integral of Eq. (3-147), the particular integral can be written by comparing with Eq. (3-46).

EXERCISES

3-6.1. A string of mass density σ and tension τ is fixed at $x = 0$. The other end at $x = L$ is terminated by a massless ring that is free to slide without friction on a vertical rod. Find the appropriate boundary condition at $x = L$ and solve the equation of motion to find the normal modes of vibration.

3-6.2. Reconsider problem 3-6.1 for the case of a terminating ring of mass m. Show that, for $m \to \infty$, the normal mode solution goes over into the case considered in the text where both ends are fixed.

3-6.3. The principal difference between a harpsicord and a piano is that in a harpsicord the strings are plucked, whereas in a piano they are struck with a hammer. Consider the case where a hammer of mass m and velocity u_0 strikes the midpoint of a stretched wire, rebounding elastically with velocity $-u_0$. The wire has a linear mass density σ and is stretched to a tension τ. Assume that the hammer has negligible width. Compare the normal modes derived for this case with the harpsicord case that can be considered to be described by Eq. (3-141).

3-6.4. Use Fourier analysis to derive the normal modes of a piano where the hammer has a finite width l.

3-6.5. Find the steady-state solution (particular integral) for a string of length L, tension τ, and damping coefficient β subjected to a driving force

$$F(x, t) = F_0 \cos \omega t,$$

where F_0 is a constant. This driving force is considered to be applied only at the point $x = a$. Show for this case that the particular integral is given by

$$y_p = \frac{2}{L} \frac{F_0}{[\sigma^2(\omega_m^2 - \omega^2) + \omega^2\beta^2]^{1/2}} \sin \frac{m\pi a}{L} \cos (\omega t - \varphi),$$

where

$$\varphi = \arctan \frac{\omega\beta}{\sigma(\omega_m^2 - \omega^2)}.$$

3-6.6. Show for the case of problem 3-6.5 that if the point $x = a$ is a node for some particular normal coordinate g_s, then g_s is not excited. Explain why this is so.

3-6.7. A driving force

$$F(x, t) = F_0 \sin \frac{n\pi x}{L} \cos \omega t$$

is applied to a stretched string of length L, tension τ, and damping coefficient β. Solve the equation of motion to find the steady-state solution.

3-7 Wave Propagation along a String

The normal mode solution to the motion of a string under tension with fixed boundary conditions at both ends was largely the work of Daniel Bernoulli. Euler and d'Alembert also solved the problem; their approach, however, was somewhat different. They envisioned the oscillatory motion of the string as a system of *traveling waves*. The equivalence of these two points of view was not established until 1807 when Fourier devised the series representation that bears his name and showed how the coefficients can be calculated by using the orthogonality properties of the sine and cosine functions.

A *wave* (which we shall define as a disturbance propagating with constant speed in the surrounding homogeneous medium) traveling in the x direction with velocity v can be represented in the most general way as a function $f(x - vt)$ of x and t. Similarly, a wave traveling in the minus x direction is represented by a function $g(x + vt)$.

As in the preceding discussion, we will examine the motion of a homogeneous string of mass density σ under a constant tension τ. The present analysis is also limited to small displacements in the y direction as before. As we saw in Section 3-6, the equation of motion for such a string is

$$\frac{\partial^2 y}{\partial t^2} = \frac{\tau}{\sigma} \frac{\partial^2 y}{\partial x^2}. \tag{3-150}$$

We wish to ascertain whether our wave functions $f(x - vt)$ and $g(x + vt)$ satisfy this equation or, since Eq. (3-150) is linear, whether the linear combination

$$y = f(x - vt) + g(x + vt) \tag{3-151}$$

satisfies Eq. (3-150). In order to streamline the algebra, we will let

$$\xi = x - vt, \qquad \eta = x + vt. \tag{3-152}$$

In terms of these variables, our solution to be tested is

$$y = f(\xi) + g(\eta). \tag{3-153}$$

We find

$$\frac{\partial y}{\partial t} = \frac{df}{d\xi} \frac{\partial \xi}{\partial t} + \frac{dg}{d\eta} \frac{\partial \eta}{\partial t} = v\left(-\frac{df}{d\xi} + \frac{dg}{d\eta}\right)$$

and

$$\frac{\partial^2 y}{\partial t^2} = v^2\left(\frac{d^2 f}{d\xi^2} + \frac{d^2 g}{d\eta^2}\right). \tag{3-154}$$

Similarly, the spatial derivative is found to be

$$\frac{\partial^2 y}{\partial x^2} = \frac{d^2 f}{d\xi^2} + \frac{d^2 g}{d\eta^2}. \tag{3-155}$$

Substituting Eqs. (3-154) and (3-155) into Eq. (3-150), we obtain

$$\left(v^2 - \frac{\tau}{\sigma}\right)\left(\frac{d^2 f}{d\xi^2} + \frac{d^2 g}{d\eta^2}\right) = 0. \tag{3-156}$$

Since $(d^2 f/d\xi^2) + (d^2 g/d\eta^2)$ cannot vanish, in general, owing to the fact that f and g are arbitrary functions, we must regard

$$v^2 = \frac{\tau}{\sigma} \tag{3-157}$$

as a possible solution for the velocity of propagation of the wave. In fact, this very arbitrariness of f and g suggests that Eq. (3-157) is the general solution, which is indeed the case. The crest of a wave traveling to the right ($+x$ direction) is shown in Fig. 3-12. At time t_0 the crest is located at $x = x_0$. The

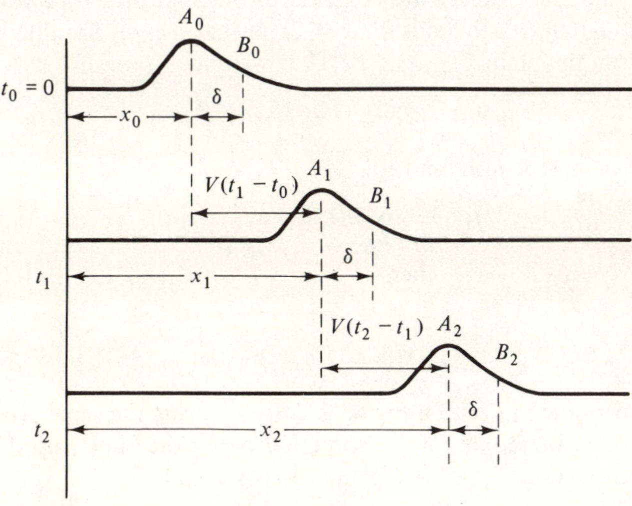

Fig. 3-12

displacement in the y direction and the lateral velocity (zero at the crest) are constant for all of the instantaneous crest points A_i located at x_i and points B_i located at $x_i + \delta$. This remark is equivalent to the statement that the wave is constant in its phase. Let us set $x_0 = t_0 = 0$ for convenience. If we represent the form of the wave at $t = t_0$ by the function $f(x)$, it is evident from our discussion in Fig. 3-12 that at time $t = t_1$ the form (constant in time) must be given by

$$f(x) = f(x - x_1) = f(x - vt_1). \tag{3-158}$$

In like manner, we find

$$f(x - x_2) = f(x - vt_2) = f(x - x_3) = f(x - vt_3), \tag{3-159}$$

and so on. In this way, we see that the arbitrary function $f(x - vt)$ does indeed represent a wave moving in the $+x$ direction. It can easily be verified that $g(x + vt)$ represents a wave moving toward lower values of x. So the general solution of Eq. (3-150) can be represented as the superposition of two waves traveling in opposite directions. This view is entirely equivalent to the normal mode or *standing wave* point of view.

The specification of the functions f and g is determined by the boundary and initial conditions. As in the previous section, we will consider both ends of the string, of length L, to be fixed. At $x = 0$ the boundary condition takes the form

$$f(-vt) = -g(vt). \tag{3-160}$$

Since the functions f and g are constant throughout the propagation of the wave, Eq. (3-160), which holds for $x = 0$, must also apply for all other values of x between the limits $0 \leq x \leq L$. Thus we find

$$y = f(x - vt) - f[-(x + vt)]. \tag{3-161}$$

At $x = L$, y must vanish, and so

$$f(L - vt) - f(-L - vt) = 0. \tag{3-162}$$

If we let $\zeta = (-L - vt)$, then $L - vt = \zeta + 2L$ and Eq. (3-162) can be written

$$f(\zeta) = f(\zeta + 2L). \tag{3-163}$$

Thus f is periodic in ζ with a period $2L$. By applying the same procedure to the function g, we see that it is also periodic with the same period.

In terms of time, Eqs. (3-162) and (3-163) yield

$$f(L - vt) = f(L - vt - 2L) = f\left[L - v\left(t + \frac{2L}{v}\right)\right] \tag{3-164}$$

from which it is obvious that f is periodic in t with a period given by

$$\tau = \frac{2L}{v}. \tag{3-165}$$

During one period the wave travels a distance $2L$. We define this period to be the *wavelength*

$$\lambda = 2L. \tag{3-166}$$

The wavelength is related to the propagation velocity v, the circular frequency ν, the angular frequency ω, and the period τ by

$$\frac{\lambda}{v} = \tau = \frac{1}{\nu} = \frac{2\pi}{\omega}. \tag{3-167}$$

The quantities τ and ν are the *fundamental* period and frequency, respectively, and correspond to the longest wavelength possible in order that the string have a node at $x = 0$ and L. We will subsequently denote these fundamental quantities by assigning them the subscript 1. The propagation velocity, as we have shown, depends only on τ and σ and is the same for all wave harmonics.

In order to establish the equivalence of the wave point of view and the normal mode analysis described in the preceding section, we shall anticipate the application of Fourier series to the problem and investigate the applicability of sinusoidal waves, such as

$$y = \sin\left[k(x - vt)\right], \tag{3-168}$$

where k is a constant known as the *wave number* and is defined by

$$k = \frac{2\pi}{\lambda} = \frac{\omega}{v}. \tag{3-169}$$

Using Eq. (3-169), Eq. (3-168) becomes

$$y = \sin \omega\left(\frac{x}{v} - t\right). \tag{3-170}$$

We have already shown that the general solution of Eq. (3-150) can be written in terms of a superposition of waves traveling in both the $+x$ and $-x$ directions. This fact, together with the knowledge that arbitrary initial conditions can be represented in terms of a Fourier series involving both sine and cosine terms, suggests that we try a wave solution of the form

$$y = A \sin \omega\left(\frac{x}{v} - t\right) + B \sin \omega\left(\frac{x}{v} + t\right) + C \cos \omega\left(\frac{x}{v} - t\right)$$

$$+ D \cos \omega\left(\frac{x}{v} + t\right). \tag{3-171}$$

We must first see whether such an assumption can be made to satisfy the boundary conditions. Setting $y = 0$ at $x = 0$ and equating the coefficients of the sine terms and cosine terms separately to zero in order to guarantee that the boundary conditions hold for all values of t, we find

$$A = B, \quad C = -D. \tag{3-172}$$

With this, Eq. (3-171) becomes

$$y = A\left[\sin \omega\left(\frac{x}{v} - t\right) + \sin \omega\left(\frac{x}{v} + t\right)\right]$$

$$+ C\left[\cos \omega\left(\frac{x}{v} - t\right) - \cos \omega\left(\frac{x}{v} + t\right)\right]. \qquad (3\text{-}173)$$

This equation reduces to zero at $x = 0$ for any value of ω. If we now require that $y = 0$ at $x = L$ for arbitrary values of A and C, we find as before that ω can only have certain values. We have already encountered such a restriction in considering the fundamental wave mode. Equations (3-165) and (3-167) can be combined to yield for the angular frequency of the fundamental mode

$$\omega_1 = \frac{2\pi v}{\lambda_1} = \frac{\pi v}{L} \qquad (3\text{-}174)$$

or using Eq. (3-157),

$$\omega_1 = \frac{\pi}{L}\sqrt{\frac{\tau}{\sigma}}, \qquad (3\text{-}175)$$

which also agrees with the frequency of the fundamental normal mode found in the preceding section. It is evident that the boundary condition at $x = L$ can also be satisfied in a more general way by

$$\omega_n = n\omega_1 = \frac{n\pi}{L}\sqrt{\frac{\tau}{\sigma}}, \qquad n = 1, 2, 3, \ldots \qquad (3\text{-}176)$$

which is identical with Eq. (3-128). The displacement corresponding to the nth wave harmonic is therefore

$$y_n(x, t) = A_n\left[\sin \omega_n\left(\frac{x}{v} - t\right) + \sin \omega_n\left(\frac{x}{v} + t\right)\right]$$

$$+ B_n\left[\cos \omega_n\left(\frac{x}{v} - t\right) - \cos \omega_n\left(\frac{x}{v} + t\right)\right]. \qquad (3\text{-}177)$$

The final step in establishing the equivalence of Eqs. (3-177) and (3-129) can be carried out by using the trigonometric identities for the sine and cosine of the sum and difference of two angles. This step is left as an exercise. The superposition of the various sinusoidal wave modes to accommodate any reasonable initial conditions follows just as in the discussion of the normal mode analysis.

It is of considerable general interest now for us to examine the effect of an abrupt change in the linear mass density on a sinusoidal wave. We shall consider an infinitely long string having a mass density σ_1 for $x < 0$ and a mass density σ_2 for all positive values of x. The tension is uniform throughout the string and is equal to τ. A wave of unit amplitude

$$y_1 = \cos \omega \left(\frac{x}{v_1} - t \right) \tag{3-178}$$

is incident on the junction from the left. We postulate the existence of a reflected wave

$$y_1' = A \cos \omega \left(\frac{x}{v_1} + t \right) \tag{3-179}$$

and a transmitted wave

$$y_2 = B \cos \omega \left(\frac{x}{v_2} - t \right). \tag{3-180}$$

We wish to determine the values of the reflected and transmitted wave amplitudes. In order to do so, we will impose boundary conditions on the junction such that the displacement y and the slope $\partial y / \partial x$ are continuous across the boundary. The first of these conditions is obvious; the second amounts to the requirement that the restoring force in the string be the same on either side of the junction. These conditions can be expressed mathematically as

$$(y_1 + y_1')_{x=0} = (y_2)_{x=0} \tag{3-181}$$

and
$$\left(\frac{\partial y_1}{\partial x} + \frac{\partial y_1'}{\partial x} \right)_{x=0} = \left(\frac{\partial y_2}{\partial x} \right)_{x=0}. \tag{3-182}$$

Substituting Eqs. (3-178) to (3-180) into these boundary conditions, we have

$$1 + A = B \tag{3-183}$$

and
$$\frac{1 - A}{v_1} = \frac{B}{v_2}. \tag{3-184}$$

Solving these equations simultaneously and using Eq. (3-157), we obtain

$$A = \frac{v_2 - v_1}{v_1 + v_2} = \frac{\sqrt{\sigma_1} - \sqrt{\sigma_2}}{\sqrt{\sigma_2} + \sqrt{\sigma_1}} \tag{3-185}$$

for the amplitude of the reflected wave and

$$B = \frac{2v_2}{v_1 + v_2} = \frac{2\sqrt{\sigma_2}}{\sqrt{\sigma_2} + \sqrt{\sigma_1}} \tag{3-186}$$

for the amplitude of the transmitted wave.

Let us determine the rate at which energy flows into and out of the boundary. We will visualize the particle of string located just at $x = 0$ and calculate the power input P to this particle from the adjacent parts of the string. This power input is merely the product of the restoring force $-\tau \, \partial y/\partial x$ and the transverse velocity $\partial y/\partial t$ of the particle. Thus

$$P = -\tau \frac{\partial y}{\partial x} \cdot \frac{\partial y}{\partial t}\bigg|_{x=0}. \tag{3-187}$$

The input from the left due to the incident and reflected waves is calculated by writing

$$y = y_1 + y_1', \tag{3-188}$$

evaluating the spatial and time derivatives at $x = 0$, and inserting the result into Eq. (3-187):

$$P = \tau \frac{\omega^2}{v_1} (1 + A)(1 - A) \cos^2 \omega t$$

$$= \tau \frac{\omega^2}{v_1} (1 - A^2) \cos^2 \omega t. \tag{3-189}$$

Averaging this quantity over a period, we find

$$\langle P \rangle = \frac{\tau \omega^2}{2 v_1} (1 - A^2). \tag{3-190}$$

Similarly, the average power transmitted into the region of positive x values is

$$\langle P_+ \rangle = \frac{\tau \omega^2}{2 v_2} B^2. \tag{3-191}$$

It can easily be shown by the use of Eqs. (3-183) and (3-184) that

$$\langle P \rangle = \langle P_+ \rangle. \tag{3-192}$$

EXERCISES

3-7.1. Reconsider the example illustrated in Fig. 3-11 from the traveling wave point of view. Superpose two wave functions, $f(x - vt)$ and $g(x + vt)$, chosen to satisfy the boundary and initial conditions, and obtain the solution given in Eq. (3-141).

3-7.2. Show that Eqs. (3-129) and (3-177) are entirely equivalent.

3-7.3. A wave traveling from the left encounters a junction between the heavy string in which it moves and a lighter one. The incoming wave divides equally into a transmitted and reflected component (that is, $A = B$). What is the ratio of the densities of the strings?

3-7.4. For a wave described by Eq. (3-178) incident from the left on a discontinuity in linear mass density located at $x = 0$, show that the reflected wave will be in phase with the incident wave if $\sigma_1 > \sigma_2$ and out of phase by

$180°$ if $\sigma_2 > \sigma_1$. How will the incident and transmitted waves compare in phase for these two cases?

3-7.5. In Fig. 3-13 we show the forces acting on an element dy of a stretched spring. The displacement of the point P due to spring stretching is η and the linear mass density of the string is σ. Write the equation of motion of the element dy and the Hooke's law (statement of the linearity of the restoring force) and, combining them, derive the wave equation for the spring.

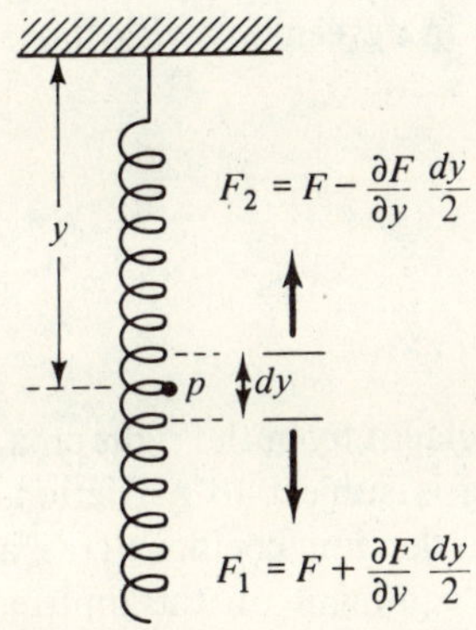

Fig. 3-13 Forces on a stretched spring.

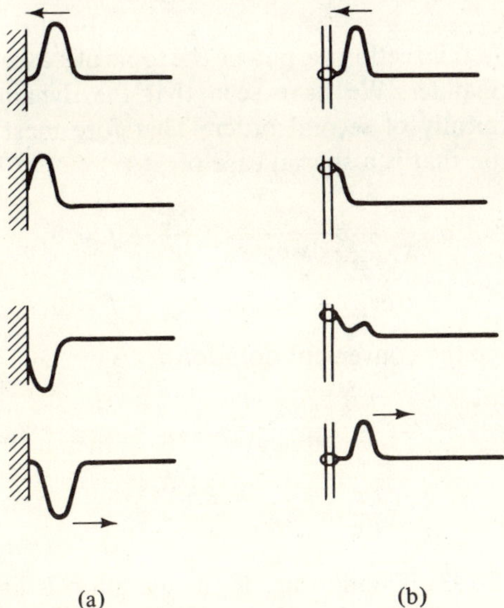

Fig. 3-14 Reflections of a wave at the termination of a string.

3-7.6. A wave in a long string is incident from the right on a fixed end as shown in Fig. 3-14(a). Show that the phase of the reflected wave is changed by 180° with respect to the incident wave. Then consider the case shown in Fig. 3-14(b) where the termination is a massless ring free to slide without friction along a vertical rod. Show that in this case the reflected wave is identical in phase with the incident wave.

3-7.7. Reconsider problem 3-7.6, replacing the wall in part (a) by a string of mass density $\sigma_1 > \sigma_2$, the mass density of the string for $x > 0$. In part (b), replace the ring and rod arrangement by a string of mass density $\sigma_1 < \sigma_2$. Show that the results are in agreement with those of problem 3-7.6 for the limiting cases

(a) $\dfrac{\sigma_1}{\sigma_2} \to \infty,$

(b) $\dfrac{\sigma_1}{\sigma_2} \to 0.$

3-7.8. Consider a wave incident from the right on a ring and rod arrangement wherein the massless ring is subject to a friction force proportional to its velocity. Formulate the reflection coefficient as a function of the friction coefficient. Examine and comment on the limiting cases of very large and very small friction.

3-8 General Remarks on Waves

In this section we pursue the topic of waves in a more general, if less physical, manner. We have seen that the dynamical equations in mechanics are typically of second order. Therefore most physical theories involve an equation that is a special case of

$$a\frac{\partial^2 z}{\partial x^2} + b\frac{\partial^2 z}{\partial x\,\partial y} + c\frac{\partial^2 z}{\partial y^2} + f = 0,$$

$$\tag{3-193}$$

or
$$az_{xx} + bz_{xy} + cz_{yy} + f = 0,$$

where we introduce the convenient notation

$$\frac{\partial^2 z}{\partial x^2} = z_{xx}, \qquad \frac{\partial^2 z}{\partial x\,\partial y} = z_{xy}, \qquad \frac{\partial^2 z}{\partial y^2} = z_{yy},$$

$$\tag{3-194}$$

$$\frac{\partial z}{\partial x} = z_x, \qquad \frac{\partial z}{\partial y} = z_y.$$

In general, Eq. (3-193) is nonlinear. If the equation is linear in its highest derivatives z_{xx}, z_{xy}, z_{yy}, it is said to be *quasilinear*. If the coefficients a, b, and c are functions of x and t only, the equation is *semilinear*; and if f is a linear function of z and its derivatives, then the equation is *linear*. Under what conditions does this equation represent a wave? By "wave" we mean a disturbance propagating at finite speed into a state in which all the field quantities are constant in time.

If we assume, for simplicity, that a, b, and c are functions of z, z_x, z_y only and do not depend on x and t, then Eq. (3-193) possesses a solution of constant state subject to the condition

$$f(u) = 0 \tag{3-195}$$

and we can consider the wave to propagate into this constant state. We denote this constant state ahead of the *wavefront* as region II. The wavefront, which we define to be the boundary between the disturbed state (region I) and the undisturbed state (region II), can be specified as

$$\varphi(x, t) = 0. \tag{3-196}$$

In the undisturbed region all derivatives of z vanish, whereas in the disturbed region there are, in general, nonvanishing derivatives of z. Thus there must exist discontinuities in certain quantities across the wavefront. We will postulate a "smooth" wave such that z, z_x, and z_t are continuous across the wavefront and discontinuity occurs in the second-order derivatives.[4] In the neighborhood of the wavefront we shall introduce curvilinear coordinates φ and ψ defined by

$$\varphi(x, t) = \text{const}$$
$$\psi(x, t) = \text{const} \tag{3-197}$$

in terms of which the wavefront $\varphi = 0$ is conveniently described. We will assume that φ_x, φ_t, ψ_x, and ψ_t are continuous across the wavefront. In terms of φ and ψ, Eq. (3-193) now reads

$$z_{\varphi\varphi}Q(\varphi, \varphi) + 2z_{\varphi\psi}Q(\varphi, \psi) + z_{\psi\psi}Q(\psi, \psi) + z_\varphi\mathcal{L}(\varphi) + z_\psi\mathcal{L}(\psi) + f = 0, \tag{3-198}$$

where
$$Q(\varphi, \psi) = a\varphi_x\psi_x + \tfrac{1}{2}b(\varphi_x\psi_t + \varphi_t\psi_x) + c\varphi_t\psi_t \tag{3-199}$$

and
$$\mathcal{L}(\varphi) = a\varphi_{xx} + b\varphi_{xt} + c\varphi_{tt}. \tag{3-200}$$

Since discontinuities only occur in crossing the wavefront, we can assume that the second-order derivative in the direction across the wavefront is discontinuous

$$[z_{\varphi\varphi}] \neq 0, \tag{3-201}$$

---

[4] If z is continuous across the wavefront and discontinuities occur only in the first- (and/or higher) order derivatives, the wave is said to be a *weak wave*. Waves in which z and its derivatives are discontinuous across the wavefront are called *strong waves*.

whereas those in the direction along the wavefront are continuous

$$[z_{\varphi\psi}] = [z_{\psi\psi}] = 0, \tag{3-202}$$

and the symbol [ ] denotes the *jump* in the indicated quantity across $\varphi = 0$. We now wish to evaluate Eq. (3-198) at points P_1 and P_2 on either side of the wavefront. We subtract the resulting equations and let P_1 and P_2 approach P as shown in Fig. 3-15. Accordingly, we obtain

$$Q(\varphi, \varphi) = 0$$

or
$$a\varphi_x^2 + b\varphi_x\varphi_t + c\varphi_t^2 = 0. \tag{3-203}$$

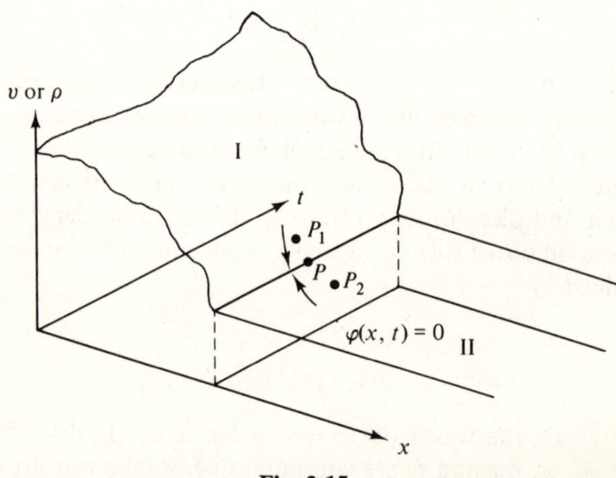

Fig. 3-15

From this result we can conclude that Eq. (3-193) describes a wave as we have defined it only if a real function φ exists that satisfies Eq. (3-203). The criterion for a real solution is

$$b^2 - 4ac > 0 \tag{3-204}$$

and $\varphi(x, t)$ as a solution of Eq. (3-203) determines the wavefront. The velocity of propagation can also be determined by using Eq. (3-203). Since $\varphi(x, t) = 0$, then $\varphi_x\, dx + \varphi_t\, dt = 0$; hence

$$c\, dx^2 - b\, dx\, dt + a\, dt^2 = 0 \tag{3-205}$$

and
$$\frac{dx}{dt} = \frac{1}{2c}\,[b \pm (b^2 - 4ac)^{1/2}]. \tag{3-206}$$

Equation (3-193) is termed *hyperbolic*, *parabolic*, or *elliptic*, depending on whether $b^2 - 4ac$ is positive, zero, or negative. An example of a mixed equation is

$$(1 - M^2)\frac{\partial^2 \phi}{\partial x^2} + \frac{\partial^2 \phi}{\partial y^2} = 0, \tag{3-207}$$

the equation for two-dimensional compressible flow around a rigid body. If the Mach number $M < 1$ for subsonic flow, then $b^2 - 4ac < 0$ and the equation is elliptic. For $M = 1$, the flow is transonic and $b^2 - 4ac = 0$ (a parabolic equation). For supersonic flow, $M > 1$ and $b^2 - 4ac > 0$, indicating the transition to a hyperbolic equation for which we expect wave solutions. Physically, the supersonic speed range is characterized by the formation of a shock wave.

The assumption of a "smooth" wavefront can be relaxed and the hyperbolic condition given in Eq. (3-204) is still found as the criterion for a wave equation. Equation (3-206) defines *characteristic base curves*. There are two such curves for the hyperbolic case, one for the parabolic, and none for the elliptic case.

We shall apply this approach, called the *method of characteristics*, to analyze hydrodynamic and magnetohydrodynamic waves in Chapter 9.

EXERCISES

3-8.1. Examine the following equations of mathematical physics and identify the wave (hyperbolic) equations.

(a) $\dfrac{\partial}{\partial x}\left(\dfrac{\partial z}{\partial x} - z\right) - \dfrac{\partial}{\partial t}\left(\dfrac{\partial z}{\partial t} - z\right) = 0.$

(b) $\dfrac{\partial^2 z}{\partial x^2} + \dfrac{\partial^2 z}{\partial t^2} = 0.$

(c) $\dfrac{\partial^2 z}{\partial x^2} = k\dfrac{\partial z}{\partial t}.$

(d) $\dfrac{\partial^2 z}{\partial x^2} + k^2 z = 0.$

(e) $\dfrac{\partial^2 z}{\partial x^2} + 2\dfrac{\partial^2 z}{\partial x\,\partial t} + \dfrac{\partial^2 z}{\partial t^2} = 0.$

(f) $\dfrac{\partial^2 z}{\partial x^2} - \dfrac{1}{v^2}\dfrac{\partial^2 z}{\partial t^2} = 0.$

3-8.2. Apply Eq. (3-206) to the wave equation for the stretched string, Eq. (3-150), to show that the velocity of propagation of the wavefront is given by Eq. (3-157).

SYSTEMS OF PARTICLES

Up to this point in our study of Newtonian mechanics we have limited the discussion to the motion of single particles. Even though massive bodies are always endowed with some degree of spatial extension, we have been able to treat them as single particles acted on by outside forces by neglecting rotational motions. In the present chapter we develop some principles that are applicable to systems of mutually interacting particles. These principles will find immediate application in the problem of two interacting particles. It will be recalled that the two-body problem for the special case of a $1/r^2$ force of attraction was the first problem solved by Newton, using his laws of mechanics. It is of interest to note that this same problem was the first to be solved by Schrödinger in connection with the dynamics of the hydrogen atom, using his wave mechanics. The nature of the dynamical problem of three or more mutually interacting bodies is also examined, and we shall see the occurrence of mathematical difficulties that have had a profound influence on the course of physics.

4-1 Conservation Laws

In Sections 2-1 and 2-4 we derived some general conservation relations for a single particle. We now wish to extend these notions for a system of mutually interacting particles. In so doing, we must distinguish between *internal* forces due to the mutual interaction of the particles that constitute the system and *external* forces that arise from some agency outside the system.

Suppose that we have a system of n particles having masses $m_1, \ldots, m_n,$

each of which is located with respect to the coordinate origin by a radius vector $r_1, \ldots, r_n$ as shown in Fig. 4-1. We shall write the external force acting on the jth particle as F_j and the internal force on the jth particle due to the presence of the kth particle as F_{jk}. In terms of these forces, the equation of motion of the jth particle under the action of the internal and external forces is

$$m_j \ddot{r}_j = F_j + \sum_{k=1}^{n} F_{jk}. \tag{4-1}$$

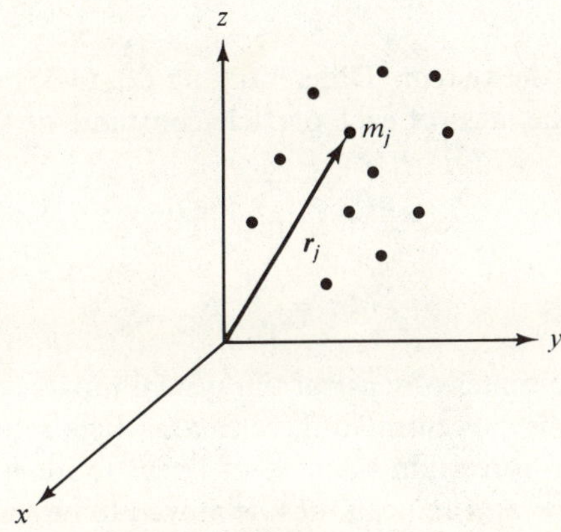

Fig. 4-1

In this equation we are assuming, as heretofore, that the mass m_j remains constant. Let us sum this equation over all n particles. We obtain

$$\sum_{j=1}^{n} m_j \ddot{r}_j = \sum_{j=1}^{n} F_j + \sum_{\substack{j=1 \\ j \neq k}}^{n} \sum_{k=1}^{n} F_{jk}, \tag{4-2}$$

where we have inserted the condition on the double sum that terms like F_{jj} may never arise, since the particle cannot exert a force on itself.

Next, we use Newton's third law of motion, which states that every action involves an *equal* and *opposite* reaction.[1] As applied to particles j and k, it implies that

$$F_{jk} = -F_{kj}. \tag{4-3}$$

This result, together with the physical requirement that $j \neq k$, causes the last term on the right-hand side of Eq. (4-2) to vanish, leaving us with the result

---

[1] We will limit the use of this "action at a distance" law to apply only to those cases where the time that it takes a light beam to pass between particles j and k is much less than the shortest time of interest.

that the motion of the system as a whole is governed solely by the external forces:

$$\sum_{j=1}^{n} m_j \ddot{\mathbf{r}}_j = \sum_{j=1}^{n} \mathbf{F}_j. \tag{4-4}$$

The radius vector to the *center of mass* of the system is defined by

$$\mathbf{r}_c = \frac{\sum m_j \mathbf{r}_j}{\sum m_j}, \tag{4-5}$$

where

$$M = \sum_{j=1}^{n} m_j \tag{4-6}$$

is the total mass of the system. Differentiating Eq. (4-5) twice with respect to time and holding the mass of each particle constant, we find

$$M\ddot{\mathbf{r}}_c = \sum_{j=1}^{n} m_j \ddot{\mathbf{r}}_j; \tag{4-7}$$

hence

$$M\ddot{\mathbf{r}}_c = \sum_{j=1}^{n} \mathbf{F}_j. \tag{4-8}$$

In other words, the center of mass of the system moves as though the entire mass of the system is concentrated there as a particle acted on by the vector sum of the external forces. In the case of linear motion, the external force vectors act as *free vectors* and can thus be moved to be summed at the center of mass. Rotational motion of the system, as we shall see, depends explicitly on the point of application of the external forces. It follows immediately from Eq. (4-8) that if the vector sum of the external force vectors is zero, then the linear momentum of the system

$$\mathbf{p} = \sum_{j=1}^{n} m_j \dot{\mathbf{r}}_j = M\dot{\mathbf{r}}_c \tag{4-9}$$

is a constant.

Turning next to the angular motion of the system, we shall define the angular momentum of the jth particle about the origin denoted by $\mathbf{L}_j$ as

$$\mathbf{L}_j = \mathbf{r}_j \times \mathbf{p}_j = m_j(\mathbf{r}_j \times \dot{\mathbf{r}}_j). \tag{4-10}$$

In the more general case, the angular momentum of the jth particle about an arbitrary point A (see Fig. 4-2) can be written

$$\mathbf{L}_{jA} = m_j(\mathbf{r}_j - \mathbf{r}_A) \times (\dot{\mathbf{r}}_j - \dot{\mathbf{r}}_A). \tag{4-11}$$

With this definition in hand, we write the linear equation of motion of the jth particle

$$\dot{\mathbf{p}}_j = \mathbf{F}_j + \sum_{\substack{k=1 \\ j \neq 1}}^{n} \mathbf{F}_{jk} \tag{4-12}$$

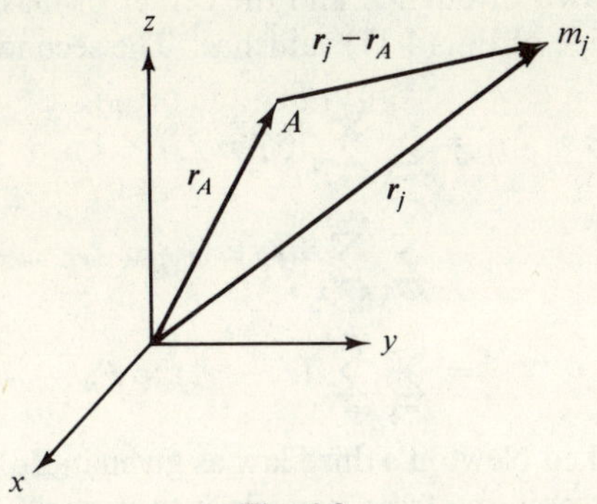

Fig. 4-2

and take its moment about the point A by cross multiplying from the left by $(r_j - r_A)$:

$$(r_j - r_A) \times \dot{p}_j = (r_j - r_A) \times F_j + \sum_k (r_j - r_A) \times F_{jk}. \qquad (4\text{-}13)$$

Using Eq. (4-11), we find

$$\dot{L}_{jA} = m_j(r_j - r_A) \times (\ddot{r}_j - \ddot{r}_A) + m_j(\dot{r}_j - \dot{r}_A) \times (\dot{r}_j - \dot{r}_A)$$

$$= m_j(r_j - r_A) \times (\ddot{r}_j - \ddot{r}_A)$$

$$= (r_j - r_A) \times \dot{p}_j - m_j(r_j - r_A) \times \ddot{r}_A$$

$$= (r_j - r_A) \times F_j + (r_j - r_A) \times \sum_k F_{jk} - m_j(r_j - r_A) \times \ddot{r}_A, \qquad (4\text{-}14)$$

where Eq. (4-12) has been used. Defining the total angular momentum and total external torque about the point A as

$$J_A = \sum_{j=1}^{n} L_{jA} \qquad (4\text{-}15)$$

and

$$N_A = \sum_{j=1}^{n} (r_j - r_A) \times F_j, \qquad (4\text{-}16)$$

respectively, we find on summing Eq. (4-14) over all n particles and using Eqs. (4-5) and (4-6),

$$\frac{dJ_A}{dt} = N_A + \sum_{j=1}^{n} \sum_{k \neq j}^{n} (r_j - r_A) \times F_{jk} - M(r_c - r_A) \times \ddot{r}_A. \qquad (4\text{-}17)$$

We see that if (1) $r_c = r_A$, the point A coincides with the center of mass; (2) $\ddot{r}_A = 0$, A is a fixed point in an inertial frame; or (3) the acceleration of A

133

is along the line drawn through A and the center of mass, then the last term on the right-hand side of Eq. (4-17) vanishes. The second term on the right is

$$\sum_{j=1}^{n}\sum_{k\neq j}^{n}(r_j - r_A)\times F_{jk} = \sum_{j=1}^{n}\sum_{k=1}^{j-1}[(r_j - r_A)\times F_{jk} + (r_k - r_A)\times F_{kj}]$$

$$= \sum_{j=1}^{n}\sum_{k=1}^{j-1}[(r_j - r_A) - (r_k - r_A)\times F_{jk}]$$

$$= \sum_{j=1}^{n}\sum_{k=1}^{j-1}(r_j - r_k)\times F_{jk}, \tag{4-18}$$

where we have invoked Newton's third law as given in Eq. (4-3). The quantity $(r_j - r_k)$ is a vector pointing from particle k to particle j. Clearly, if F_{jk} lies along this line, as is true for central forces, or if the F_{jk} are zero, then the second term on the right of Eq. (4-17) also vanishes and we are left with

$$\dot{J}_A = N_A. \tag{4-19}$$

If the vector sum of the moments of all the external forces is zero or, more simply, if the external forces are zero, then the total angular momentum of the system is a constant of the motion.

In many cases of practical interest, the total force acting on particle j

$$\mathscr{F}_j = F_j + \sum_{k=1}^{n} F_{jk} \tag{4-20}$$

depends only on the radius vectors of the particles in the system:

$$\mathscr{F}_j = \mathscr{F}_j(r_1, \ldots, r_n). \tag{4-21}$$

In principle, a potential function V may then exist such that[2]

$$\mathscr{F}_j = -\nabla_j V \tag{4-22}$$

if the force field is a conservative one. The equation of motion of the jth particle subject to such forces is

$$\dot{p}_j = -\nabla_j V. \tag{4-23}$$

Taking the dot product of $p_j/m_j = \dot{r}_j$ with both sides of this equation, we find

$$\frac{d}{dt}\left(\frac{p_j^2}{2m_j}\right) + \nabla_j V \cdot \dot{r}_j = 0. \tag{4-24}$$

Summing this equation over all n particles gives

$$\frac{d}{dt}\sum_{j=1}^{n}\left(\frac{p_j^2}{2m_j}\right) + \sum_{j=1}^{n}\nabla_j V \cdot \dot{r}_j = 0. \tag{4-25}$$

---

[2] The nature of such potentials for systems of particles subject to various forces of mutual interaction is discussed in Chapter 8.

The first term of this equation is the time derivative of the total kinetic energy of the system

$$\frac{dT}{dt} = \frac{d}{dt} \sum_{j=1}^{n} \left(\frac{P_j^2}{2m_j} \right) = \frac{d}{dt} \frac{P^2}{2M}, \tag{4-26}$$

and the second is dV/dt:

$$\frac{dV}{dt} = \sum_{j=1}^{n} \boldsymbol{\nabla}_j V \cdot \dot{\boldsymbol{r}}_j = \boldsymbol{\nabla} V \cdot \dot{\boldsymbol{r}} = \frac{d}{dt} \boldsymbol{\nabla} V \cdot \boldsymbol{r}. \tag{4-27}$$

Therefore Eq. (4-25) can be written

$$\frac{d}{dt}(T + V) = \frac{dE}{dt} = 0 \tag{4-28}$$

or

$$E = \text{const}, \tag{4-29}$$

where E is the total energy of the system.

If the internal forces are conservative but the external forces are not, then the total energy of the system will vary according to

$$\frac{d}{dt}(T + V^{(i)}) = \sum_{j=1}^{n} \boldsymbol{F}_j \cdot \dot{\boldsymbol{r}}_j, \tag{4-30}$$

where

$$V^{(i)} = -\sum_{j=1}^{n} \int_{\infty}^{r_j} \boldsymbol{F}_{jk} \cdot d\boldsymbol{r}_j \tag{4-31}$$

is the potential energy corresponding to the field of the internal forces.

It is worthwhile to note that the total kinetic energy T of a system of particles can be written as the sum of the kinetic energy of translation of the center of mass plus the kinetic energy of the particles relative to the center of mass. This situation is easily shown.

Let us define two vectors $\boldsymbol{r}_c$ and $\boldsymbol{r}_j$, depicted in Fig. 4-3, as the radius vector to a point C and a vector from C to the jth particle, respectively. In terms of these vectors, the total kinetic energy of the system is

$$T = \sum_{j=1}^{n} T_j = \sum_{j=1}^{n} \frac{m_j}{2} \dot{r}_j^2 = \sum_{j=1}^{n} \left[\frac{m_j}{2} (\dot{\boldsymbol{r}}_c + \dot{\boldsymbol{r}}_j') \cdot (\dot{\boldsymbol{r}}_c + \dot{\boldsymbol{r}}_j') \right]$$

$$= \frac{\dot{r}_c^2}{2} \sum_{j=1}^{n} m_j + \dot{\boldsymbol{r}}_c \cdot \sum_{j=1}^{n} m_j \dot{\boldsymbol{r}}_j' + \sum_{j=1}^{n} \frac{m_j}{2} \dot{r}_j'^2. \tag{4-32}$$

If we then specify the point C to be the center of mass of the system, the term $\sum_{j=1}^{n} m_j \dot{\boldsymbol{r}}_j'$ must vanish and we are left with

$$T = \frac{M}{2} \dot{r}_c^2 + \sum_{j=1}^{n} \frac{m_j}{2} \dot{r}_j'^2 \tag{4-33}$$

as promised.

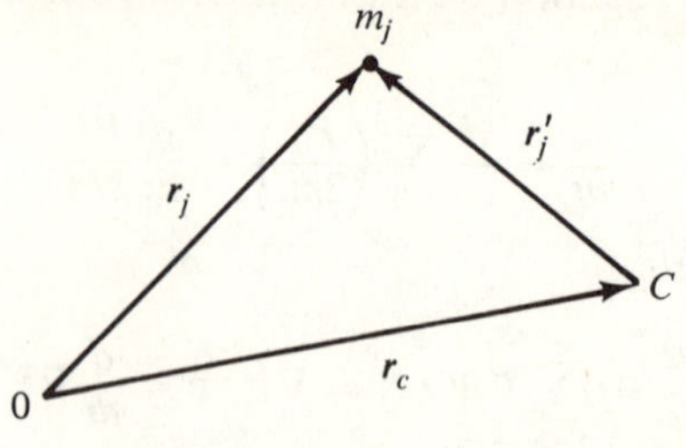

Fig. 4-3

EXERCISES

4-1.1. An amount of energy E is released when a gun of mass m_1 fires a projectile of mass m_2. Consider the two cases in which (1) the gun carriage rests on a frictionless surface and (2) the gun carriage is firmly fixed in the earth. Calculate the projectile velocity and the fraction of the energy lost to recoil for each case.

4-1.2. A boy who weighs 80 lb is standing on one end of a 20-ft long, 200-lb raft that has its other end just against the shore. The boy walks to the other end of the raft and then stops. How far is he from the shore?

4-1.3. The distance between the centers of the earth and the moon is about 60 earth radii. The center of mass of the earth–moon system is located within the earth at a point about $0.23R_e$ below the surface. What is the ratio of the lunar mass to that of the earth?

4-1.4. A block of soft wood of mass m_1 is suspended by a string of length l. A bullet of mass m_2 is fired from a gun and lodges in the block. The pendulum arrangement is then observed to deflect until a maximum angle θ is reached. Find the speed of the bullet.

4-1.5. A massless, elastic hacksaw blade has one end embedded in a wall. The free end is known to experience a 2-cm deflection when a 500-g mass is placed on the tip. A blob of putty having a mass of 200 g is dropped onto the blade tip, where it sticks with no energy loss. The maximum excursion of the blade in the oscillation that ensues is observed to coincide with the initial horizontal position of the blade. From what height was the putty dropped?

4-2 The Two-Body Problem

Here we are concerned with the problem of two interacting particles. The radius vectors in terms of which we shall describe the problem are shown in Fig. 4-4. The only forces assumed to act are F_1 on particle 1 and

136

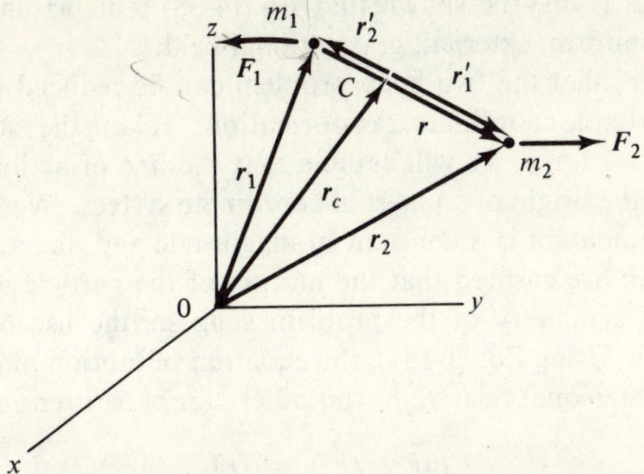

Fig. 4-4 Schematic definition of internal coordinates.

F_2 on particle 2. These forces need not be central. In terms of the cartesian system centered at o, the equations of motion of the two particles are

$$m_1\ddot{r}_1 = F_1, \qquad m_2\ddot{r}_2 = F_2. \tag{4-34}$$

Since

$$r_2 = r_1 + r, \tag{4-35}$$

then

$$m_2\ddot{r}_2 = m_2\ddot{r}_1 + m_2\ddot{r}. \tag{4-36}$$

Using Eq. (4-34), we find

$$F_2 = \frac{m_2}{m_1} F_1 + m_2\ddot{r}. \tag{4-37}$$

Since $F_2 = -F_1$ by Newton's third law, Eq. (4-37) can be expressed as

$$\frac{m_1 m_2}{m_1 + m_2} \ddot{r} = F_2, \tag{4-38}$$

which is a very useful result. Equation (4-36) says, in effect, that the motion of particle 2 relative to particle 1 is equivalent to the situation in which particle 1 is fixed (given infinite mass) and m_2 is replaced by

$$\mu = \frac{m_1 m_2}{m_1 + m_2}, \tag{4-39}$$

which is called the *reduced mass* of the system. Note that, in deriving this result, we have implicitly assumed that the particle masses remain constant when we wrote Eq. (4-36). We have also neglected the influence of external forces, although it may be shown that Eq. (4-38) remains unaltered in the presence of a uniform, external, gravitational field.

We see, then, that the two-body problem can be reduced to a one-body problem by a suitable coordinate transformation. Taking the interaction force $\mathbf{F}_2$ to be a central force, we will assume that the line of action of the force passes through the origin of an inertial coordinate system. We have seen that the angular momentum is a constant in magnitude and direction for such a force, and so we are ensured that the motion of the particle will be planar. The cylindrical symmetry of the problem suggests the use of plane polar coordinates r, θ. Using Eq. (1-131), the equation of motion of either particle (usually the lighter one) relative to the other can be written in scalar form:

$$\mu(\ddot{r} - r\dot{\theta}^2) = f(r), \qquad (4\text{-}40)$$

$$\mu(r\ddot{\theta} + 2\dot{r}\dot{\theta}) = 0, \qquad (4\text{-}41)$$

where μ is the reduced mass of the system and

$$\mathbf{F} = f(r)\hat{\mathbf{r}} \qquad (4\text{-}42)$$

is the force of interaction. Equation (4-41) is equivalent to

$$\mu \frac{d}{dt}(r^2\dot{\theta}) = 0. \qquad (4\text{-}43)$$

This equation can be integrated directly to yield

$$\mu r^2 \dot{\theta} = J = \text{const}, \qquad (4\text{-}44)$$

which is merely the angular momentum.

We can now use Eq. (4-44) to eliminate $\dot{\theta}$ from Eq. (4-40). The radial equation of motion then takes the form

$$\ddot{r} - \frac{J^2}{\mu^2 r^3} = \frac{f(r)}{\mu}, \qquad (4\text{-}45)$$

which involves only r and t. If desirable, Eqs. (4-44) and (4-45) could be solved for $\theta(t)$ and $r(t)$. It is more useful to eliminate t and write an equation in terms of r and θ from which the equation of the particle orbit can be derived. Using Eq. (4-44), $\ddot{r}$ can be written

$$\ddot{r} = \frac{J^2}{\mu^2 r^4}\left[\frac{d^2 r}{d\theta^2} - \frac{2}{r}\left(\frac{dr}{d\theta}\right)^2\right]. \qquad (4\text{-}46)$$

The differential equation for the particle orbit is then

$$\frac{d^2 r}{d\theta^2} - \frac{2}{r}\left(\frac{dr}{d\theta}\right)^2 - r = \frac{\mu r^4}{J^2} f(r). \qquad (4\text{-}47)$$

This equation is nonlinear. However, the simple substitution of $u = 1/r$ for the dependent variable with

$$\frac{dr}{d\theta} = -\frac{1}{u^2}\frac{du}{d\theta} \tag{4-48}$$

and

$$\frac{d^2r}{d\theta^2} = \frac{2}{u^3}\left(\frac{du}{d\theta}\right)^2 - \frac{1}{u^2}\frac{d^2u}{d\theta^2} \tag{4-49}$$

transforms Eq. (4-47) into the simpler form

$$\frac{d^2u}{d\theta^2} + u = -\frac{\mu}{J^2}\frac{f(1/u)}{u^2}. \tag{4-50}$$

At this point we will specify the force law to be an inverse-square attractive force

$$f(r) = -kr^{-2} \rightarrow f\left(\frac{1}{u}\right) = -ku^2, \tag{4-51}$$

for which the right-hand side of Eq. (4-50) is a constant:

$$\frac{d^2u}{d\theta^2} + u = \frac{k\mu}{J^2}. \tag{4-52}$$

This is the equation of an undamped oscillator with a constant driving term, and we can write the solution by inspection as

$$u = A\cos(\theta + \varphi) + \frac{k\mu}{J^2}, \tag{4-53}$$

or in terms of r

$$r = \left[A\cos(\theta + \varphi) + \frac{k\mu}{J^2}\right]^{-1}, \tag{4-54}$$

where A and φ are the constants of integration. In order to interpret Eq. (4-54), we recall from analytic geometry the general definition of a conic section: the curve followed by a point moving on a plane in such a way that the ratio of its distance from a fixed point to its distance from a fixed line remains constant. The fixed line is called the *directrix*, the fixed point is the *focus*, and the constant ratio is the *eccentricity*, ϵ. In terms of Fig. 4-5, the focus is at O, and the vertical line CB, located a horizontal distance p from the focus, is the directrix. We can define the conic section $\mathscr{C}$ by

$$\epsilon = \frac{OP}{PC} = \frac{r}{p - OA} = \frac{r}{p - r\cos\theta}, \tag{4-55}$$

or solving for r

$$r = \left(\frac{1}{\epsilon p} + \frac{1}{p}\cos\theta\right)^{-1}. \tag{4-56}$$

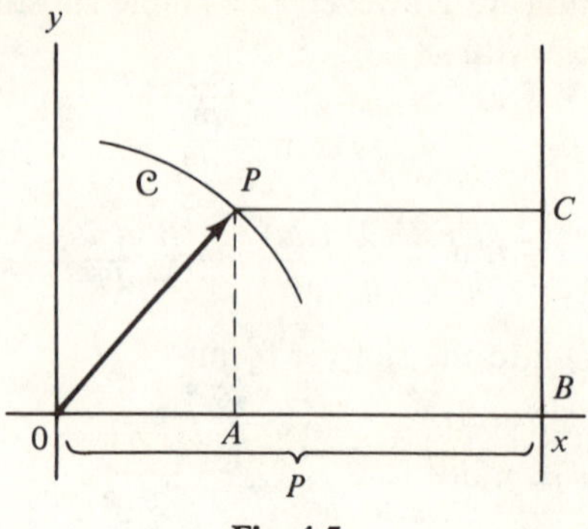

Fig. 4-5

Comparing Eqs. (4-54) and (4-56), we see that a central, inverse-square attractive force leads to particle orbits that are conic sections. For convenience, we can set $\varphi = 0$ in Eq. (4-54), thereby putting the solution into standard conic form:

$$r = \left(\frac{k\mu}{J^2} + A \cos \theta\right)^{-1}. \tag{4-57}$$

The arbitrary constant A is now determined by the boundary conditions of the particular problem.

It is also interesting to look at the problem from the energy point of view. In addition to the angular momentum integral expressed in Eq. (4-44), we can write an energy integral

$$E = T + V, \tag{4-58}$$

where E is a constant. Written explicitly, this equation becomes

$$\frac{\mu}{2}(\dot{r}^2 + r^2\dot{\theta}^2) - \int_{\infty}^{r} f(r)\, dr = E. \tag{4-59}$$

Since

$$\dot{r} = \frac{dr}{d\theta}\,\dot{\theta}, \tag{4-60}$$

Eq. (4-59) becomes

$$\frac{J^2}{2\mu r^4}\left[\left(\frac{dr}{d\theta}\right)^2 + r^2\right] - \int_{\infty}^{r} f(r)\, dr = E, \tag{4-61}$$

where Eq. (4-44) has been used to eliminate $\dot{\theta}$. This is the equation for the orbit. Specifying $f(r) = -k/r^2$, the potential energy becomes $-k/r$ and Eq. (4-61) becomes

$$\left(\frac{dr}{d\theta}\right)^2 + r^2 = \frac{2\mu r^4}{J^2}\left(E + \frac{k}{r}\right). \tag{4-62}$$

140

Solving for the derivative and separating variables, we find, on carrying out the integration,

$$\theta = \arcsin\left[\frac{\mu k r - J^2}{r(2J^2\mu E + \mu^2 k^2)^{1/2}}\right] - \phi,\qquad(4\text{-}63)$$

where ϕ is the constant of integration. Choosing $\phi = 3\pi/2$ and solving for r, we have

$$r = \frac{J^2/\mu k}{1 + [(2J^2 E/\mu k^2) + 1]^{1/2}\cos\theta}.\qquad(4\text{-}64)$$

Comparing Eqs. (4-54), (4-56), and (4-64), we see that they are mutually compatible if the eccentricity is given by

$$\epsilon = \sqrt{\frac{2J^2 E}{\mu k^2} + 1}\qquad(4\text{-}65)$$

and the integration constant A in Eq. (4-54) is specified to be

$$A = \frac{\mu k \epsilon}{J^2}.\qquad(4\text{-}66)$$

The nature of the orbit—that is, whether it is a circle, ellipse, parabola, or hyperbola—depends on the value of the eccentricity ϵ. The total energy E of the particle can be written in terms of J and ϵ:

$$E = \frac{\mu k^2}{2J^2}(\epsilon^2 - 1).\qquad(4\text{-}67)$$

For circular orbits, $\epsilon = 0$ and the energy of a particle in a circular orbit is

$$E_c = -\frac{\mu k^2}{2J^2}.\qquad(4\text{-}68)$$

For elliptical orbits, the eccentricity lies in the range

$$0 < \epsilon < 1,\qquad(4\text{-}69)$$

and the corresponding range of allowable total energies is

$$E_c < E_e < 0.\qquad(4\text{-}70)$$

Parabolic orbits, like circular orbits, are a singular case, corresponding as they do to $\epsilon = 1$:

$$E_p = 0.\qquad(4\text{-}71)$$

The hyperbolic orbits, $\epsilon > 1$, describe particles that are too energetic to be bound:

$$E_h > 0.\qquad(4\text{-}72)$$

This description can be further clarified by constructing an energy diagram. For our purpose, it is convenient to visualize the problem in one dimension and to express motion in the θ direction in terms of J by means of the angular momentum integral Eq. (4-44):

$$\frac{\mu}{2} r^2 \dot{\theta}^2 = \frac{J^2}{2\mu r^2}. \tag{4-73}$$

This term, no longer containing a derivative, can then be lumped with the potential $V = -k/r$ to form an *effective potential*

$$V_{\text{eff}} = \frac{J^2}{2\mu r^2} - \frac{k}{r}. \tag{4-74}$$

In Fig. 4-6 we have plotted the two terms of the effective potential and their sum, shown as a solid curve for arbitrary values of J, k, and μ. For $\epsilon = 0$ (circular orbit), the coefficient of the cosine term in Eq. (4-54) vanishes [see Eq. (4-66)] and

$$r_c = \frac{J^2}{\mu k}, \tag{4-75}$$

as can also be verified by solving for the location of the minimum of V_{eff}. For a higher-energy particle with E still less than zero, the orbit will be elliptic and radial turning points will be solutions of the equation

$$V_{\text{eff}} = E = \frac{J^2}{2\mu r^2} - \frac{k}{r}. \tag{4-76}$$

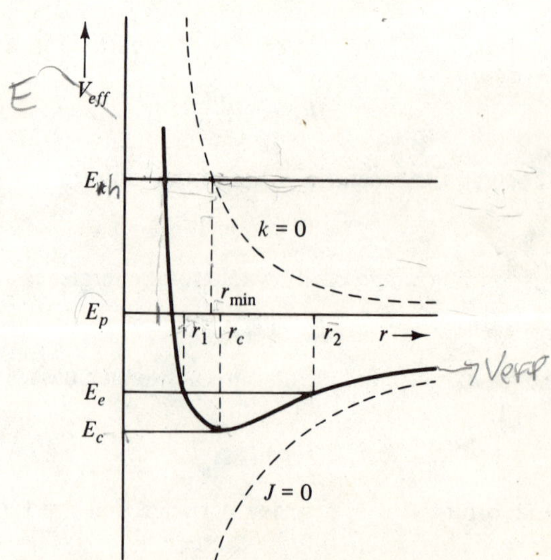

Fig. 4-6 Effective potential for inverse-square central force.

It is left as an exercise to show that these turning points are given by

$$\frac{1}{r_1} = \frac{\mu k}{J^2} + \sqrt{\left(\frac{\mu k}{J^2}\right)^2 + \frac{2\mu E}{J^2}},$$ (4-77)

$$\frac{1}{r_2} = \frac{\mu k}{J^2} - \sqrt{\left(\frac{\mu k}{J^2}\right)^2 + \frac{2\mu E}{J^2}}.$$ (4-78)

As the particle energy is increased until a critical value $E = 0$ is reached, the second turning point moves out to infinity and the first becomes

$$r_p = \frac{J^2}{2\mu k}$$ (4-79)

and the orbit becomes parabolic. For $E > 0$, the orbit is hyperbolic and the closest distance of approach is given by Eq. (4-77).

It is interesting, for the case of elliptical orbits, to look at the properties of the orbit in terms of the major axis $2a$. This, the longest diameter of the orbit, is simply the sum of the radial extrema r_1 and r_2:

$$2a = r_1 + r_2.$$ (4-80)

For the purpose of this calculation, it is convenient to multiply Eq. (4-74) by r^2 before solving for the roots. We then obtain r_1 and r_2 as

$$r_1 = -\frac{k}{2E} - \sqrt{\left(\frac{k}{2E}\right)^2 + \frac{J^2}{2\mu E}},$$ (4-81)

$$r_2 = -\frac{k}{2E} + \sqrt{\left(\frac{k}{2E}\right)^2 + \frac{J^2}{2\mu E}}.$$ (4-82)

So we see that the major axis can be expressed as

$$2a = -\frac{k}{E}.$$ (4-83)

Then it can be shown, using Eqs. (4-64) and (4-65), that the equation of the orbit can be written

$$r = \frac{a(1 - \epsilon^2)}{1 + \epsilon \cos \theta}.$$ (4-84)

It is of interest to note that the energy of a particle traveling in an elliptical orbit in an inverse-square field depends only on the major axis of the orbit. This fact is the basis of our ability to express the energy of bound particles in quantum mechanics in terms of the *principal quantum number*.

For circular orbits, $r \to a$ in the limit as $\epsilon \to 0$. From Eq. (4-83) we see that the total energy

$$T + V = T - \frac{k}{a} = -\frac{k}{2a};$$
(4-85)

thus

$$T = \frac{k}{2a} = \frac{\mu}{2}\, a^2 \dot{\theta}^2.$$
(4-86)

The first two of Kepler's laws of planetary motion were set down in 1609. They are

1. *The orbit of a planet is an ellipse with the sun located at one focus.*
2. *The radius vector drawn from the sun to a planet sweeps out equal areas of the orbit in equal times.*

In 1619 Kepler presented a third law.

3. *The square of the period of a planet's orbit is proportional to the cube of the major axis.*

These laws were derived empirically from the data of Tycho Brahe's observations. Newton was able to show over 50 years later that these laws are consistent only with an inverse-square attractive force between the sun and its planets.

The first of Kepler's laws has been established by the derivation of Eq. (4-54). In order to verify the second, consider Fig. 4-7, which shows a particle of mass μ moving along the curve $\mathscr{C}$ under the influence of a central force seated at O. The area of the infinitesimal triangle OBB' is

$$dA = \tfrac{1}{2} r^2\, d\theta,$$
(4-87)

and so

$$r^2 \dot{\theta} = 2\frac{dA}{dt} = \frac{J}{\mu},$$
(4-88)

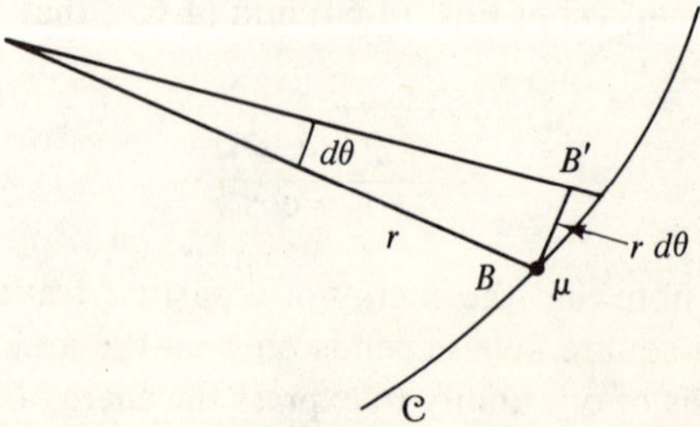

Fig. 4-7

a constant as required by Kepler's second law. In order to see that this law of constant areal velocity can only arise for an inverse-square law of attraction, substitute the standard form of a conic section written in terms of $u = 1/r$

$$u = \frac{1}{\epsilon p} + \frac{1}{p} \cos \theta \tag{4-89}$$

into Eq. (4-50) and solve for $f(r) = f(1/u)$. The result is

$$f(r) = -\frac{J^2}{\mu \epsilon p r^2} = -\frac{k}{r^2}. \tag{4-90}$$

Let us now look at Kepler's third law. We can write the period of motion of a planet in its orbit as

$$\tau = \frac{\pi ab}{dA/dt} = \frac{2\pi ab\mu}{J}, \tag{4-91}$$

where πab is the total area of the orbit, a and b are the semimajor and minor axes of the orbit, respectively, and we have made use of Eq. (4-88). The quantity $\epsilon p = b^2/a$ for an ellipse. Using this relation and eliminating J by means of Eq. (4-90), the period can be expressed as

$$\tau^2 = \frac{4\pi^2 \mu a^3}{k} \tag{4-92}$$

in accordance with Kepler's third law.

Most of the planets move in orbits that are very nearly circular. The eccentricity of the earth's orbit is only about 1/60. Such orbits can be treated as *perturbed* circular orbits, a point that will be discussed in Chapter 12.

Hyperbolic orbits are of interest in studying the nature of the interaction of two particles during a *scattering* process. If a light particle having $E > 0$ is caused to interact via an inverse-square central force with a heavy particle, the light particle will follow a hyperbolic path. In analyzing an atomic collision, the main item of observational datum is the scattering or deflection angle $\Theta = \pi - 2\alpha$ between the incoming and outgoing asymptotes of the hyperbola. This situation is shown in Fig. 4-8, drawn for a repulsive force. The orbit is determined by the interaction constant k and by

$$J = p_0 s \tag{4-93}$$

and
$$E = T_0 = \frac{p_0^2}{2\mu}, \tag{4-94}$$

where p_0 is the initial momentum of the incoming particle. The equation of the hyperbola in polar coordinates is

$$r = \frac{a(\epsilon^2 - 1)}{-1 + \epsilon \cos \theta}. \tag{4-95}$$

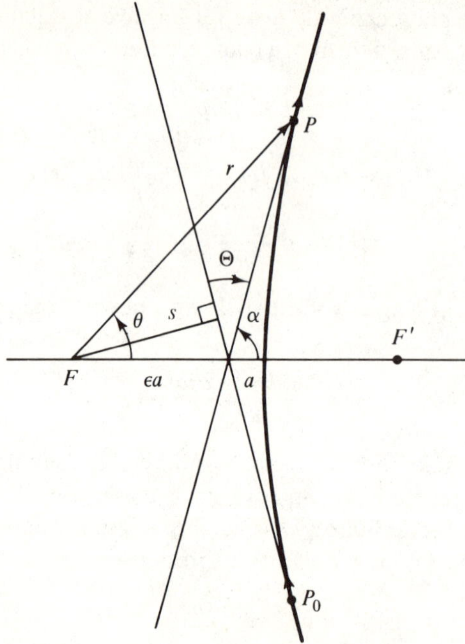

Fig. 4-8 Schematic representation of Rutherford scattering.

It can be seen from Fig. 4-8 that as $r \to \infty$, $\theta \to \alpha$. Applying this limit to Eq. (4-95), we find

$$\cos \alpha = \frac{1}{\epsilon}.$$

(4-96)

Using Eqs. (4-65) and (4-96), we see that

$$\tan \frac{\Theta}{2} = \cot \alpha = (\epsilon^2 - 1)^{-1/2} = \left(\frac{\mu k^2}{2J^2 E} \right)^{1/2},$$

(4-97)

or with Eqs. (4-93) and (4-94)

$$\tan \frac{\Theta}{2} = \frac{k}{2Es}.$$

(4-98)

In a typical scattering experiment a beam of particles, all moving in the same direction, is directed on a foil target and the *distribution* of scattering angles among the emerging beam particles is measured. The impact parameter s is no longer a useful input parameter, since we do not know it.

We will define a *differential scattering cross section $d\sigma$* in the following way. If the fraction of the total number of incident particles N scattered into the angular range Θ to $\Theta + d\Theta$ is dN/N, then we define

$$\frac{dN}{N} = n \, d\sigma, \tag{4-99}$$

where n is the number of scattering centers per unit area of the foil. The differential cross section $d\sigma$ can also be given in terms of Fig. 4-9 as

$$d\sigma = 2\pi s \, ds. \tag{4-100}$$

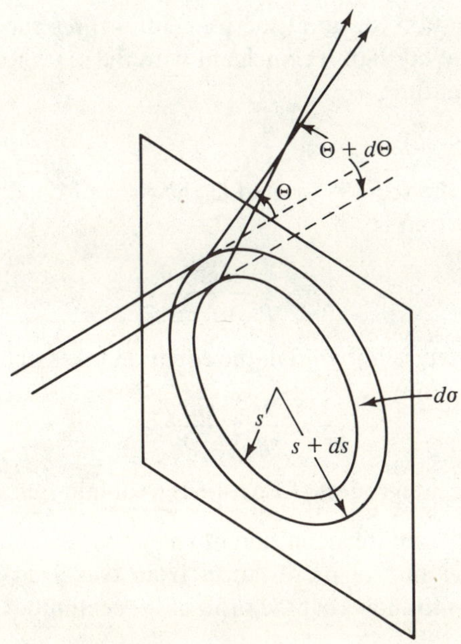

Fig. 4-9 Differential cross section for scattering into the angular range Θ to $\Theta + d\Theta$.

Eliminating s between Eqs. (4-98) and (4-100), we find for $d\sigma$:

$$d\sigma = \frac{\pi}{2} \left(\frac{k}{2E} \right)^2 \sin \Theta \, \csc^4 \frac{\Theta}{2} \, d\Theta, \tag{4-101}$$

where a minus sign has been used in the calculation of ds in order to specify a repulsive force. The scattering distribution is then

$$dN = N(\Theta) \, d\Theta = \frac{\pi N n}{2} \left(\frac{k}{2E} \right)^2 \sin \Theta \, \csc^4 \frac{\Theta}{2} \, d\Theta. \tag{4-102}$$

The interaction between a proton, say, and a heavy atomic nucleus is quite properly a problem for quantum mechanics. The result of the quantum mechanical calculation is, as luck would have it, just the same as we obtain classically. If this were not so, then Rutherford's alpha scattering experiment that established the model of the nuclear atom would not have been properly interpreted and the history of atomic physics might have been quite different.[3]

EXERCISES

4-2.1. Prove that Eq. (4-38) is valid if the two particles are located in a uniform gravitational field.

4-2.2. Find the equation for an attractive central force such that all circular orbits with centers of curvature coincident with the seat of the force have the same angular momentum.

4-2.3. Derive Eq. (4-46).

4-2.4. Show that if the focus is located at the right of the directrix, the equation of the conic section is

$$\frac{1}{r} = \frac{1}{\epsilon p} - \frac{\cos \theta}{p},$$

and that if the directrix is horizontal, the equation takes one of the two forms

$$\frac{1}{r} = \frac{1}{\epsilon p} \pm \frac{\sin \theta}{p}.$$

4-2.5. Carry out the integration of Eq. (4-62) to obtain Eqs. (4-63) and (4-64).

4-2.6. Use Eq. (4-84) and the definition of an ellipse: the locus of a point that moves such that the sum of its distances from two fixed points (the focii) remains a constant—to show that the ratio of the semiminor to the semimajor axis is

$$\frac{b}{a} = \sqrt{1 - \epsilon^2}$$

and the perigee (closest distance of approach of a particle to the force center) and apogee (greatest distance from the force center) are $a(1 - \epsilon)$ and $a(1 + \epsilon)$, respectively.

4-2.7. Verify Eq. (4-75) by comparing with the minimum of the V_{eff} curve as given by Eq. (4-74).

4-2.8. Show that the radial turning points for V_{eff} are given by Eqs. (4-77) and (4-78).

---

[3] See J. Norwood, *Twentieth Century Physics*. Englewood Cliffs, N.J.: Prentice-Hall, Inc., 1976.

4-2.9. In Yukawa's theory of nuclear forces the potential between two nuclei has the form

$$V = -\frac{k}{r}\, e^{-\alpha r},$$

where k is a positive quantity. Derive the force corresponding to this potential and comment on the qualitative difference between this force and the inverse-square force. Show that, to first order in the exponential, the turning points for elliptic orbits are given by Eqs. (4-77) and (4-78) with E replaced by $E - k\alpha$. Evaluate J and E for a circular orbit of radius a.

4-2.10. Sketch the effective potential for an inverse-cube attractive force and discuss the various types of motion possible under such a force. Solve Eq. (4-50) and write the solution for $\mu k/J^2 < 1$, $\mu k/J^2 > 1$, and $\mu k/J^2 = 1$. Sketch typical orbits for each case. Evaluate the range of J and E for each of these motions in terms of the integration constants.

4-2.11. If the mass of the sun were suddenly reduced to half its present value, what changes would take place in earth's orbit, assumed to be previously circular?

4-2.12. Two particles move about each other in circular orbits under an inverse-square force of attraction. The motion is suddenly halted and the particles are released to fall toward each other. How long does it take for them to collide in terms of the period of their circular motion?

4-2.13. A comet moves along a parabolic path in the plane of earth's orbit, which we shall assume to be circular. Prove that the maximum time that the comet can remain inside earth's orbit is $2/(3\pi)$ year.

4-2.14. An artificial earth satellite is determined to have a perigee of 380 km above the earth's surface, at which point its speed is 29,635 km/hr. Determine its apogee and the period of its orbit.

4-2.15. A space probe is initially in an elliptical orbit about the earth with perigee r_1 and apogee r_2. At a certain point in the orbit a thruster is fired for a sufficient time to give the probe an added velocity increment δv in order to allow the probe to escape the influence of earth's gravity with a final velocity V_0 relative to the earth. At what point in the orbit will this increment be a minimum? Calculate δv for this case as a function of the parameters of the elliptical orbit.

4-2.16. Calculate the Rutherford scattering cross section for the case of an attractive inverse-square force and show that the same result is obtained as in the case of repulsive scattering.

4-2.17. A particle is projected at a smooth hard sphere of radius R as shown in Fig. 4-10. The impact parameter for the scattering is $s < R$. Assuming specular reflection—that is, equal angles of incidence and reflection—

(a) find the scattering angle Θ as a function of s.
(b) calculate the differential scattering cross section $d\sigma$.
(c) integrate $d\sigma$ to show that

$$\sigma = \int \sigma(\Theta)\, d\Theta = \pi R^2$$

as one might expect.

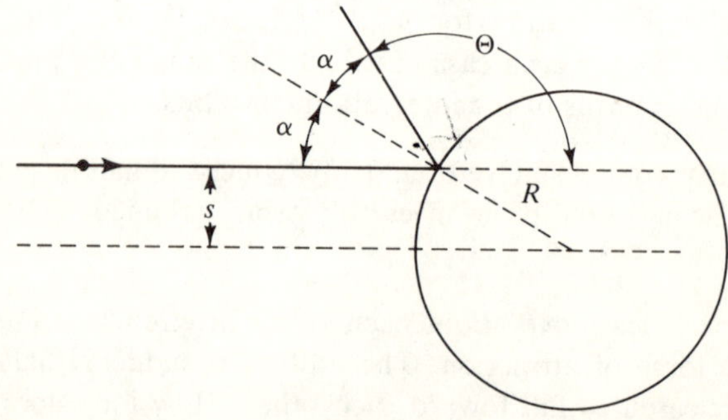

Fig. 4-10

4-3 The *N*-Body Problem

We saw in the preceding section that the two-body problem can be resolved into two equivalent one-body problems by considering the motion of one body relative to the other and the motion of the center of mass of the two-body system. A review of our procedure, resulting in the definition of the reduced mass of the system, suggests that no such analysis can be applied to systems of more than two bodies and, unfortunately, this is true. With all our sophistication, an analytical analysis of even the three-body problem is beyond our means. Instead, numerical solutions of great complexity must be carried out for specific initial conditions. These results, because of the approximations that must be made, apply only for a limited period of time.

Let us examine, as the simplest example, the problem of three bodies interacting under mutual gravitational attraction. The masses of the three particles are m_1, m_2, m_3, and for notational convenience we will specify a cartesian coordinate system wherein the positions of the bodies are specified by the coordinates (q_1, q_2, q_3), (q_4, q_5, q_6), (q_7, q_8, q_9), respectively. The potential energy of the system will then be given by

$$V = -\gamma\{m_2 m_3[(q_4 - q_7)^2 + (q_5 - q_8)^2 + (q_6 - q_9)^2]^{-1/2}$$
$$+ m_3 m_1[(q_7 - q_1)^2 + (q_8 - q_2)^2 + (q_9 - q_3)^2]^{-1/2}$$
$$+ m_1 m_2[(q_1 - q_4)^2 + (q_2 - q_5)^2 + (q_3 - q_6)^2]^{-1/2}\} \quad (4\text{-}103)$$

and the scalar equations of motion will be

$$m_1\ddot{q}_i = -\frac{\partial V}{\partial q_i}, \qquad i = 1, 2, 3, \qquad\qquad (4\text{-}104)$$

$$m_2\ddot{q}_j = -\frac{\partial V}{\partial q_j}, \qquad j = 4, 5, 6, \qquad\qquad (4\text{-}105)$$

$$m_3\ddot{q}_k = -\frac{\partial V}{\partial q_k}. \qquad k = 7, 8, 9. \qquad\qquad (4\text{-}106)$$

This system consists of nine second-order coupled equations and is therefore of order 18. In 1772 Lagrange was able to show that this problem can be reduced to a 6th-order system. It is done by exploiting the inherent constants of motion of the three-body system—namely, the constancy of the momentum of the center of gravity (which reduces the problem to 12th order), the constancy of the angular momentum of the system (which reduces the system to 8th order), and the energy integral (which lowers the order by two more units to 6th order). The details of these reductions are given by Whittaker.[4]

We saw in Eq. (4-33) that the kinetic energy of a system of particles can be written as the sum of the translational kinetic energy of the center of mass of the system and the internal kinetic energy relative to the center of mass. As a result of the definition of the center of mass given in Eq. (4-5), the total linear momentum of the system will be equal to the linear momentum of the center of mass; that is, the linear momentum of the center of mass is zero. The total angular momentum will be given by

$$\boldsymbol{J} = M(\boldsymbol{r}_c \times \dot{\boldsymbol{r}}_c) + \sum_{k=1}^{n} m_k(\boldsymbol{r}'_k \times \dot{\boldsymbol{r}}'_k) \qquad\qquad (4\text{-}107)$$

in which the internal angular momentum with respect to the center of mass depends only on the internal coordinates and not on the location of the origin about which $\boldsymbol{J}$ is taken. The relative positions of the particles with respect to each other similarly depend only on internal coordinates, as do the relative

---

[4] E. T. Whittaker, *A Treatise on the Analytical Dynamics of Particles and Rigid Bodies* (4th ed.). Cambridge: Cambridge University Press, 1961.

velocities; thus the internal forces and the corresponding potential energy, if definable, should also be expected to depend only on the internal coordinates.

Even though it looks as if we have effected a complete separation of the N-body problem, such is not the case in general. The motion of the center of mass of the system is separate only if the external force F is independent of the internal motion of the system. Usually it is not; the internal motions will often depend on F and on the motion of the center of mass. There are many examples in nature, however, in which the internal motions of a system of particles are *nearly* independent of external forces. A baseball, for example, maintains its identity throughout the violence done to it during a baseball game. This is also true of the atomic nucleus composed of nucleons (protons and neutrons), atoms composed of electrons and a nucleus, molecules composed of atoms, as well as the baseball itself, composed of molecules. In all such cases, the internal forces are much stronger than the external forces, and the internal equations of motion (see problem 4-3.3) depend essentially on internal forces alone. From an external point of view, the system then behaves as a particle that has (in addition to its external energy, momentum, and angular momentum) an internal (or intrinsic) energy and angular momentum (usually known as *spin*).

EXERCISES

4-3.1. Prove that the total linear momentum of an N-body system is equal to its external momentum—that is, to the momentum of its center of mass.

4-3.2. Derive Eq. (4-107) to show that cross terms involving both internal and center-of-mass coordinates drop out.

4-3.3. Use Eq. (4-1) and the definition of the internal coordinates implied by Fig. 4-3 to show that the internal equations of motion have the form

$$m_j \ddot{r}'_j = F_j + \sum_{k=1}^{N} F_{jk} - m_j \ddot{r}_c. \tag{4-108}$$

4-4 Collisions

Strictly speaking, any interaction between two or more particles that lasts for a finite time can be called a collision. If the forces of interaction satisfy Newton's third law, then the total linear momentum of the particles will be unchanged by the collision. Furthermore, if the mutual forces are central ones, the angular momentum of the system will be conserved through the collision process as was shown in Section 4-1. The total energy is also conserved, and for those cases in which the potential energy is the same before and after the collision, the kinetic energy will be conserved.

Consider the simplest case of a collision between two particles in which the kinetic energy and the linear momentum are conserved. The subscripts 1 and 2 distinguish the two particles, and the subscripts i and f denote a quantity evaluated before or after the collision, respectively. The conservation relations are expressed by

$$\boldsymbol{p}_{1i} + \boldsymbol{p}_{2i} = \boldsymbol{p}_{1f} + \boldsymbol{p}_{2f} \tag{4-109}$$

and

$$\frac{p_{1i}^2}{2m_1} + \frac{p_{2i}^2}{2m_2} = \frac{p_{1f}^2}{2m_1} + \frac{p_{2f}^2}{2m_2}. \tag{4-110}$$

This amounts to a set of four scalar equations relating four momenta, each having three scalar components, plus the ratio of the two masses. If 9 out of these 13 quantities are given, then the other 4 can be found. For instance, we might be given the masses, the initial momenta, and the direction of motion of one of the particles after the collision and asked to find the final momentum of the other particle and the energy of the particle with the known final direction. Another typical situation might call for sufficient information on the initial and final energies and momenta to be gathered in order to ascertain the mass of one of the particles.

Let us examine next an equally general case that often obtains in laboratory scattering experiments where particle 2, referred to as the *target* particle, is initially at rest. After the collision the incident particle, number 1, is scattered through an angle θ and the target particle is scattered through an angle φ as shown in Fig. 4-11. The conservation relations that describe this process are

$$p_{1i} = p_{1f} \cos \theta + p_{2f} \cos \varphi, \tag{4-111}$$

$$0 = p_{1f} \sin \theta - p_{2f} \sin \varphi, \tag{4-112}$$

and

$$\frac{p_{1i}^2}{2m_1} = \frac{p_{1f}^2}{2m_1} + \frac{p_{2f}^2}{2m_2}. \tag{4-113}$$

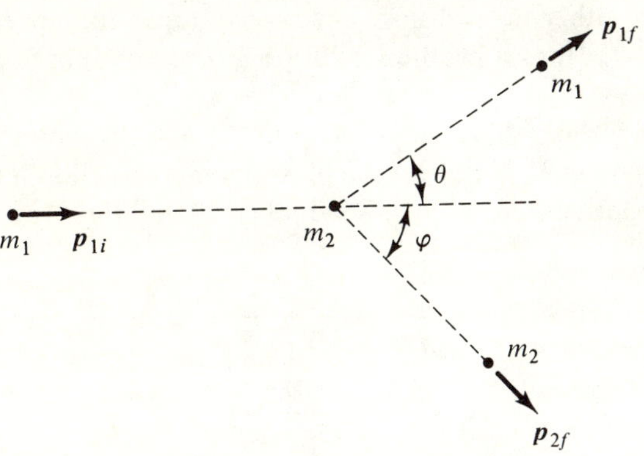

Fig. 4-11

In choosing a coordinate system wherein one of the particles is at rest, we have reduced the dimensionality of the problem from three to two. Instead of a four-equation system relating 13 quantities, we have a three-equation system relating five quantities: p_{1f}/p_{1i}, p_{2f}/p_{1i}, θ, φ, m_1/m_2, only two of which need be known. Suppose, for example, that the masses, the initial momentum of the incident particle, and the scattering angle θ are known. Then the final momenta and the scattering angle of the target are found to be

$$\frac{p_{1f}}{p_{1i}} = \frac{m_1}{m_1 + m_2} \cos\theta \pm \left[\left(\frac{m_1}{m_1 + m_2}\right)^2 \cos^2\theta + \frac{m_2 - m_1}{m_1 + m_2}\right]^{1/2} \quad (4\text{-}114)$$

$$\frac{p_{2f}}{p_{1i}} = \left[\left(\frac{p_{1f}}{p_{1i}}\right)^2 - 2\frac{p_{1f}}{p_{1i}} \cos\theta + 1\right]^{1/2}, \quad (4\text{-}115)$$

and
$$\cos\varphi = \frac{p_{1i}}{p_{2f}} - \frac{p_{1f}}{p_{2f}} \cos\theta. \quad (4\text{-}116)$$

The quantity in square brackets in Eq. (4-114) must have a value such that p_{1f}/p_{1i} is neither negative nor imaginary. It is left as an exercise to show that if $m_1 > m_2$, this stipulation implies a maximum value of θ given by

$$\cos^2\theta_{\max} = 1 - \left(\frac{m_2}{m_1}\right)^2, \qquad 0 \le \theta_{\max} \le \frac{\pi}{2}. \quad (4\text{-}117)$$

Scattering angles θ less than $\theta_{\max}$ can arise for either of two values of p_{1f}/p_{1i}, depending on how nearly head-on the collision is. As can be seen from Eqs. (4-115) and (4-116), there will also be different values of p_{2f}/p_{1i} and φ for these two cases. By way of clarification, let us examine the case of $\theta = 0$. Equations (4-114)–(4-116) become

$$\frac{p_{1f}}{p_{1i}} = \frac{m_1 \pm m_2}{m_1 + m_2}, \quad (4\text{-}118)$$

$$\frac{p_{2f}}{p_{1i}} = \sqrt{2(1 \mp 1)} \, \frac{m_2}{m_1 + m_2}, \quad (4\text{-}119)$$

and $\varphi = 0$ for either no collision [corresponding to the upper signs in Eqs. (4-118) and (4-119)] or a head-on collision (corresponding to a choice of the lower signs).

If $m_2 > m_1$, then all values of θ from 0 (no collision) to π (a head-on collision) are possible. Note that the plus sign must be chosen in Eq. (4-114); otherwise a negative value of p_{1f}/p_{1c} would result and this is not physical. For $\theta = \pi$, we find

$$\frac{p_{1f}}{p_{1i}} = \frac{m_2 - m_1}{m_1 + m_2}, \quad (4\text{-}120)$$

$$\frac{p_{2f}}{p_{1i}} = \frac{2m_2}{m_1 + m_2}, \quad (4\text{-}121)$$

$$\varphi = 0. \quad (4\text{-}122)$$

In the limit as $m_2/m_1 \to \infty$, as obtains, for example, when a tennis ball is bounced off a practice wall firmly anchored to the earth, we see that $p_{1f}/p_{1i} \to 1$ and $p_{2f}/p_{1f} \to 2$. The student is often disturbed by this result; however, since $p_{2f} = m_2 v_{2f}$ and $m_2 \to \infty$, then obviously $v_{2f} \to 0$. In terms of kinetic energy transfer, we find

$$\frac{T_{2f}}{T_{1i}} = \frac{4m_1 m_2}{(m_1 + m_2)^2}, \tag{4-123}$$

which is symmetric with respect to the masses and also holds true for $m_1 > m_2$. Clearly, T_{2f} tends to zero for large discrepancies between the masses. Equation (4-123) is also useful in another sense. Suppose that m_1 is unknown but that T_{1i} can be calculated. If m_2 is known and T_{2f} can be measured for a head-on collision, then Eq. (4-123) can be solved to find m_1:

$$\frac{m_1}{m_2} = 2\frac{T_{1i}}{T_{2f}} - 1 \pm \left[\left(2\frac{T_{1i}}{T_{2f}} - 1\right)^2 - 1\right]^{1/2}. \tag{4-124}$$

This equation serves to determine m_1 to within a choice of two values. If a different scattering angle or a target with a different mass is chosen and the experiment rerun, then m_1 is determined uniquely. It is essentially the method by which the neutron was shown to be a particle and its mass measured by Sir James Chadwick in 1932.

We have seen that if we know the initial momenta and masses and one of the scattering angles or the initial incident and first target energies, along with the target mass and a scattering angle, the rest of the information can be gleaned by manipulating the conservation relations. In order to find both scattering angles, however, the initial linear and angular momenta, as well as the force law acting between the particles, must be known. We showed in Section 4-2 how the scattering angle Θ, with respect to a coordinate system fixed in the target particle, can be calculated for the case of an inverse-square central force. We would now like to transform these results to a laboratory reference frame wherein the motions of the particles are labeled by coordinates r_1, r_2 relative to a fixed origin with respect to which the target particle (2) is initially at rest. We shall carry out this transformation in two steps. First, we introduce a center-of-mass coordinate system in which the particle positions are given by vectors r_1', r_2' with respect to the center of mass of the system

$$r_1' = r_1 - r_c, \qquad r_2' = r_2 - r_c \tag{4-125}$$

as shown in Fig. 4-4. By the definition of the center of mass

$$r_c = \frac{m_1 r_1 + m_2 r_2}{m_1 + m_2} \tag{4-126}$$

and the definition of the relative coordinate

$$r = r_1 - r_2, \tag{4-127}$$

the laboratory coordinates can be written

$$r_1 = r_c + \frac{m_2}{m_1 + m_2}\, r \tag{4-128}$$

and

$$r_2 = r_c - \frac{m_1}{m_1 + m_2}\, r. \tag{4-129}$$

So for the transformation between the center-of-mass coordinates and the relative coordinate r we have

$$r_1' = \frac{\mu}{m_1}\, r, \qquad r_2' = -\frac{\mu}{m_2}\, r. \tag{4-130}$$

As we have seen, the total linear momentum of the system is zero in terms of center-of-mass coordinates; hence the initial and final momenta of the two particles are always equal and opposite. For this reason, the sum of the scattering angles in this system is always equal to π.

The object of our inquiry is to relate the scattering angle Θ in the relative coordinate frame, previously found, to the scattering angle θ in the laboratory reference frame. Since, by Eq. (4-130), the velocity of the incident particle in the center-of-mass frame is always parallel to the relative velocity of the two particles

$$\dot{r}_1' = \frac{\mu}{m_1}\, \dot{r}, \tag{4-131}$$

it follows that the scattering angle of the incident particle in the center-of-mass system is the same as in the relative system

$$\theta' = \Theta. \tag{4-132}$$

Differentiating Eq. (4-124), we find the relation between the incident particle velocities in the center-of-mass and laboratory frames:

$$\dot{r}_1 = \dot{r}_1' + \dot{r}_c. \tag{4-133}$$

The relationship between the velocity vectors $\dot{r}_1$, $\dot{r}_1'$, and $\dot{r}_c$ evaluated after the collision and the scattering angles θ and $\theta' = \Theta$ is shown in Fig. 4-12. Using this figure, we find

$$\tan \theta = \frac{\dot{r}_{1f}' \sin \Theta}{\dot{r}_{1f}' \cos \Theta + \dot{r}_c}. \tag{4-134}$$

The constant velocity of the center of mass $\dot{r}_c$ can be expressed in terms of the laboratory frame by differentiating Eq. (4-126) and evaluating the results for the time before the collision occurs. Thus we set $\dot{r}_{2i} = 0$ and find

$$\dot{r}_c = \frac{\mu}{m_2}\, \dot{r}_{1i} = \frac{\mu}{m_2}\, \dot{r}_i, \tag{4-135}$$

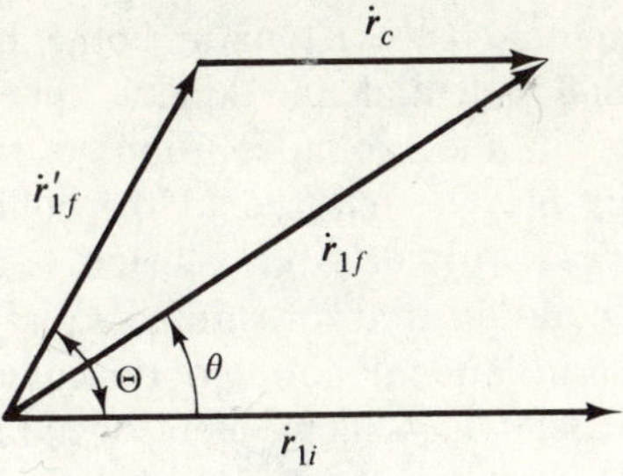

Fig. 4-12

where we have substituted $\dot{r}_{1i} = \dot{r}_i$, since the relative velocity and the laboratory velocity of the incident particle are initially the same because of the fact that the target particle is initially stationary in the laboratory frame. Evaluating Eq. (4-131) after the collision

$$\dot{r}'_{1f} = \frac{\mu}{m_1}\,\dot{r}_f, \tag{4-136}$$

we can eliminate the reduced mass between Eqs. (4-135) and (4-136) to find

$$\frac{\dot{r}_c}{\dot{r}'_{1f}} = \frac{m_1\dot{r}_i}{m_2\dot{r}_f} = \frac{m_1}{m_2}, \tag{4-137}$$

where we have taken account of the fact that the relative speeds are the same before and after the collision if the collision is an elastic one. Using Eq. (4-137), we therefore find for $\tan\theta$

$$\tan\theta = \frac{\sin\Theta}{\cos\Theta + (m_1/m_2)}. \tag{4-138}$$

In much the same way, a relation for φ, the scattering angle of the target particle, can be obtained.

We see that if $m_1 \gg m_2$, then θ must be small regardless of the value of Θ. It is left as an exercise to show that if $m_1 = m_2$, then

$$\theta = \tfrac{1}{2}\Theta. \tag{4-139}$$

By virtue of the results of problem 4-4.4, the scattering angle for particle 2 will in this case be

$$\varphi = \tfrac{1}{2}(\pi - \Theta). \tag{4-140}$$

If $m_2 \gg m_1$, then the relative one-body problem and the laboratory two-body problem reduce to the same thing and

$$\theta \approx \Theta. \tag{4-141}$$

Up to this point we have only considered elastic collisions in which the energy of motion is conserved. Because the conservation laws are also valid

157

in quantum mechanics, the results obtained apply to atomic and subatomic particles as well as to billiard balls. Atoms and other microscopic constituents of matter have internal potential and kinetic energy associated with the motions of their parts; in undergoing collisions of sufficient violence, some of this internal energy may be released to the collision or some collision energy may be absorbed. An inelastic collision is said to be *endoergic* if kinetic energy of the collision is absorbed and *exoergic* if internal energy is released to add to the translational motion of the colliding particles. Furthermore, the colliding particles may change identity and number. A proton may collide with a nucleus and be absorbed with or without the release of a neutron. It is also possible for three or more particles to be formed in a two-particle collision. In all cases, conservation of momentum applies; energy is also conserved if the changes in the internal energy of the particles are properly taken into account.

Let us consider an example in which the number of the particles does not change but the type may. An incident particle of mass m_1 collides with a stationary target particle of mass m_2 and two other particles of mass m_3 and m_4 are produced and fly off with scatter angles θ_3 and θ_4, measured with respect to the initial direction of the incident particle. A kinetic energy Q is produced in the collision that may be greater than zero for an exoergic collision or less than zero for an endoergic collision. Our momentum and energy conservation relations then take the form

$$p_1 = p_3 \cos \theta_3 + p_4 \cos \theta_4, \qquad (4\text{-}142)$$

$$0 = p_3 \sin \theta_3 - p_4 \sin \theta_4, \qquad (4\text{-}143)$$

$$\frac{p_1^2}{2m_1} + Q = \frac{p_3^2}{2m_3} + \frac{p_4^2}{2m_4}. \qquad (4\text{-}144)$$

We thus have three equations in terms of six quantities ($p_1, p_3, p_4, \theta_3, \theta_4, Q$), assuming the masses to be known. A common situation involves calculating Q from a knowledge of p_1 and measurement of, say, p_3 and θ_3. Eliminating θ_4 between Eqs. (4-142) and (4-143) and substituting the resulting expression for p_4 into Eq. (4-144), we find

$$Q = T_3 + \frac{p_4^2}{2m_4} - T_1 = T_3\left(1 + \frac{m_3}{m_4}\right) - T_1\left(1 - \frac{m_1}{m_4}\right)$$

$$- 2\left(\frac{m_1 m_3 T_1 T_3}{m_4^2}\right)^{1/2} \cos \theta_3. \qquad (4\text{-}145)$$

This equation, like all others in Newtonian mechanics, is only valid for velocities much less than c, the speed of light. For velocities approaching that of light, the relationship between p and T previously used is no longer valid and we must rewrite Eq. (4-144) as

$$T_1 + Q = T_3 + T_4, \tag{4-146}$$

where
$$T = \sqrt{m^2c^4 + p^2c^2} - mc^2, \tag{4-147}$$

which can be shown to approach $p^2/2m$ for $p \ll mc$.

EXERCISES

4-4.1. Verify Eq. (4-114).

4-4.2. If $m_1 > m_2$, show that the maximum value of the incident particle scattering angle is given by Eq. (4-117).

4-4.3. If $m_1 > m_2$ and the incident particle is known to experience a maximum scattering angle, find p_{1f}/p_{1i}, p_{2f}/p_{1i}, and φ.

4-4.4. Show that for $m_1 = m_2$

$$\frac{p_{1f}}{p_{1i}} = \cos\theta, \qquad \frac{p_{2f}}{p_{1i}} = \sin\theta, \qquad \theta + \varphi = \frac{\pi}{2}.$$

4-4.5. Verify Eq. (4-123) and solve for the ratio of the masses to obtain Eq. (4-124).

4-4.6. Show that if $m_1 = m_2$, then $\theta = \frac{1}{2}\Theta$.

4-4.7. Prove that in the low velocity limit—that is, $p \ll mc$—Eq. (4-147) reduces to the familiar Newtonian relation, $T = p^2/2m$.

4-4.8. Assume a completely inelastic collision in which a particle of mass m_1 and speed v_1 collides with a wad of putty of mass m_2 and sticks in it. How much kinetic energy is converted to thermal energy in such a collision?

4-4.9. Rewrite Eq. (4-101) for the Rutherford scattering cross section in terms of θ, the laboratory scattering angle, for the case $m_1 = m_2$.

4-5 Motions Involving Variable Mass

Newton's second law of motion says that the rate of change of the momentum mu of a body is equal to the force F applied to the body—that is[5]

$$\frac{d}{dt}(mu) = F. \tag{4-148}$$

---

[5] We shall restrict ourselves to one-dimensional motion for this discussion.

In the usual case in which the mass of the body remains constant, the left-hand side reduces to the product of mass times acceleration. If the mass is allowed to vary, however, the second law must be used in the more general form given in Eq. (4-148). Considerable caution is required in formulating the problem, as the following example will show.

Consider a box filled with sand that is sliding down a smooth plane of inclination angle φ. The box is leaking sand through a hole at a constant rate. We wish to formulate an equation to describe the motion of the box and the sand. At time t let m be the instantaneous mass of the box plus the sand remaining in it. At a time dt later the mass of the box has changed from m to $m - dm$ and the velocity has changed from u to $u + du$. The momentum of the sliding box has therefore changed from mu to $(m - dm)(u + du)$. The momentum carried away by the sand that has leaked out is $u\,dm$. Since the total change in the momentum of the system over the time interval dt is equal to $F\,dt$, where $F = mg \sin \varphi$, we find

$$u\,dm + (m - dm)(u + du) - mu$$

$$= u\,dm + mu - u\,dm + m\,du - dm\,du - mu$$

$$= mg \sin \varphi\, dt. \tag{4-149}$$

Neglecting products of infinitesimals, Eq. (4-149) reduces, after dividing by dt, to

$$m \frac{du}{dt} = mg \sin \varphi \tag{4-150}$$

for the motion of the box and sand independent of whether leakage occurs or not. The key to understanding this problem is to notice that the motion of the box is not changed by the leakage of the sand, since this event imparts no momentum to the box.

A very different result would have been obtained if we had assumed, for example, that the sand was discharged in such a way as always to leave the ejected sand at rest. Such would be the case if the element dm is given a velocity $-u$ equal to the instantaneous velocity of the box and opposite in direction. In this rather artificial example the increment dm will have zero momentum at the end of the time interval dt. It is left as an exercise to show that the equation of motion of the box in this case is given by

$$m \frac{du}{dt} = u \frac{dm}{dt} + mg \sin \varphi, \tag{4-151}$$

where the first term on the right-hand side describes the rate at which momentum is being imparted to the box by the ejection of the sand.

A more practical example of this type of behavior is the rocket. Here the assumption is that burned gas products are ejected through a nozzle at the rear of the rocket with a speed v relative to the rocket. If the rocket plus its unburned fuel has a momentum mu at time t, then at time $t + dt$ the momentum of the rocket will be $(m - dm)(u + du)$ and the momentum of the mass dm ejected through the nozzle in the interval dt will be $(u - v)\,dm$. Thus we find for the change in the momentum of the system

$$dp = (u - v)\,dm + (m - dm)(u + du) - mu = F\,dt,$$

where F embodies any and all external forces (gravity, air resistance, pressure drops across the rocket nozzle, etc.) acting on the rocket. To first order in infinitesimals, we find for the equation of motion of the rocket

$$m\frac{du}{dt} = v\frac{dm}{dt} + F. \tag{4-152}$$

In this last quarter of the twentieth century, when chemically fueled rockets are being used to explore the solar system, it is interesting to integrate Eq. (4-152) for some cases of interest. We will assume a rocket operated in space over regions where changes in the gravitational potential are small, thereby justifying the neglect of the external force term.

Setting $F = 0$ and multiplying the equation by dt/m, we have

$$du = v\frac{dm}{m}. \tag{4-153}$$

If the initial conditions are $u(0) = u_0$ and $m(0) = m_0$, then, noting that dm and du have opposite signs, we find upon integrating

$$u - u_0 = v\ln\frac{m_0}{m}. \tag{4-154}$$

The rocket plus its payload but without fuel has a mass M_0. The fuel has a mass M_f. Therefore after a time τ_b to "burnout," the mass of the rocket will be $m(\tau_b) = M_0$, at which time a maximum velocity will have been reached. So

$$u_{\max} - u_0 = v\ln\left(1 + \frac{M_f}{M_0}\right). \tag{4-155}$$

It is this logarithmic dependence on mass that requires enormous rockets to deliver small payloads. Let us look at the problem from an energy point of view. If the power production is assumed to be a constant fraction of the mass of the rocket less fuel—that is,

$$P = \alpha M_0 = \tfrac{1}{2}\dot{m}v^2, \tag{4-156}$$

and the rocket burns for a time τ_b, then, since

$$M_f = \dot{m}\tau_b, \tag{4-157}$$

we find

$$u_{\max} - u_0 = v \ln\left(1 + \frac{V^2}{v^2}\right), \tag{4-158}$$

where
$$V = \sqrt{2\alpha\tau_b} \tag{4-159}$$

is a second characteristic velocity that is a function of the burning time and the power-to-mass ratio of the system. If V is fixed, then the optimum value of the exhaust velocity v in order to maximize $\Delta u = u_{\max} - u_0$ is $v \approx 0.505\, V$, for which $\Delta u \approx 0.805\, V$ and $M_f/M_0 \approx 3.92$. If we take for α the moderately optimistic figure of $\alpha = 200\ \text{W/kg}$ and $\tau_b = 100$ days (a reasonable time for interplanetary flight), then v is about $3 \cdot 10^4$ m/sec and $\Delta u \approx 4.7 \cdot 10^4$ m/sec. The acceleration of such a rocket would be quite low, $\Delta u/\tau_b \approx 5.5 \cdot 10^{-3}$ m/sec^2, and it could not ascend from the surface of the earth but would need to be placed or assembled in orbit by high-thrust rockets. If the exhaust velocity is limited to the value attainable with chemical rockets, $v \approx 10^3$ m/sec, then multistaging or very large mass ratios must be employed in order to attain high speeds. Clearly, if large loads are to be transported between planets in an economical way, a high-exhaust velocity, low-thrust drive is required.

EXERCISES

4-5.1. Show that Eq. (4-151) is the equation of motion of a sandbox that slides without friction down a plane inclined at an angle φ in such a way that sand is expelled from the box with a velocity equal to the instantaneous velocity of the box but in the opposite direction.

4-5.2. Plot $\Delta u/V$ as a function of v/V to show that the curve has a single maximum at $v/V \approx 0.505$.

4-5.3. A rocket is projected vertically upward from rest in a uniform gravitational field. The exhaust velocity is 10^3 m/sec and $M_f/M_0 = 6$. Find the maximum altitude attained by the rocket.

4-5.4. A two-stage rocket capable of accelerating a payload M_p to a speed u in deep space, starting from rest, is to be designed. Structural factors impose the limit that the empty rocket less fuel and payload cannot weigh less than 10% of its fuel capacity. Find the optimum distribution of mass between the two stages so that the total initial weight of the rocket is a minimum.

4-5.5. Derive an expression for the time that a rocket with fuel mass M_f, final mass M_0, and exhaust velocity v can hover in a uniform gravitational field having acceleration g.

4-6 Suspended Cables

All bodies found in nature are *deformable* to a certain extent. In many cases, such as those to be examined in the next chapter, it is possible to treat the bodies as being perfectly rigid, since they undergo negligible deformation under the action of forces of the magnitude usually encountered. In other cases, the forces will be large enough to cause deformation. If the forces are not too large, the deformation that they cause will disappear when the force is removed and the body will return to its former shape. We say that such bodies are *elastic*. In other cases, such as a string lying on a table, moderate forces produce permanent deformation quite easily. Such bodies are said to be *freely deformable*.

In this section we treat a freely deformable cable hanging in static equilibrium from two fixed points under the action of a uniform gravitational field. We shall limit ourselves to those cases in which the curve of the cable lies in a single plane. In analyzing a deformable body, we assume that it consists of a large number of rigid elements joined together. The forces on each element are the external force due to gravity and the tension force acting along the cable between adjacent elements. We shall treat two cases that differ in the distribution of weight along the cable.

First, we examine the case in which the weight of the cable itself is negligible and the load is distributed uniformly along a horizontal direction. This is, of course, the case of a suspension bridge (see Fig. 4-13). The support points are on the same level. In part (b) we show a short section of the cable taken near the middle of the span such that the left-hand end is horizontal. The forces acting on segment AB are the tension T_0 at A (horizontal to the left), the tension T at B directed tangent to the cable at an upward angle θ, and the total weight W of the section of the roadbed which acts at the midpoint. If λ is the weight per unit length of the roadbed, the equilibrium conditions on segment AB are

$$T \sin \theta = \lambda x, \tag{4-160}$$

$$T \cos \theta = T_0. \tag{4-161}$$

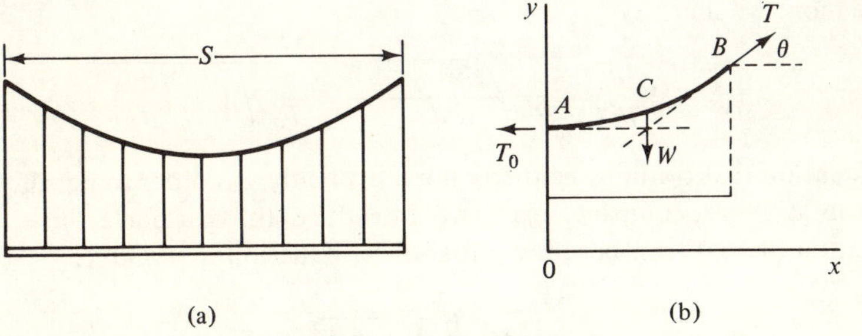

Fig. 4-13 The suspension bridge.

Dividing Eq. (4-160) by Eq. (4-161) and noting that $\tan\theta = dy/dx$, we have

$$\frac{dy}{dx} = \frac{\lambda x}{T_0},\tag{4-162}$$

which we integrate to obtain

$$y = \frac{\lambda}{2T_0}\,x^2 + \beta,\tag{4-163}$$

the equation of a parabola. The tension T at any point along the cable can be found by squaring Eqs. (4-160) and (4-161) and adding them. This step yields

$$T = \sqrt{T_0^2 + \lambda^2 x^2}.\tag{4-164}$$

If the span of the bridge between its support towers is S, then the maximum tension in the support cable, experienced at the points of support, will be

$$T_{\max} = \tfrac{1}{2}\sqrt{4T_0^2 + \lambda^2 S^2}.\tag{4-165}$$

Consider next the case of a heavy cable hanging from two level support points under its own weight. In the previous case, the weight was distributed horizontally; here it is distributed uniformly along the cable itself. Figure 4-13(b) still applies, but the weight W, instead of being given by λx, is given by μs, where μ is the weight of the cable per unit length and s is a coordinate measured along the cable from point A. The equations of static equilibrium are

$$T\sin\theta = \mu s,\tag{4-166}$$

$$T\cos\theta = T_0.\tag{4-167}$$

As before, we divide the upper equation by the lower one to obtain

$$\frac{\mu s}{T_0} = \tan\theta = \frac{dy}{dx},\tag{4-168}$$

which is called the intrinsic equation of the *catenary*. In order to express this equation in terms of x and y, we again use $\tan\theta = dy/dx$ and differentiating once more we find

$$\frac{d^2y}{dx^2} = \frac{\mu}{T_0}\frac{ds}{dx} = \frac{\mu}{T_0}\frac{\sqrt{dx^2 + dy^2}}{dx} = \frac{\mu}{T_0}\sqrt{1 + \left(\frac{dy}{dx}\right)^2},\tag{4-169}$$

an equation that contains neither x nor y explicitly. In order to put the equation in a more compact form, we introduce the parameter $p = dy/dx$. Equation (4-169) then becomes a first-order equation in p and x:

$$\frac{dp}{dx} = \frac{\mu}{T_0}\sqrt{1 + p^2}.\tag{4-170}$$

The variables can easily be separated; after integrating, the result is

$$\operatorname{arcsinh} p = \frac{\mu x}{T_0}, \tag{4-171}$$

the constant of integration being zero, since $p = dy/dx = 0$ at $x = 0$. Taking the sinh of both sides of Eq. (4-171)

$$p = \frac{dy}{dx} = \sinh \frac{\mu x}{T_0} \tag{4-172}$$

and integrating, we find

$$y = \frac{T_0}{\mu} \cosh \frac{\mu x}{T_0} \tag{4-173}$$

in which the constant of integration is zero if $y(0) = T_0/\mu$. It is occasionally convenient to express the variable s as a function of x, which can be done by using Eqs. (4-168) and (4-173). Differentiating Eq. (4-173) and substituting for dy/dx from Eq. (4-168), we have

$$s = \frac{T_0}{\mu} \sinh \frac{\mu x}{T_0}. \tag{4-174}$$

The tension at any point of the cable can be found by using Eqs. (4-167), (4-173), and (4-174). Since $\cos \theta = dx/ds$, we obtain

$$T = T_0 \frac{ds}{dx} = T_0 \cosh \frac{\mu x}{T_0} = \mu y. \tag{4-175}$$

The tension at any point of the cable is therefore proportional to the height of that point above a horizontal line located a distance T_0/μ below the lowest part of the cable. In terms of s, the tension is

$$T = \sqrt{T_0^2 + \mu^2 s^2}. \tag{4-176}$$

The maximum value of T occurs at the support points where

$$|x| = \frac{S}{2}, \qquad y = h + \frac{T_0}{\mu}, \tag{4-177}$$

S is the span, and h is the *sag*, defined as the vertical distance between the lowest part of the cable and the support points. At this point the tension in the cable is

$$T_{\max} = T_0 \cosh \frac{\mu S}{2T_0} = \mu h + T_0. \tag{4-178}$$

EXERCISES

4-6.1. Carry out a series expansion of Eq. (4-173) to show that in the case of a tightly stretched cable (large T_0/μ) the catenary becomes a parabola to a first approximation.

4-6.2. Derive Eq. (4-176).

4-6.3. A string of total weight w and length l is suspended from two points at the same level and a weight W is attached to the lowest point of the string. If the string makes an angle φ with the horizontal at either side of its lowest point, find the equation of the catenary assumed by either branch of the string. Find the maximum tension in the string also.

4-6.4. A uniform chain of mass m per unit length hangs vertically so that its lower end just touches a horizontal table. If it is released at the top, show that at the time a length x of chain has fallen, the force on the table is equivalent to the weight of a length $3x$ of the chain.

4-6.5. A uniform heavy cable hangs from two fixed points at the same level. If θ is the angle between the tangent to the cable and the horizontal at any point, show that

$$y = \frac{T_0}{\mu} \sec \theta.$$

RIGID BODY MECHANICS

We shall define a *rigid body* as a system of particles, the relative internal velocities of which are all zero—that is, a system in which the spatial relationship of the particles is constant. Our emphasis will be on the macroscopic manifestations of such a system, and we will not consider the "graininess" due to microscopic structure except as a derivational device.

In order to describe the motion of a rigid body, we will require two coordinate systems: an inertial frame (the laboratory system) and a coordinate system fixed in the rigid body. Therefore six quantities must be specified so as to fix the position of the body.[1] Since rigid bodies are, by definition, not subject to distortion, any finite motion must be expressible as the linear translational motion of some point located in the body plus a rotation about that point. *Chasles' theorem* is even more general; it states that the line of translation can be made to coincide with the axis of rotation.[2] This means, in effect, that the most general displacement of a rigid body is a *screw* displacement in which the ratio of the distance of translation to the angle of rotation defines the *pitch* of the screw. If the axis of the screw passes through the center of mass of the body, then the separation of the motion into linear and angular components allows us to write the kinetic energy of the body and its angular momentum as the sum of motions *of* and *about* the center of mass as in Eqs. (4-33) and (4-107).

---

[1] For extended bodies, one must specify not only *location* but also *orientation*.

[2] For proof of this theorem, see E. T. Whittaker, *A Treatise on the Analytical Dynamics of Particles and Rigid Bodies*. Cambridge: Cambridge University Press, 1961, pp. 4–5.

5-1 Rigid Bodies in Static Equilibrium

The concept of the center of mass of an extended body is a useful one. Earlier we saw that the radius vector describing the center of mass of a system of particles can be given by

$$Mr_c = \sum_{j=1}^{n} m_j r_j, \qquad (5\text{-}1)$$

where $M = \sum m_j$ is the total mass of the system and m_j and r_j are the masses and radius vectors describing the individual constituent particles. In the limit of a continuous extended body, $n \to \infty$ and the summation goes over into an integral:

$$Mr_c = \iiint_{V} \rho r \, d\tau, \qquad (5\text{-}2)$$

where ρ is the mass density of the body—in general, a function of r—and the integral is taken to include the volume of the body. The calculation of centers of mass can usually be simplified by recognizing the symmetries and taking advantage of them in arranging the calculation. For instance, it can easily be shown that if a body is symmetrical with respect to a plane, then the center of mass must lie in that plane.[3] This theorem has several obvious corollaries. If a body is symmetrical with respect to two planes, the centroid lies somewhere along the line of intersection of the planes. If the body is symmetrical with respect to three planes having only one common point, that point must correspond to the centroid.

As an example of the use of symmetries to save calculation, consider a triangular lamina of uniform density. It can easily be shown that the centroid of such a body lies at the intersection of its medians. In Fig. 5-1 we show a

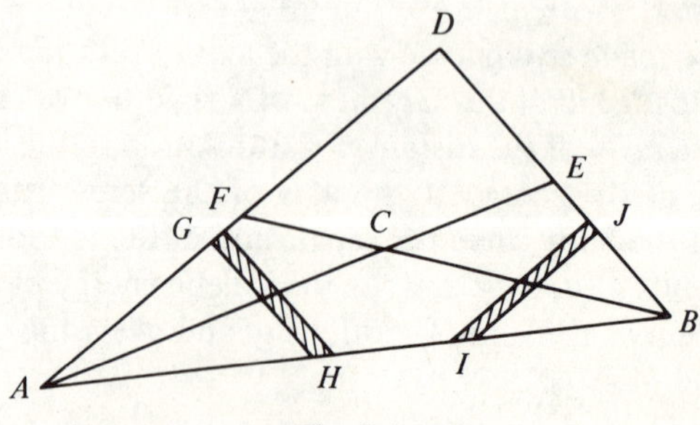

Fig. 5-1

---

[3] Our symmetry assumption extends to the mass density ρ, which must be a constant or spherically symmetrical. In this case, the center of mass of the body coincides with the *centroid* of its volume.

triangle ABD. The point E bisects the line DB; hence AE is a median. If we take as our mass element $dm = \rho\,d\tau$, the strip GH parallel to DB, we see that GH is, in effect, a uniform one-dimensional rod that must have its centroid at the midpoint. The midpoint of GH always lies on the median AE, and so the centroid of the triangle must lie somewhere along the line AE. If we then apply the same logic to the median FB and mass element IJ, we see that the centroid must lie along FB. Since FB and AE only intersect at the point C, this point is the center of mass of the triangle.

If a body is composed of a number of parts, each of whose centers of mass is known, then these composite centers of mass can be regarded as particles having masses corresponding to their corresponding parts, and Eq. (5-1) can be used to calculate the center of mass of the entire body. For example, suppose that we wish to find the center of mass of a mallet consisting of a cubical head of side a and a cylindrical handle of length b as shown in Fig. 5-2.

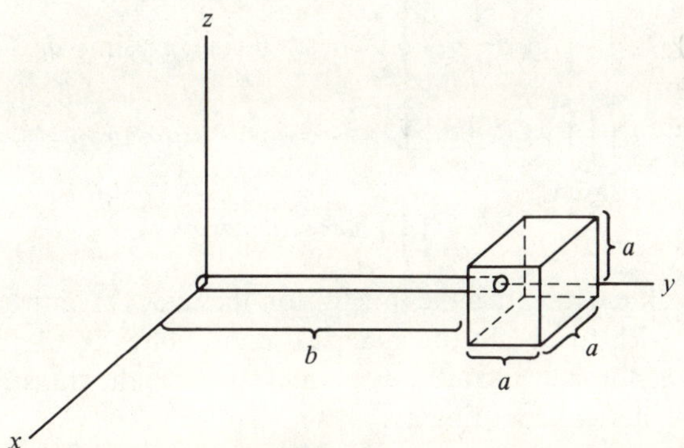

Fig. 5-2

If the handle has a linear mass density λ and the head has a volume mass density ρ, we see by inspection that the center of mass of the handle is located at $(0, b/2, 0)$ and that the center of mass of the head is located at $(0, b + a/2, 0)$. Hence, by Eq. (5-1), the center of mass of the mallet will be given by

$$y_c = \frac{\lambda b \cdot b/2 + \rho a^3(b + a/2)}{\lambda b + \rho a^3}. \tag{5-3}$$

Another class of configurations that can be evaluated by symmetry and summation is represented by bodies possessing some sort of cavity. Suppose that we consider the example shown in Fig. 1-23 of a sphere of radius R containing a spherical cavity of radius $R/2$ tangent to the surface of the sphere. We can readily find the centroid of a uniform solid sphere. The centroid of the hollowed sphere can therefore be written

$$Mx_c = M_1 x_{c_1} - M_2 x_{c_2}, \tag{5-4}$$

where M_1 and M_2 are the mass of a solid sphere of radius R and $R/2$, respectively, and

$$x_{c_1} = 0, \qquad x_{c_2} = \frac{R}{2} \tag{5-5}$$

are their respective centroids as derived by inspection from Fig. 1-23. It is left for the reader to show that the centroid of the hollowed sphere shown in this figure is located at a distance of $R/14$ to the left of the origin.

In the most general case, we must, of course, deal directly with the volume integral. Any of several coordinate systems can be used in the integration; generally, however, we will want the resulting coordinates of the center of mass to be expressed in the cartesian system. For instance, if spherical coordinates are indicated, we find, using Table 1-1,

$$M x_c = \iiint \rho x \, d\tau = \iiint \rho(r \cos \theta \sin \varphi) r^2 \sin \varphi \, dr \, d\theta \, d\varphi, \tag{5-6}$$

$$M y_c = \iiint \rho y \, d\tau = \iiint \rho(r \sin \theta \sin \varphi) r^2 \sin \varphi \, dr \, d\theta \, d\varphi, \tag{5-7}$$

$$M z_c = \iiint \rho z \, d\tau = \iiint \rho(r \cos \varphi) r^2 \sin \varphi \, dr \, d\theta \, d\varphi, \tag{5-8}$$

where

$$M = \iiint \rho r^2 \sin \varphi \, dr \, d\theta \, d\varphi. \tag{5-9}$$

In all these volume integrals the integration includes the entire volume of the body. If we wish to obtain the spherical coordinates of the center of mass r_c, θ_c, φ_c, we can first evaluate x_c, y_c, z_c and then use the transformation from cartesian to spherical coordinates:

$$r_c = x_c^2 + y_c^2 + z_c^2, \tag{5-10}$$

$$\theta_c = \arctan \frac{y_c}{x_c}, \tag{5-11}$$

$$\varphi_c = \arctan \frac{\sqrt{x_c^2 + y_c^2}}{z_c}. \tag{5-12}$$

Having discussed the center of mass and its calculation, let us examine the static equilibrium of a rigid body. The equations of motion of a rigid body are

$$\dot{p}_c = \sum_{j=1}^{n} F_j, \tag{5-13}$$

where p_c is the linear momentum of the center of mass and

$$\dot{j} = \sum_{j=1}^{n} N_j, \tag{5-14}$$

which were derived in the last chapter as Eqs. (4-8) and (4-19). Equation (5-13) describes the linear motion of the center of mass in terms of the external forces acting on the body. Equation (5-14) describes the rotary motion of the body with respect to a point in terms of the total external torque about that point.[4]

For a rigid body in a state of static equilibrium, the left-hand sides of Eqs. (5-13) and (5-14) must vanish and we find

$$\sum_{j=1}^{n} \boldsymbol{F}_j = 0 \tag{5-15}$$

$$\sum_{j=1}^{n} \boldsymbol{N}_j = 0 \tag{5-16}$$

as the equilibrium conditions. These equations also include *stationary equilibrium* states wherein the linear momentum and the angular momentum are constant but nonzero. The specification of static equilibrium is made by means of the initial conditions

$$\boldsymbol{p}_c|_{t=0} = \boldsymbol{J}|_{t=0} = 0. \tag{5-17}$$

Next, we wish to show that if the torque about some point O is known to be zero, then the torque about any other point O' is also zero if Eqs. (5-15) and (5-16) are satisfied. If $\boldsymbol{r}_j$ is a radius vector from some convenient origin to the point of application of the force $\boldsymbol{F}_j$ and $\boldsymbol{r}_0$ and $\boldsymbol{r}_{0'}$ are radius vectors from the same origin to the points O and O', respectively, then the torque about O' is given by

$$\begin{aligned}
\sum_{j} \boldsymbol{N}_{j0'} &= \sum_{j} (\boldsymbol{r}_j - \boldsymbol{r}_{0'}) \times \boldsymbol{F}_j \\
&= \sum_{j} (\boldsymbol{r}_j - \boldsymbol{r}_0 + \boldsymbol{r}_0 - \boldsymbol{r}_{0'}) \times \boldsymbol{F}_j \\
&= \sum_{j} (\boldsymbol{r}_j - \boldsymbol{r}_0) \times \boldsymbol{F}_j + (\boldsymbol{r}_0 - \boldsymbol{r}_{0'}) \times \boldsymbol{F}_j \\
&= \sum_{j} \boldsymbol{N}_{j0} + (\boldsymbol{r}_0 - \boldsymbol{r}_{0'}) \times \sum_{j} \boldsymbol{F}_j. \tag{5-18}
\end{aligned}$$

Thus if $\sum_j \boldsymbol{N}_{j0}$ and $\sum_j \boldsymbol{F}_j$ are both equal to zero, it follows that the torque about any point O' must also vanish.

In calculating the rate of change of momentum due to a force $\boldsymbol{F}$ applied to a rigid body, we need only know the magnitude and direction of the force. In calculating the torque due to $\boldsymbol{F}$, we must know not only the magnitude and

---

[4] The reference point may be chosen as the center of mass, a fixed point in an intertial frame, such as the laboratory frame, or a point accelerating along the line joining it to the center of mass, such as the point of contact between a rolling wheel and the plane on which it rolls.

direction of F but also the point at which the force acts on the body. If we draw a line, which we shall call the *line of action*, through this point having the direction of F, then the torque about a given point will be independent of the location of the point of application of the force along this line. This situation can easily be shown by using Fig. 5-3. We wish to compare the torques about O due to application of the force F at points A and B in line with the direction of F. Since $r_1 \times F$ and $r_2 \times F$ both have the same directionality, we can prove that the torques are the same if their magnitudes are identical. Applying the law of sines to triangle OAB, we find

$$\frac{r_1}{\sin(\pi - \theta_2)} = \frac{r_1}{\sin\theta_2} = \frac{r_2}{\sin\theta_1}$$

or
$$r_1 \sin\theta_1 = r_2 \sin\theta_2;$$

thus

$$|F||r_1| \sin\theta_1 = |F||r_2| \sin\theta_2 \tag{5-19}$$

and we have established that the torque is indeed independent of the point of application of the force along the line of action.

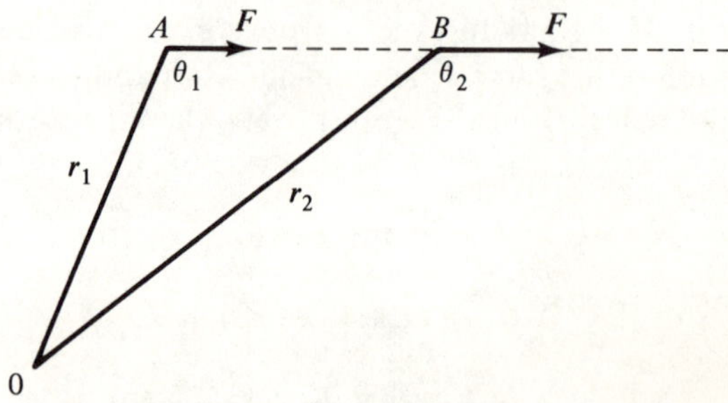

Fig. 5-3

By the distributive property of the vector product as expressed in Eq. (1-53), we see that the torque due to a force can be calculated by summing the torques due to each of the component forces. If F_j is a system of forces acting at points r_j and F is the vector sum or resultant of this system acting at a point r, then

$$F = \sum_j F_j \tag{5-20}$$

and
$$(r - r_0) \times F = \sum_j (r_j - r_0) \times F_j, \tag{5-21}$$

where r_0 is any point about which the torques are to be calculated.

A system of forces may have a null resultant

$$\sum_j F_j = 0 \qquad (5\text{-}22)$$

without having a null resultant torque if the forces are directed along parallel but distinct lines of action. Such an arrangement is called a *couple*. As we can see from Eq. (5-18), a couple exerts the same net torque about every point and can thus be characterized by a single vector giving the total torque. The simplest system of this type is a pair of forces, equal in magnitude but opposite in direction, acting at points A and B and separated by a vector r such that the torque about any point is given by

$$N = r \times F. \qquad (5\text{-}23)$$

This arrangement is shown in Fig. 5-4.

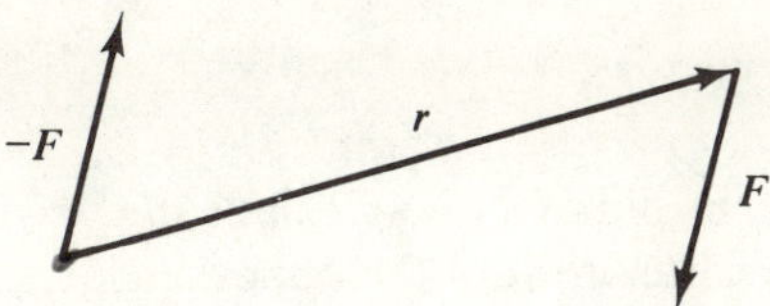

Fig. 5-4 The simplest couple.

It is clear that the simplest system equivalent to any system of forces must be composed of a single force acting through some arbitrary point plus a couple. Let the vector sum of the forces be F and the total torque about the arbitrarily chosen point A (normally taken to be the center of mass) be N. If we then allow F to act at A and add a couple whose torque is N, we have a system whose effect on the body will be wholly equivalent to that of the original system of forces. Since the simplest couple is composed of two forces, one of which may be chosen to act at an arbitrary point, we can let one of the couple forces act at A and add it to F as shown in Fig. 5-5. We see that the system of forces F_j can be reduced to a system containing two forces, $\mathscr{F} = F + F_c$ acting at A and $-F_c$ acting at B. In general, any system of forces can be reduced to an equivalent system containing no more than two forces.

If the couple lies in the same plane as the single nonzero force—that is, $F \perp N$—then the system can be reduced to an equivalent system containing a single force. This statement is easily proven. The torque N is equivalent to a couple composed of two equal and opposite forces F_c and $-F_c$. Only the vector product $N = r \times F_c$ is fixed; the vectors F_c and r are separately arbitrary. So we can choose $F = F$ and elect to apply the vector $-F_c$ at A, the point of application of F. Thus the single force vector is canceled and

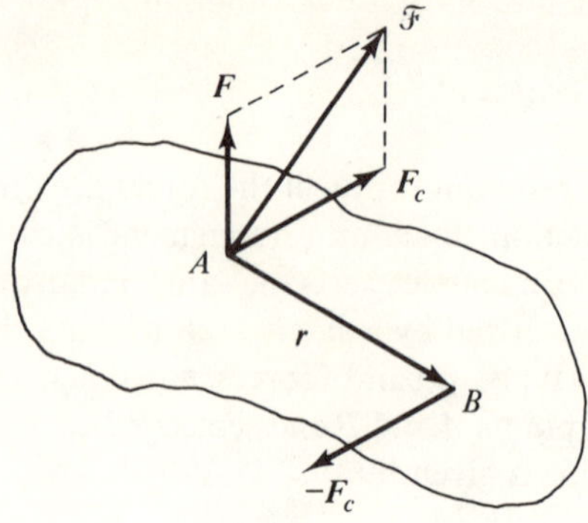

Fig. 5-5

only the vector $\boldsymbol{F}_c$ at B remains. The point B is located with respect to the point A by a vector $\boldsymbol{r}$ such that $\boldsymbol{N} = \boldsymbol{r} \times \boldsymbol{F}_c$.

EXERCISES

5-1.1. Prove that the centroid of a triangle lies $\frac{2}{3}$ of the way along any median as measured from the vertex toward the opposite side.

5-1.2. Locate the center of mass of a uniform wire bent into the shape of the arc of a circle of radius a. The angle subtended by the arc at the vertex is 2ϕ.

5-1.3. Use the results of problems 5-1.1 and 5-1.2 to find the center of mass of a thin sheet in the shape of a sector of a circle of radius a. The vertex angle is 2ϕ.

5-1.4. Show that the centroid of the uniform hollowed sphere shown in Fig. 1-23 is located a distance $R/14$ to the left of the origin.

5-1.5. Find the center of mass of a uniform segment of a sphere. The solid angle subtended is Ω and the radius of the sphere is a. Show that your result approaches the expected values in the limits as $\Omega \to 0$ (a uniform rod) and $\Omega \to 4\pi$ (a sphere). The solid angle is defined as the area of any subtended spherical surface divided by the square of its radius.

5-1.6. Find the center of mass of a uniform hemispherical shell of outer radius a and inner radius b. Show that, for very thin shells, the center of mass approaches $a/2$.

5-1.7. A homogeneous hemisphere of mass M and radius R rests with its curved surface in contact with a smooth horizontal surface. One edge is tied to a point on the horizontal surface by a light, inextensible string of length $l < R$. Find the tension in the string.

174

5-1.8. Prove that any system of forces acting on a rigid body can be reduced to a couple plus a single force normal to the plane of the couple.

5-2 Rotating Coordinate Systems

We have seen that the linear motion of an extended rigid body can be described in terms of the motion of a particle located at the center of mass. Therefore our main task will be to describe rotational motion. In order to do so, we must first derive the relations between fixed and rotating coordinate systems.

Consider two cartesian reference systems, one a fixed or inertial system and the other an arbitrary system that may exhibit linear and rotational motion with respect to the fixed system. In Fig. 5-6 we define the radius

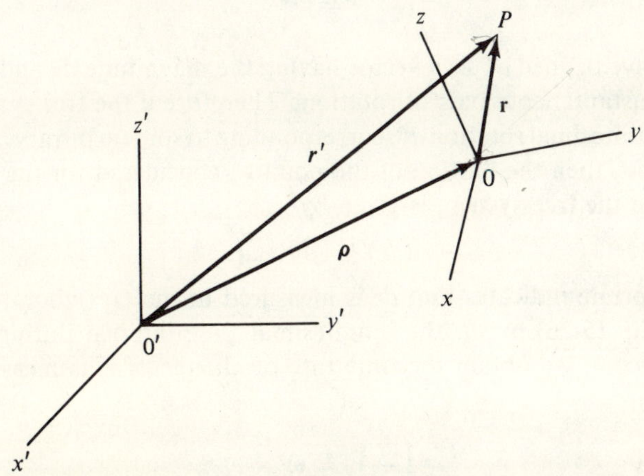

Fig. 5-6

vectors by which a point P can be specified in terms of these coordinates. The fixed system is described by coordinates (x', y', z') and the rotating system is given by coordinates (x, y, z). In terms of the point P, we see that

$$r' = \rho + r, \tag{5-24}$$

where r' and r are the radius vectors to P in the fixed and free systems, respectively, and ρ locates O with respect to O'.

An arbitrary *infinitesimal* displacement of O with respect to O' can always be represented as a pure rotation about some axis that we shall term the *instantaneous axis of rotation*. Referring to Fig. 5-7, we see that dr can be expressed as

$$dr = d\theta \times r \tag{5-25}$$

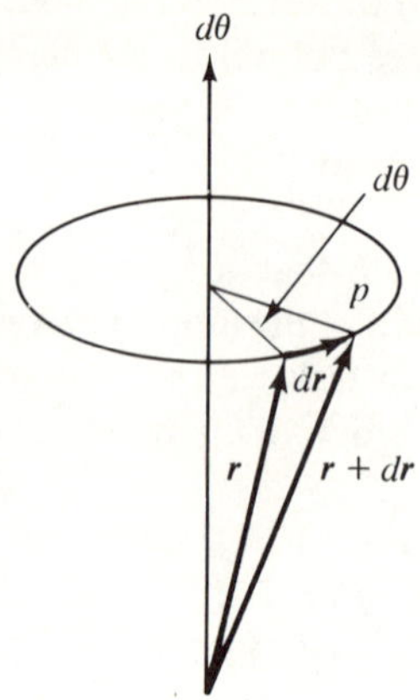

Fig. 5-7

when we have defined $d\boldsymbol{\theta}$ as a vector having the magnitude $d\theta$ and a direction along the instantaneous axis of rotation. Therefore if the free system experiences an infinitesimal rotation $d\boldsymbol{\theta}$, corresponding to some arbitrary infinitesimal displacement, then the motion of the point P, considered for the moment to be at rest in the free system, is given by

$$(d\boldsymbol{r})' = d\boldsymbol{\theta} \times \boldsymbol{r}, \tag{5-26}$$

where the prime indicates that $d\boldsymbol{r}$ is measured in the O' (laboratory) frame. Dividing Eq. (5-26) by dt, the infinitesimal time interval during which the rotation occurs, we obtain the time rate of change of $\boldsymbol{r}$ as measured in the fixed system

$$\left(\frac{d\boldsymbol{r}}{dt}\right)' = \boldsymbol{\omega} \times \boldsymbol{r}. \tag{5-27}$$

The vector quantity $\boldsymbol{\omega}$, defined by

$$\boldsymbol{\omega} = \frac{d\boldsymbol{\theta}}{dt}, \tag{5-28}$$

is called the *angular frequency* of the rotation and is directed along the axis of the rotation. If we then generalize the situation by allowing the point P to have a velocity $d\boldsymbol{r}/dt$ in the O system, $(d\boldsymbol{r}/dt)'$ will be given by

$$(\dot{\boldsymbol{r}})' = \dot{\boldsymbol{r}} + \boldsymbol{\omega} \times \boldsymbol{r}, \tag{5-29}$$

where we have used the raised dot notation to denote the time derivatives. Even though we chose the radius vector $\boldsymbol{r}$ for the derivation of Eq. (5-29), the form of this expression is equally valid for any arbitrary vector $\boldsymbol{A}$. Hence the rate of change of $\boldsymbol{A}$ measured in the laboratory system will be

$$(\dot{\boldsymbol{A}})' = \dot{\boldsymbol{A}} + \boldsymbol{\omega} \times \boldsymbol{A}. \tag{5-30}$$

We can now use Eqs. (5-24) and (5-29) to obtain an expression for the velocity of the point P as measured in the laboratory frame. Differentiating Eq. (5-24) with respect to time, we find

$$(\dot{r}')' = (\dot{\rho})' + (\dot{r})', \tag{5-31}$$

and so

$$(\dot{r}')' = (\dot{\rho})' + \dot{r} + \omega \times r. \tag{5-32}$$

This result can be written in a more compact notation as

$$u' = U + u + \omega \times r, \tag{5-33}$$

where u' is the velocity of P relative to O', U is the velocity of O relative to O', u is the velocity of P relative to O, ω is the angular frequency of the free system, and $\omega \times r$ is the velocity of P arising from the rotation of the free system.

EXERCISE

5-2.1. Show that the angular acceleration $\alpha = d\omega/dt$ is the same in both the fixed and rotating systems.

5-3 The Inertia Tensor

Let us examine a rigid body that we shall consider to be composed of n particles of mass m_λ, where the index λ runs from 1 to n. Assume that the body rotates with an instantaneous angular frequency ω about a point fixed with respect to the unprimed (body) coordinate system. For generality, we assume that this point has a linear velocity U with respect to the laboratory system. The instantaneous velocity of the λth particle can then be specified by using Eq. (5-33). Because of the rigidity, however, the velocity of any particle relative to the body system vanishes. Thus

$$u = \frac{dr}{dt} = 0, \tag{5-34}$$

and we find

$$u_\lambda = U + \omega \times r_\lambda. \tag{5-35}$$

The prime has been deleted from u_λ for notational simplicity, since all velocities are now measured in the fixed system. This deletion should cause no confusion because all velocities are zero with respect to the rotating frame. The kinetic energy of the λth particle of the body is

$$T_\lambda = \tfrac{1}{2}m_\lambda u_\lambda^2. \tag{5-36}$$

Summing over all n particles and using Eq. (5-35), we find for the total kinetic energy

$$T = \tfrac{1}{2} \sum_\lambda m_\lambda (U + \boldsymbol{\omega} \times r_\lambda) \cdot (U + \boldsymbol{\omega} \times r_\lambda) \tag{5-37}$$

or expanding the dot product,

$$T = \tfrac{1}{2} \sum_\lambda m_\lambda U^2 + \sum_\lambda m_\lambda U \cdot \boldsymbol{\omega} \times r_\lambda + \tfrac{1}{2} \sum_\lambda m_\lambda (\boldsymbol{\omega} \times r_\lambda)^2. \tag{5-38}$$

We can now effect a considerable simplification by choosing to locate the origin of the body coordinate frame at the center of mass of the body. The procedure is as follows. First, we rewrite the cross-term in Eq. (5-38) by taking U and $\boldsymbol{\omega}$ outside the summation:

$$\sum_\lambda m_\lambda U \cdot \boldsymbol{\omega} \times r_\lambda = U \cdot \boldsymbol{\omega} \times \sum_\lambda m_\lambda r_\lambda. \tag{5-39}$$

We recognize immediately that

$$\sum_\lambda m_\lambda r_\lambda = M r_c. \tag{5-40}$$

By choosing the origin of the coordinate system to coincide with the center of mass, this term vanishes and the kinetic energy of the body is

$$\begin{aligned}
T &= \tfrac{1}{2} \sum_\lambda m_\lambda U^2 + \tfrac{1}{2} \sum_\lambda m_\lambda (\boldsymbol{\omega} \times r_\lambda)^2 \\
&= \tfrac{1}{2} M U^2 + \tfrac{1}{2} \sum_\lambda m_\lambda (\boldsymbol{\omega} \times r_\lambda)^2 \\
&= T_t + T_r,
\end{aligned} \tag{5-41}$$

where T_t and T_r are the translational and rotational kinetic energies, respectively. Thus we have effected the separation of the kinetic energy into two independent parts as indicated in the introduction to this chapter.

The rotational term is of special interest to us. Using Eq. (1-65), we can rewrite $(\boldsymbol{\omega} \times r_\lambda)^2$ as

$$(\boldsymbol{\omega} \times r_\lambda)^2 = (\boldsymbol{\omega} \times r_\lambda) \cdot (\boldsymbol{\omega} \times r_\lambda) = \omega^2 r_\lambda^2 - (\boldsymbol{\omega} \cdot r_\lambda)^2, \tag{5-42}$$

and so

$$T_r = \tfrac{1}{2} \sum_\lambda m_\lambda [\omega^2 r_\lambda^2 - (\boldsymbol{\omega} \cdot r_\lambda)^2]. \tag{5-43}$$

For reasons that will become apparent momentarily, we choose to label our coordinates (x_1, x_2, x_3) rather than (x, y, z) as formerly. The radius vector r_λ in the rotating frame is then expressible as $(x_{\lambda,1}, x_{\lambda,2}, x_{\lambda,3})$, and the ith components of the vectors $\boldsymbol{\omega}$ and r_λ are ω_i and $x_{\lambda,i}$, respectively. In this notation the rotational kinetic energy becomes

$$T_r = \tfrac{1}{2} \sum_\lambda m_\lambda \left[\left(\sum_i \omega_i^2 \right) \left(\sum_k x_{\lambda,k}^2 \right) - \left(\sum_i \omega_i x_{\lambda,i} \right) \left(\sum_j \omega_j x_{\lambda,j} \right) \right]. \tag{5-44}$$

This expression can be simplified by using the Kronecker delta to write $\omega_i = \sum_i \omega_j \, \delta_{ij}$, and we find

$$T_r = \tfrac{1}{2} \sum_\lambda \sum_{i,j} m_\lambda \left[\omega_i \omega_j \, \delta_{ij} \left(\sum_k x_{\lambda,k}^2 \right) - \omega_i \omega_j x_{\lambda,i} x_{\lambda,j} \right]$$

$$= \tfrac{1}{2} \sum_{i,j} \omega_i \omega_j \sum_\lambda m_\lambda \left(\delta_{ij} \sum_k x_{\lambda,k}^2 - x_{\lambda,i} x_{\lambda,j} \right)$$

$$= \tfrac{1}{2} \sum_{i,j} I_{ij} \omega_i \omega_j, \tag{5-45}$$

where we define

$$I_{ij} \equiv \sum_\lambda m_\lambda \left[\delta_{ij} \sum_k x_{\lambda,k}^2 - x_{\lambda,i} x_{\lambda,j} \right], \tag{5-46}$$

or written as a 3×3 array and reverting to x, y, z for the coordinates

$$\mathsf{I} = \begin{pmatrix} \displaystyle\sum_\lambda m_\lambda (y_\lambda^2 + z_\lambda^2) & -\displaystyle\sum_\lambda m_\lambda x_\lambda y_\lambda & -\displaystyle\sum_\lambda m_\lambda x_\lambda z_\lambda \\[2ex] -\displaystyle\sum_\lambda m_\lambda y_\lambda x_\lambda & \displaystyle\sum_\lambda m_\lambda (x_\lambda^2 + z_\lambda^2) & -\displaystyle\sum_\lambda m_\lambda y_\lambda z_\lambda \\[2ex] -\displaystyle\sum_\lambda m_\lambda z_\lambda x_\lambda & -\displaystyle\sum_\lambda m_\lambda z_\lambda y_\lambda & \displaystyle\sum_\lambda m_\lambda (x_\lambda^2 + y_\lambda^2) \end{pmatrix}. \tag{5-47}$$

If we then let $m_\lambda \to 0$ and the number of particles in the body $n \to \infty$, the summation over λ goes over into an integral and we find the expression for I_{ij} appropriate to a body with a mass density $\rho(r)$:

$$I_{ij} = \iiint_{V} \rho(r) \left(\delta_{ij} \sum_k x_k^2 - x_i x_j \right) d\tau. \tag{5-48}$$

The quantity I turns out to be a tensor and is known as the *inertia tensor*.[5] The diagonal elements I_{11}, I_{22}, I_{33} are called the *moments of inertia* about the x_1, x_2, and x_3 axes, respectively, and $-I_{12}, -I_{13}, -I_{21}, -I_{23}, -I_{31}, -I_{32}$ are called the *products of inertia*. We note immediately that the inertia tensor is symmetric—that is,

$$I_{ij} = I_{ji}, \tag{5-49}$$

and so there are only six independent elements. If the rotation axis is made to coincide with one of the coordinate axes, say x_1, then the equation for the rotational energy goes over into the simple form

$$T_r = \tfrac{1}{2} I_{11} \omega_1^2 = \tfrac{1}{2} I \omega^2 \tag{5-50}$$

encountered in elementary treatments.

---

[5] The truth of our assumption that I is a tensor depends on the establishment of its behavior under a coordinate transformation. This process is carried out later in this section.

Having examined the formulation of the energy of rotation, let us investigate the angular momentum. Again we select a point at rest in the body coordinate system as our reference point. With respect to this point, the angular momentum of the body is

$$J = \sum_{\lambda} r_{\lambda} \times p_{\lambda}, \tag{5-51}$$

where p_{λ} is the linear momentum of the λth particle. If the axis of rotation lies within the body, then any point along this axis will be at rest in the laboratory frame as well as the body frame and will constitute a convenient choice of reference point; if no part of the body is at rest in the laboratory frame, then the reference point is chosen to be the center of mass.

Referred to the rotating (body) coordinate system, the linear momentum p_{λ} is

$$p_{\lambda} = m_{\lambda} u_{\lambda} = m_{\lambda} \omega \times r_{\lambda}, \tag{5-52}$$

and so

$$J = \sum_{\lambda} m_{\lambda} r_{\lambda} \times (\omega \times r_{\lambda}). \tag{5-53}$$

Using the identity for the vector triple product, Eq. (1-64), the expression for J can be rewritten

$$J = \sum_{\lambda} m_{\lambda} [r_{\lambda}^2 \omega - r_{\lambda}(r_{\lambda} \cdot \omega)]. \tag{5-54}$$

Using the same notation as previously employed to describe the energy, we find for the ith component of the angular momentum vector

$$\begin{aligned}
J_i &= \sum_{\lambda} m_{\lambda} \left[\omega_i \sum_{j} x_{\lambda,k}^2 - x_{\lambda,i} \sum_{k} x_{\lambda,j} \omega_j \right] \\
&= \sum_{\lambda} m_{\lambda} \sum_{j} \left(\omega_j \, \delta_{ij} \sum_{k} x_{\lambda,k}^2 - x_{\lambda,i} x_{\lambda,j} \omega_j \right) \\
&= \sum_{j} \omega_j \sum_{\lambda} m_{\lambda} \left(\delta_{ij} \sum_{k} x_{\lambda,k}^2 - x_{\lambda,i} x_{\lambda,j} \right) \\
&= \sum_{j} I_{ij} \omega_j, \tag{5-55}
\end{aligned}$$

or in tensor notation

$$J = I \cdot \omega. \tag{5-56}$$

Then multiplying J_i by $\tfrac{1}{2}\omega_i$ and summing over i give

$$\tfrac{1}{2} \sum_{i} \omega_i J_i = \tfrac{1}{2} \sum_{i,j} I_{ij} \omega_i \omega_j = T_r, \tag{5-57}$$

or

$$T_r = \tfrac{1}{2} \omega \cdot J. \tag{5-58}$$

In tensor form this equation is

$$T_r = \tfrac{1}{2}\boldsymbol{\omega}\cdot\mathsf{I}\cdot\boldsymbol{\omega}. \tag{5-59}$$

From Eq. (5-56) we can abstract an important law of tensor analysis—namely, that the dot product of a second-rank tensor and a first-rank tensor (vector) yields a first-rank tensor; the dot product of two first-rank tensors, as we already know, results in a tensor of zero rank (scalar).

As shown in the simple example given in Eq. (5-50), the expression for T_r is much more compact if the body coordinate system is chosen appropriately. Intuitive choice is obviously limited. A more fruitful approach results from formulating the problem such that the off-diagonal elements of the inertia tensor vanish. This procedure can be seen as follows. If I_{ij} can be cast in the form

$$I_{ij} = I_i\,\delta_{ij} \tag{5-60}$$

or, equivalently,

$$\mathsf{I} = \begin{pmatrix} I_1 & 0 & 0 \\ 0 & I_2 & 0 \\ 0 & 0 & I_3 \end{pmatrix}, \tag{5-61}$$

then J_i would assume the form

$$J_i = \sum_j I_i\,\delta_{ij}\,\omega_j = I_i\omega_i \tag{5-62}$$

and the energy would be

$$T_r = \tfrac{1}{2}\sum_{i,j} I_i\,\delta_{ij}\,\omega_i\omega_j = \tfrac{1}{2}\sum_i I_i\omega_i^2, \tag{5-63}$$

both quite simple. Our task is to find the set of body axes for which the off-diagonal elements of I all vanish. Such axes are called the *principal axes of inertia.* If the axis of rotation coincides with a principal axis, then both $\boldsymbol{\omega}$ and $\boldsymbol{J}$ will be directed along this axis. If I is the corresponding moment of inertia, then

$$\boldsymbol{J} = I\boldsymbol{\omega} \tag{5-64}$$

obtains. Comparing Eqs. (5-64) and (5-55), the components of J can be written

$$\begin{aligned}
J_1 &= I\omega_1 = I_{11}\omega_1 + I_{12}\omega_2 + I_{13}\omega_3 \\
J_2 &= I\omega_2 = I_{21}\omega_1 + I_{22}\omega_2 + I_{23}\omega_3 \\
J_3 &= I\omega_3 = I_{31}\omega_1 + I_{32}\omega_2 + I_{33}\omega_3
\end{aligned} \tag{5-65}$$

or

$$\begin{aligned}
(I_{11} - I)\omega_1 + I_{12}\omega_2 + I_{13}\omega_3 &= 0 \\
I_{21}\omega_1 + (I_{22} - I)\omega_2 + I_{23}\omega_3 &= 0 \\
I_{31}\omega_1 + I_{32}\omega_2 + (I_{33} - I)\omega_3 &= 0
\end{aligned} \tag{5-66}$$

a simultaneous set of homogeneous equations. Cramer's method for solving simultaneous algebraic equations teaches us that the determinant of the coefficients must vanish in order that a nontrivial solution exist:

$$\begin{vmatrix} I_{11} - I & I_{12} & I_{13} \\ I_{21} & I_{22} - I & I_{23} \\ I_{31} & I_{32} & I_{33} - I \end{vmatrix} = 0. \qquad (5\text{-}67)$$

The expansion of this determinant leads to a cubic equation for I. Each of the three roots I_1, I_2, I_3, called the *principal moments of inertia*, corresponds to the moments of inertia about one of the three principal axes. If the rigid body rotates about one of these axes, say x_1, Eq. (5-64) becomes

$$\boldsymbol{J} = I_1\boldsymbol{\omega} \qquad (5\text{-}68)$$

with $\boldsymbol{J}$ and $\boldsymbol{\omega}$ directed along the x_1 axis. The direction of this axis is then found by substituting I_1 for I in Eq. (5-66) and determining the ratios of the components of $\boldsymbol{\omega}$—that is, $\omega_1:\omega_2:\omega_3$. This step yields the direction cosines of the x_1 axis. The directions of the x_2 axis and the x_3 axis are found in the same manner. These principal axes are orthogonal.

In most practical instances, the body of interest will have a regular shape and the principal axes can be determined intuitively from an examination of its symmetry. For example, if the body had cylindrical symmetry, one of the principal axes will coincide with the symmetry axis. The other two axes will lie in a plane perpendicular to the symmetry axis and may be directed arbitrarily. Obviously, in this case, if I_1 is the moment of inertia for rotation about the symmetry axis, then $I_2 = I_3$. If the symmetry is spherical rather than cylindrical, then $I_1 = I_2 = I_3$.

To illustrate, consider a cube, l on a side. In problem 5-3.1 the reader is asked to show that if the origin of the body coordinates is chosen at one corner, with axes directed along the adjacent edges of the cube, then the inertia tensor is

$$\mathsf{I} = Ml^2 \begin{pmatrix} \tfrac{2}{3} & -\tfrac{1}{4} & -\tfrac{1}{4} \\ -\tfrac{1}{4} & \tfrac{2}{3} & -\tfrac{1}{4} \\ -\tfrac{1}{4} & -\tfrac{1}{4} & \tfrac{2}{3} \end{pmatrix}. \qquad (5\text{-}69)$$

The axes chosen are clearly not principal ones. If, for instance, the cube rotates about the x_3 axis, then $\boldsymbol{\omega} = \omega_3\hat{k}$ and, using Eq. (5-65), we can write

$$J_1 = -\tfrac{1}{4}Ml^2\omega_3,$$

$$J_2 = -\tfrac{1}{4}Ml^2\omega_3, \qquad (5\text{-}70)$$

$$J_3 = \tfrac{2}{3}Ml^2\omega_3.$$

Obviously, J is not parallel to ω for this choice of axes. In order to find the principal moments of inertia, we write the determinant equation

$$\begin{vmatrix} \tfrac{2}{3}Ml^2 - I & -\tfrac{1}{4}Ml^2 & -\tfrac{1}{4}Ml^2 \\ -\tfrac{1}{4}Ml^2 & \tfrac{2}{3}Ml^2 - I & -\tfrac{1}{4}Ml^2 \\ -\tfrac{1}{4}Ml^2 & -\tfrac{1}{4}Ml^2 & \tfrac{2}{3}Ml^2 - I \end{vmatrix} = 0, \qquad (5\text{-}71)$$

and to avoid the complication of a frontal assault on the corresponding cubic equation, we use the fact that the value of a determinant is unaffected by adding or subtracting any row from any other row (which also holds true for columns). Subtract the top row from the middle row, thereby obtaining

$$\begin{vmatrix} \tfrac{2}{3}Ml^2 - I & -\tfrac{1}{4}Ml^2 & -\tfrac{1}{4}Ml^2 \\ -\tfrac{11}{12}Ml^2 + I & \tfrac{11}{12}Ml^2 - I & 0 \\ -\tfrac{1}{4}Ml^2 & -\tfrac{1}{4}Ml^2 & \tfrac{2}{3}Ml^2 - I \end{vmatrix}$$

$$= (\tfrac{11}{12}Ml^2 - I)\begin{vmatrix} \tfrac{2}{3}Ml^2 - I & -\tfrac{1}{4}Ml^2 & -\tfrac{1}{4}Ml^2 \\ -1 & 1 & 0 \\ -\tfrac{1}{4}Ml^2 & -\tfrac{1}{4}Ml^2 & \tfrac{2}{3}Ml^2 - I \end{vmatrix}$$

$$= (\tfrac{11}{12}Ml^2 - I)[(\tfrac{2}{3}Ml^2 - I)^2 - \tfrac{1}{8}M^2l^4 - \tfrac{1}{4}Ml^2(\tfrac{2}{3}Ml^2 - I)] = 0. \quad (5\text{-}72)$$

The quadratic term in brackets can readily be solved, and we find for the three roots

$$I_1 = \tfrac{1}{6}Ml^2, \qquad I_2 = I_3 = \tfrac{11}{12}Ml^2. \qquad (5\text{-}73)$$

Since two of the roots are equal, the principal axis associated with the other, I_1, must be a symmetry axis. In order to find its direction, we substitute $I_1 = (Ml^2)/6$ for I in Eq. (5-66):

$$(\tfrac{2}{3} - \tfrac{1}{6})\omega_{11} - \tfrac{1}{4}\omega_{21} - \tfrac{1}{4}\omega_{31} = 0,$$

$$-\tfrac{1}{4}\omega_{11} + (\tfrac{2}{3} - \tfrac{1}{6})\omega_{21} - \tfrac{1}{4}\omega_{31} = 0, \qquad (5\text{-}74)$$

$$-\tfrac{1}{4}\omega_{11} - \tfrac{1}{4}\omega_{21} + (\tfrac{2}{3} - \tfrac{1}{6})\omega_{31} = 0,$$

where the factor Ml^2 common to all terms has been factored out. The second subscript 1 on the ω_i is included to indicate that we are considering the principal axis associated with I_1. These equations yield the result that

$$\omega_{11} = \omega_{21} = \omega_{31}; \qquad (5\text{-}75)$$

hence the ratios are all one to one and this principal axis must correspond to the diagonal of the cube. Since $I_2 = I_3$, the orientation of the principal axes associated with these moments is arbitrary and may lie anywhere in a plane normal to the diagonal.

We saw earlier that in order to separate the kinetic energy of a body into translational and rotational terms, it is necessary to situate the origin of the body coordinate frame at the center of mass. For certain shaped bodies, computation of the inertia tensor may be very inconvenient in such a system. Therefore let us consider a body coordinate system whose origin does not coincide with the center of mass and label the axes X_i. We shall require that this system have the same orientation as the center-of-mass system. If such a body frame can be found, in terms of which the inertia tensor can conveniently be calculated, then, as we shall see, a transformation can be made to find the inertia tensor in terms of the center-of-mass system.

The elements of the inertia tensor in the X_i frame can be written

$$I_{ij}^0 = \sum_\lambda m_\lambda \left(\delta_{ij} \sum_k X_{\lambda,k}^2 - X_{\lambda,i} X_{\lambda,j} \right). \tag{5-76}$$

In Fig. 5-8 we show the vectors involved. Since

$$\boldsymbol{R} = \boldsymbol{\rho} + \boldsymbol{r} \tag{5-77}$$

with components

$$X_i = \rho_i + x_i, \tag{5-78}$$

we can rewrite I_{ij}^0 in terms of x_i and ρ_i as

$$I_{ij}^0 = \sum_\lambda m_\lambda \left[\delta_{ij} \sum_k (x_{\lambda,k} + \rho_k)^2 - (x_{\lambda,i} + \rho_i)(x_{\lambda,j} + \rho_j) \right]$$

$$= \sum_\lambda m_\lambda \left(\delta_{ij} \sum_k x_{\lambda,k}^2 - x_{\lambda,i} x_{\lambda,j} \right)$$

$$+ \sum_\lambda m_\lambda \left[\delta_{ij} \sum_k (2x_{\lambda,k}\rho_k + \rho_k^2) - (\rho_i x_{\lambda,j} + \rho_j x_{\lambda,i} + \rho_i \rho_j) \right]. \tag{5-79}$$

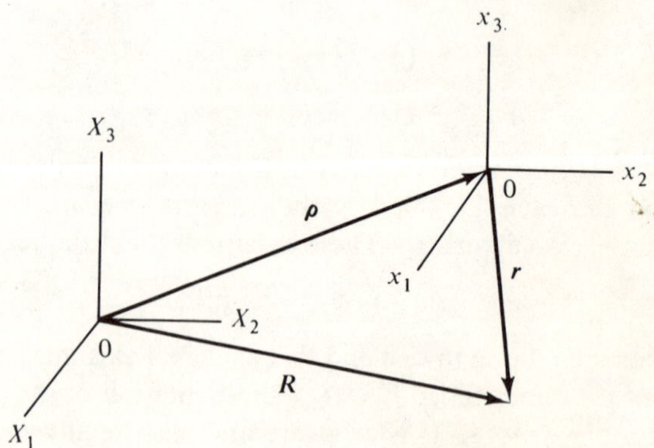

Fig. 5-8

The first term is simply I_{ij}. Rearranging the second term, we have

$$I_{ij}^0 = I_{ij} + \sum_\lambda m_\lambda \left(\delta_{ij} \sum_k \rho_k^2 - \rho_i \rho_j \right)$$

$$+ \sum_\lambda m_\lambda \left(2\delta_{ij} \sum_k x_{\lambda,k} \rho_k - \rho_i x_{\lambda,j} - \rho_j x_{\lambda,i} \right). \tag{5-80}$$

Since the x_i system has as its origin the center of mass, $\sum_\lambda m_\lambda \mathbf{r}_\lambda = 0$, and all terms like $\sum_\lambda m_\lambda x_{\lambda,k}$ also vanish. The last summation is entirely composed of such terms, and so

$$I_{ij}^0 = I_{ij} + \sum_\lambda m_\lambda \left(\delta_{ij} \sum_k \rho_k^2 - \rho_i \rho_j \right)$$

$$= I_{ij} + M(\rho^2 \delta_{ij} - \rho_i \rho_j). \tag{5-81}$$

Solving for I_{ij}, the inertia tensor corresponding to the center-of-mass coordinates, we have

$$I_{ij} = I_{ij}^0 - M(\rho^2 \delta_{ij} - \rho_i \rho_j). \tag{5-82}$$

This equation is the most general statement of the parallel axis theorem, usually given in scalar form in elementary texts. It is left as a problem for the reader to show that the difference between corresponding diagonal elements of I and I^0 is merely the mass of the body times the square of the distance between the parallel axes of rotation.

In order to tidy up loose ends, let us close this rather lengthy section by examining the properties of the tensor under coordinate transformation. We choose as our starting point the equation relating the angular momentum and angular frequency vectors, Eq. (5-55), which we write in terms of the indices k and l:

$$J_k = \sum_l I_{kl} \omega_l. \tag{5-83}$$

Since this is a vector equation in good standing, an entirely analogous expression must obtain in any coordinate system rotated with respect to the system in which Eq. (5-83) applies:

$$J_i' = \sum_j I_{ij}' \omega_j'. \tag{5-84}$$

Here the primed quantities refer to the rotated frame. We saw in Section 1-2 that vectors obey an orthogonal transformation [Eqs. (1-7) and (1-8)]; thus we can write

$$J_k = \sum_m \lambda_{mk} J_m', \tag{5-85}$$

and

$$\omega_l = \sum_j \lambda_{jl} \omega_j'. \tag{5-86}$$

Substituting for ω_l and J_k in Eq. (5-83) from these transformation relations gives

$$\sum_m \lambda_{mk} J'_m = \sum_l I_{kl} \sum_j \lambda_{jl} \omega'_j. \tag{5-87}$$

Multiply both sides of this equation by λ_{ik} and sum over k. This step gives, with appropriate rearrangement,

$$\sum_m \left(\sum_k \lambda_{ik} \lambda_{mk} \right) J'_m = \sum_j \left(\sum_{k,l} \lambda_{ik} \lambda_{jl} I_{kl} \right) \omega'_j. \tag{5-88}$$

The parenthetical term on the left-hand side is, by Eq. (1-14), simply δ_{im}. So performing the summation over m, we have

$$J'_i = \sum_j \left(\sum_{k,l} \lambda_{ik} \lambda_{jl} I_{kl} \right) \omega'_j \tag{5-89}$$

and by comparison with Eq. (5-84) we see that

$$I'_{ij} = \sum_{k,l} \lambda_{ik} \lambda_{jl} I_{kl}. \tag{5-90}$$

This transformation rule, derived for the inertia tensor, enjoys general validity for *all* second-rank tensors. This equation can be written in tensor form as

$$\mathsf{I}' = \lambda \mathsf{I} \tilde{\lambda}, \tag{5-91}$$

where we have transposed the second orthogonal matrix, thereby inverting the order of its multiplication with I. Since, for orthogonal matrices, the transpose of the matrix is equal to its inverse, Eq. (5-91) can also be written

$$\mathsf{I}' = \lambda \mathsf{I} \lambda^{-1}. \tag{5-92}$$

This type of transformation is called a *similarity transformation*. A second-rank tensor is defined as an object that obeys such a transformation. We recall that tensors of zero rank (scalars) obey

$$A' = A \tag{5-93}$$

and tensors of the first rank (vectors) obey

$$A'_i = \sum_l \lambda_{il} A_l. \tag{5-94}$$

From this and Eq. (5-90) we can presume that a tensor A of arbitrary rank obeys

$$A'_{ijk\ldots} = \sum_{l,m,n\ldots} \lambda_{il} \lambda_{jm} \lambda_{kn} \ldots A_{lmn\ldots} \tag{5-95}$$

as is indeed the case.

EXERCISES

5-3.1. Use Eq. (5-48) to calculate the inertia tensor for a cube, l on a side, of density ρ and mass m. The origin of the cartesian coordinate system is located at one corner of the cube, and the three adjacent edges lie along the coordinate axes. From this arrangement, show that

$$
\mathsf{I} = ml^2 \begin{pmatrix} \frac{2}{3} & -\frac{1}{4} & -\frac{1}{4} \\ -\frac{1}{4} & \frac{2}{3} & -\frac{1}{4} \\ -\frac{1}{4} & -\frac{1}{4} & \frac{2}{3} \end{pmatrix}.
$$

5-3.2. Calculate the moments of inertia I_1, I_2, I_3 for a homogeneous ellipsoid of mass m and minor axes $a < b < c$. Show by setting $a = b = c$ in the resulting relations that, for a sphere of radius a rotating about a diameter, the moment of inertia is $I = \frac{2}{5}ma^2$.

5-3.3. Consider a thin homogeneous rod of length l and mass m rotating about an axis located at one end and aligned perpendicular to the rod. Find the point at which, if all the mass were concentrated, the moment of inertia would be the same as the real moment of inertia. The distance of this point from the axis of rotation is called the *radius of gyration*.

5-3.4. Find the moment of inertia of a homogeneous thin disk of radius a and mass m rotating about an axis perpendicular to the flat surface and passing through the center.

5-3.5. You are given two cylinders having the same diameter and length and equal masses. It is known that one of them is homogeneous and that the other is a thin hollow shell. Excluding any destructive testing, describe in detail how you would distinguish between them.

5-3.6. Show that the difference between corresponding diagonal elements of the inertia tensors I and I^0 for the same body referred to two coordinate systems having the same orientation, the origin of one frame being at the center of mass, is equal to the mass of the body time the square of the distance between the parallel axes of rotation.

5-3.7. Prove that the sum of the moments of inertia of a plane lamina about any two perpendicular axes in the plane of the lamina is equal to the moment of inertia about an axis through their point of intersection perpendicular to the lamina. (*Hint:* If it can be established from a point element of mass dm, then the proof follows immediately by integration.)

5-3.8. Use the *perpendicular axis theorem* established in problem 5-3.7 to calculate the moment of inertia of a uniform thin ring of radius a and mass m about an axis in the plane of the ring passing through its center—that is, a diameter. What is the moment of inertia of the ring if we now consider an axis tangent to the ring?

5-3.9. Reconsider the cube described in problem 5-3.1. Using the inertia tensor calculated there and the parallel axis theorem, Eq. (5-82), calculate a new inertia tensor referred to a coordinate system with the same orientation but with its origin at the center of mass.

5-3.10. Show that the inertia tensor for the cube calculated in problem 5-3.9 for the center-of-mass system is invariant in regard to the orientation of the coordinate axes.

5-4 Euler's Equations of Motion for Rigid Bodies

The motion of a rigid body was shown in Section 5-1 to be governed by

$$\frac{d\mathbf{p}}{dt} = \mathbf{F} \tag{5-96}$$

and

$$\frac{d\mathbf{J}}{dt} = \mathbf{N}, \tag{5-97}$$

where

$$\mathbf{p} = m\mathbf{u} \tag{5-98}$$

and

$$\mathbf{J} = \mathbf{I} \cdot \boldsymbol{\omega}. \tag{5-99}$$

Here $\mathbf{F}$ is the total external force applied to the body, $\mathbf{N}$ is the total torque about some convenient point O, $\mathbf{u}$ is the velocity of the center of mass, and $\mathbf{I}$ and $\boldsymbol{\omega}$ are the inertia tensor and the angular frequency about O.

The rotational equations are basically much more complicated than the linear ones because of the fact that $\mathbf{p}$ and $\mathbf{u}$ are related by a scalar, the mass m, and thus are always parallel, whereas $\mathbf{J}$ and $\boldsymbol{\omega}$ are related by a tensor, $\mathbf{I}$, and are therefore not, in general, parallel. A more serious difference arises from the fact that $\mathbf{I}$ is not constant with reference to the laboratory system but changes as the body rotates. The most serious complication in describing the rotational motion occurs because no symmetrical set of three coordinates (such as a cartesian set, x, y, z) exists with which to describe the orientation of the body in space. This last point may not be obvious; a little experimentation with a solid body, however, will convince the reader that this is so. To illustrate, suppose that we choose a zero position for the body and try to specify the orientation by specifying rotation angles $\varphi_x, \varphi_y, \varphi_z$ about three perpendicular axes. Take, for example, the position specified by $\varphi_x = \pi/2$, $\varphi_y = \pi/2, \varphi_z = 0$. If we first rotate through a right angle about the x axis and then about the y axis, the result will be different than if we had first rotated about the y axis and then about the x axis.

The problems arising from the fact that I varies as the body rotates can be circumvented by referring Eq. (5-97) to a coordinate frame fixed in the body. If we let "$d*/dt$" denote the time derivative with reference to the body frame, then, by Eq. (5-30), Eq. (5-97) becomes

$$\frac{d*J}{dt} + \boldsymbol{\omega} \times J = N. \tag{5-100}$$

Since I remains constant in the body frame, we can use Eq. (5-99) to obtain

$$I \cdot \frac{d\boldsymbol{\omega}}{dt} + \boldsymbol{\omega} \times (I \cdot \boldsymbol{\omega}) = N, \tag{5-101}$$

where we have used the results of problem 5-2.1—that is,

$$\frac{d*\boldsymbol{\omega}}{dt} = \frac{d\boldsymbol{\omega}}{dt}. \tag{5-102}$$

For maximum convenience, we choose the body axes to be principal ones. Equation (5-101) can then be written in component form

$$I_1\dot{\omega}_1 + (I_3 - I_2)\omega_3\omega_2 = N_1,$$
$$I_2\dot{\omega}_2 + (I_1 - I_3)\omega_1\omega_3 = N_2, \tag{5-103}$$
$$I_3\dot{\omega}_3 + (I_2 - I_1)\omega_2\omega_1 = N_3.$$

These are Euler's equations of motion for a rigid body. If one point in the body is held fixed, that point is selected as the origin of the body frame. If there are no constraints on the motion, the center of mass will be chosen as the origin.

The energy theorem can be derived from Euler's equations in the following manner. First, we take the dot product of Eq. (5-101) with $\boldsymbol{\omega}$:

$$\boldsymbol{\omega} \cdot I \cdot \frac{d\boldsymbol{\omega}}{dt} = \boldsymbol{\omega} \cdot N. \tag{5-104}$$

Since I is a symmetric tensor and therefore equal to its transpose, the order of multiplication of the left-hand side is immaterial and we find

$$\boldsymbol{\omega} \cdot I \cdot \frac{d\boldsymbol{\omega}}{dt} = \frac{d\boldsymbol{\omega}}{dt} \cdot I \cdot \boldsymbol{\omega} = \frac{1}{2}\frac{d*}{dt}(\boldsymbol{\omega} \cdot I \cdot \boldsymbol{\omega}) = \frac{dT_r}{dt} \tag{5-105}$$

by Eq. (5-59). Here we have used the fact that d/dt and $d*/dt$ both produce the same result when applied to a scalar. Comparing Eqs. (5-104) and (5-105), we obtain the energy theorem

$$\frac{dT_r}{dt} = \boldsymbol{\omega} \cdot N. \tag{5-106}$$

If $d\boldsymbol{\omega}/dt = 0$, we see from Eq. (5-101) that

$$\boldsymbol{\omega} \times (\mathbf{I} \cdot \boldsymbol{\omega}) = \mathbf{N} \tag{5-107}$$

for a body rotating with constant angular frequency. If the torque is to vanish, then $\mathbf{I} \cdot \boldsymbol{\omega}$ must be parallel to $\boldsymbol{\omega}$; this result holds only if $\boldsymbol{\omega}$ is directed along a principal axis of the body. In other words, if a body, such as a wheel, is to spin freely without exerting forces and torques on its bearings, then it must not only be statically balanced (center of mass on the axis of rotation), it must also be dynamically balanced, which in our terms means that the axis of rotation must be a principal axis of the inertia tensor.

In order to solve Eqs. (5-103) for $\boldsymbol{\omega}(t)$, we would need to know the components of the torque along the principal axes of the body. Such is not usually the case unless $N = 0$. Let us examine this case—that of a freely rotating symmetrical body. We choose x_3 as the symmetry axis such that $I_1 = I_2$. The third of Eqs. (5-103) will then reduce to

$$I_3\dot{\omega}_3 = 0 \rightarrow \omega_3 = \text{const}. \tag{5-108}$$

The first two equations can be written

$$\dot{\omega}_1 + \alpha\omega_3\omega_2 = 0 \tag{5-109}$$

and

$$\dot{\omega}_2 - \alpha\omega_3\omega_1 = 0, \tag{5-110}$$

where

$$\alpha \equiv \frac{I_3 - I_1}{I_1}. \tag{5-111}$$

Equations (5-109) and (5-110) are coupled first-order linear equations. Differentiating the equations with respect to time and eliminating $\dot{\omega}_1$ and $\dot{\omega}_2$ by use of the original equations, we obtain a pair of uncoupled second-order equations

$$\ddot{\omega}_1 + (\alpha\omega_3)^2\omega_1 = 0 \tag{5-112}$$

$$\ddot{\omega}_2 + (\alpha\omega_3)^2\omega_2 = 0 \tag{5-113}$$

that we recognize as the equations of a simple harmonic oscillator with angular frequency $\alpha\omega_3$. The solutions can be conveniently written in the form

$$\omega_1 = C \cos(\alpha\omega_3 t + \varphi), \tag{5-114}$$

$$\omega_2 = C \sin(\alpha\omega_3 t + \varphi). \tag{5-115}$$

Thus the angular frequency vector $\boldsymbol{\omega}$ is seen to precess in a circle of radius C about the x_3 axis with angular frequency $\alpha\omega_3$. This precession is in the same sense as ω_3 if $I_3 > I_1$—that is, $\alpha > 0$—and in the opposite sense otherwise. The magnitude of $\boldsymbol{\omega}$ is

$$\omega = \sqrt{\omega_3^2 + C^2} \tag{5-116}$$

and is constant. The constants ω_3, C, and φ are determined by the initial conditions.

The instantaneous axis of rotation as determined by the vector $\boldsymbol{\omega}$ traces out a cone (referred to as the *body cone*) as it precesses about the axis of symmetry. The half angle of this cone, ϕ_b, is specified by

$$\phi_b = \arctan \frac{C}{\omega_3}. \tag{5-117}$$

The constants C and ω_3 can be written in terms of ϕ_b as

$$C = \omega \sin \phi_b, \qquad \omega_3 = \omega \cos \phi_b. \tag{5-118}$$

In order to describe the motion in the laboratory frame, we must locate $\boldsymbol{\omega}$ relative to $\boldsymbol{J}$, since, by Eq. (5-97), $\boldsymbol{J}$ is constant for $N = 0$. The angle between these two vectors, ϕ_l, is given by

$$\cos \phi_l = \frac{\boldsymbol{\omega} \cdot \boldsymbol{J}}{\omega J} = \frac{\boldsymbol{\omega} \cdot \mathsf{I} \cdot \boldsymbol{\omega}}{\omega J} = \frac{2T_r}{\omega J}. \tag{5-119}$$

Since by Eq. (5-106) T_r is constant, this angle ϕ_l is constant, and the axis of rotation traces out a cone in space (the *laboratory cone*). The line of contact between the laboratory and the body cones at any instant of time is simply the instantaneous axis of rotation. The picture is one in which the body cone rolls without slipping around the laboratory cone as shown in Fig. 5-9.

In 1834 Poinsot obtained a geometrical representation of the motion of a freely rotating body in terms of the constants of the motion and a geometrical

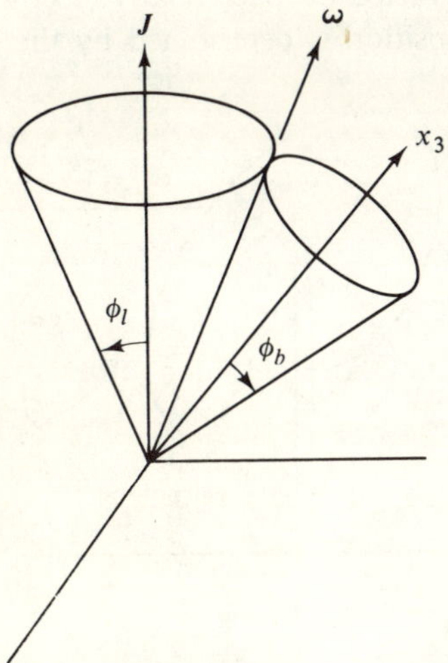

Fig. 5-9

analogy known as the *inertia ellipsoid*. The rotational kinetic energy in terms of principal axes [see Eq. (5-63)] is

$$T_r = \tfrac{1}{2}I_1\omega_1^2 + \tfrac{1}{2}I_2\omega_2^2 + \tfrac{1}{2}I_3\omega_3^2. \tag{5-120}$$

Since $I_i > 0$, it defines an ellipsoidal surface. If we define a vector

$$r = \frac{a}{\sqrt{2T_r}}\,\boldsymbol{\omega}, \tag{5-121}$$

where a is a constant, then

$$T_r = \tfrac{1}{2}\boldsymbol{\omega}\cdot\mathsf{I}\cdot\boldsymbol{\omega} = \text{const} \tag{5-122}$$

can be written in terms of r as

$$\boldsymbol{r}\cdot\mathsf{I}\cdot\boldsymbol{r} = a^2, \tag{5-123}$$

which is the equation of the inertia ellipsoid. This ellipsoid is fixed in the body and rotates with it. The normal to the ellipsoid at the point r is parallel to the vector

$$\nabla(\boldsymbol{r}\cdot\mathsf{I}\cdot\boldsymbol{r}) = 2\frac{r}{\omega}\boldsymbol{J}. \tag{5-124}$$

The tangent plane to the ellipsoid at the point r is therefore perpendicular to the constant vector $\boldsymbol{J} = \mathsf{I}\cdot\boldsymbol{\omega}$. This construction is shown in Fig. 5-10. The perpendicular distance l from the origin to the tangent plane is

$$l = \frac{\boldsymbol{r}\cdot\boldsymbol{J}}{J} = \frac{r}{\omega}\frac{\boldsymbol{\omega}\cdot\mathsf{I}\cdot\boldsymbol{\omega}}{J} = \frac{a\sqrt{2T_r}}{J} = \text{const}. \tag{5-125}$$

So the tangent plane is fixed in space relative to the origin and is called the *invariable plane*. Its position is determined by the initial conditions. Since

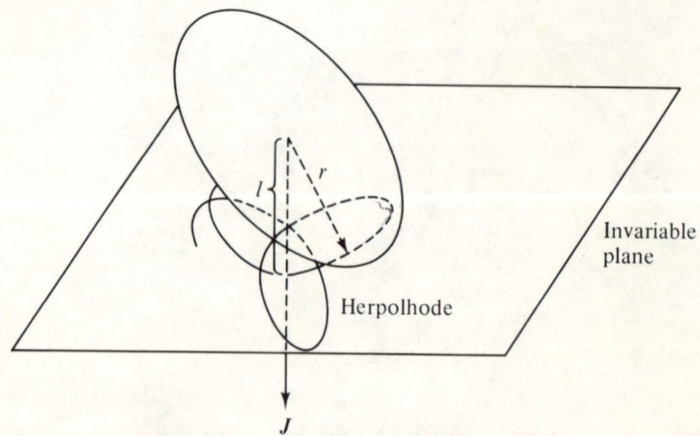

Fig. 5-10

the point of contact between the ellipsoid and the plane lies on the instantaneous axis of rotation, the ellipsoid rolls on the plane without slipping. The angular frequency at any instant has the magnitude

$$\omega = \frac{\sqrt{2T_r}}{a}\, r. \qquad (5\text{-}126)$$

As the ellipsoid rolls on the invariable plane with its center at rest at the origin, the point of contact traces out a curve on the ellipsoid called the *polhode* and a curve on the invariable plane called the *herpolhode.* The polhode is a closed curve; the herpolhode is, in general, not closed but consists of lobes within an annular envelope in the invariable plane. In the case of a symmetrical body, the polhodes are circles about the symmetry axis and the herpolhodes are circles in the invariable plane. In this case, r and ω (but not $\boldsymbol{\omega}$) are constant during the motion. The link between Poinsot's description and the body and laboratory cone picture is found in the fact that the polhode and herpolhode are the intersections of the body cone and laboratory cone with the inertia ellipsoid and invariable plane, respectively.

If $I_1 \approx I_3$, then the precession frequency $\alpha\omega_3 \ll \omega_3$. As an interesting example of this situation, the earth is known to be slightly flattened near the Poles because of the rotation such that we may think of it as approximated by an oblate spheroid with $I_3 \gtrsim I_1$. If the earth is a sufficiently good approximation to a rigid body, then we should find $\alpha\omega_3 \approx \omega_3/300$. The observed precession frequency is rather irregular, averaging about 50% less than that predicted above. This deviation is caused by the fact that the earth is *not* rigid (much of its interior is known to be molten) and the shape is not exactly that of an oblate spheroid but is, in fact, somewhat pear shaped, as a careful study of the moon's orbit has revealed.

EXERCISES

5-4.1. Show for a freely rotating symmetrical body that $\boldsymbol{J}$, $\boldsymbol{\omega}$, and the x_3 axis are coplanar.

5-4.2. Demonstrate that if the freely rotating symmetrical body is a prolate spheroid—that is, $I_3 < I_1$—the body cone rolls on the *outside* of the laboratory cone as shown in Fig. 5-9, whereas for the case of an oblate spheroid ($I_3 > I_1$), the x_3 axis lies between $\boldsymbol{J}$ and $\boldsymbol{\omega}$, and so the body cone rolls on the inside of the laboratory cone.

5-4.3. Prove the statement that if two principal moments of inertia are equal, then the polhode and herpolhode are both circles. In doing so, use only information derived from the Poinsot treatment of the freely rotating body.

5-4.4. Find the radii of the circles on the invariable plane that serve as the envelope of the herpolhode in the general case $I_1 \neq I_2 \neq I_3$.

MOTION IN TERMS
OF A NONINERTIAL FRAME

In Section 1-7 it was shown that Newtonian mechanics applied only in certain frames of reference that are unaccelerated with respect to one another. Such reference frames are called *inertial* frames. To restrict ourselves strictly to describing motion in terms of inertial frames can, in some cases, give rise to very complex equations, and it becomes a matter of convenience to see how the motion can be referred to a noninertial frame.

As the most outstanding example of our need for such a description, consider mechanical experiments carried out on or near the surface of the earth. Clearly, we would wish to choose a laboratory reference system fixed with respect to the earth. We are aware, however, that the earth experiences a complicated motion involving rotation about its own axis, rotation about the sun, and so on, and thus is accelerated with respect to Newton's "absolute" inertial frame at rest with respect to the "fixed" stars. Therefore the earth's coordinate system is a *noninertial* one, and even though we may ignore this point and still obtain sufficient accuracy for many purposes, the fact remains that many important effects do indeed result from the noninertial nature of the terrestrial frame.

In the previous chapter on rigid bodies we made extensive use of a noninertial frame, the body frame, without being specific about the underlying nature of such frames. The reader is encouraged at this stage to review Section 5-2 on rotating coordinate systems.

6-1 Centrifugal and Coriolis Forces

We have seen that Newton's second law $F = ma$ is valid only in an inertial frame. Now that we are going to examine mechanics from a more general point of view, we will use the notation of Section 5-2 and label our inertial frame by a prime superscript. Thus

$$F = ma' = m(\dot{u}')'. \tag{6-1}$$

Here the term $a' = (\dot{u}')'$ indicates that the differentiation of u' must be carried out with respect to the fixed system. In order to evaluate this term, we will differentiate Eq. (5-33) and impose the restriction of constant angular frequency ($\dot{\omega} = 0$).[1] Thus we find

$$F = m\dot{U}' + m(\dot{u})' + m\omega \times (\dot{r})'. \tag{6-2}$$

The second term on the right-hand side of Eq. (6-2) can be evaluated by substituting u for A in Eq. (5-30). This step yields

$$(\dot{u})' = \dot{u} + \omega \times u = a + \omega \times u, \tag{6-3}$$

where a is the acceleration in the rotating system. The last term on the right-hand side of Eq. (6-2) can be sorted out with the help of Eq. (5-29):

$$\omega \times (\dot{r})' = \omega \times \dot{r} + \omega \times (\omega \times r) = \omega \times u + \omega \times (\omega \times r). \tag{6-4}$$

Combining these results, Newton's second law, referred to coordinates rotating with constant angular frequency, is

$$F = ma' + m\dot{U}' + ma + m\omega \times (\omega \times r) + 2m(\omega \times u), \tag{6-5}$$

where $\dot{U}'$ is the acceleration of the origin of the noninertial coordinate system with respect to the inertial system. In most practical cases, this acceleration is negligible, and the noninertial system will have only rotation and possibly a uniform velocity. So $U = $ const or zero and we shall set $\dot{U}' = 0$. Therefore, to an observer in the rotating system, the effective force on a particle of mass m is given by

$$F_{\text{eff}} = ma = ma' - m\omega \times (\omega \times r) - 2m(\omega \times u). \tag{6-6}$$

The first term, ma', is the usual right-hand term from Newton's second law. It is modified for the observer in the rotating frame by the second and third terms. The quantity $-m\omega \times (\omega \times r)$ is called the *centrifugal force.* This term reduces to $-m\omega^2 r$ if $\omega \cdot r = 0$. The minus sign implies that the centrifugal force is directed outward from the center of rotation. The last term on the right-hand side of Eq. (6-6) is called the *Coriolis force.* It arises from the *motion* of the particle in the noninertial frame. The vector nature of the centrifugal force is depicted in Fig. 6-1. Such a simple picture of the Coriolis force is not possible, depending as it does on the arbitrary vector u.

---

[1] This restriction imposes no great limitation on the problems to be considered in this chapter.

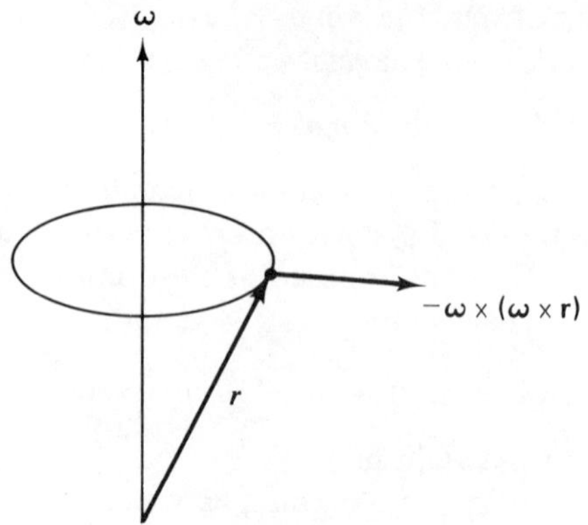

Fig. 6-1

Let us look more deeply into the physical meaning of these two forces. It is vitally important for the reader to realize that the centrifugal and Coriolis forces are not basic physical forces in the sense of the gravitational or electrostatic forces; they are introduced in an artificial manner in order to "patch up" Newtonian mechanics so as to be able to write an equation bearing a family resemblance to Newton's second law that will be valid in a noninertial frame. In other words, $F = ma'$ is valid *only* in an inertial frame. If we wish to write an equation of similar form—$F_{\text{eff}} = ma$ in a noninertial frame—then according to our findings we can formulate such an equation in terms of the real force as

$$F_{\text{eff}} = F + F_{\text{cent}} + F_{\text{Cor}}. \tag{6-7}$$

Suppose, as an example, that a body rotates in a circle about a fixed center of attraction much as the earth rotates about the sun. The only real force is the force (gravitational in the case of the earth and the sun) toward the center. An observer on the rotating body notes that the body does not fall toward the center. In order to reconcile this observation with the requirement that the net radial force vanish—that is, that the circular orbit be maintained—the observer postulates an additional force, the centrifugal force. This is an artificial construct that arises solely from our wish to extend Newton's second law to a noninertial system. The same comments apply to the Coriolis force; this force is necessary to describe motion relative to the rotating body.

The concept of centrifugal and Coriolis forces is very useful. The motion of a particle moving relative to a rotating body from the point of view of an observer in the "fixed star" frame would be highly complicated. By introducing the fictitious forces, however, the motion can be described in terms of an equation resembling Newton's equation in such a way as to be useful to us as noninertial observers.

196

EXERCISE

6-1.1. Write an equation of the same form as Eq. (6-6) to describe the dynamics of a particle relative to a noninertial frame that may rotate with a nonconstant angular frequency and be accelerated with respect to the fixed stars.

6-2 Motion On or Near the Earth's Surface

The motion of the earth with respect to the fixed star frame is dominated by the term describing the earth's daily rotation on its axis. Other terms relating to rotation about the sun, motion of the solar system in the galaxy, and so forth, are all quite small by comparison. We can, therefore, to a reasonable approximation consider a terrestrial frame of reference to be in pure rotation at constant ω with respect to the "absolute" inertial frame.

The equation of motion, relative to an inertial frame, for a particle subject to a gravitational force mg is, of course,

$$ma' = mg. \tag{6-8}$$

If we then refer the motion to the terrestrial frame and measure the position vector from the center of the earth, we find, using Eq. (6-6),

$$ma = m[g - \omega \times (\omega \times r)] - 2m\omega \times u$$

$$= mg_{\text{eff}} - 2m\omega \times u. \tag{6-9}$$

Here we have chosen to combine the gravitational and centrifugal terms, since both are proportional to the mass of the particle and depend only on its position. For this reason, we have defined an effective gravitational acceleration g_{eff} as

$$g_{\text{eff}} = g - \omega \times (\omega \times r). \tag{6-10}$$

The gravitational force on a body at rest on the earth's surface is mg_{eff}. In Fig. 6-2 we have drawn this vector for a particle in the Northern and Southern Hemispheres. A falling body released from rest near the earth's surface will initially fall with an acceleration g_{eff} until the Coriolis force begins to deflect it. A liquid, such as an ocean, comes to equilibrium with its surface normal to g_{eff}, which explains the (approximately) oblate spheroidal shape of the earth.

It is of interest to calculate the effect of the Coriolis force on a particle moving in various directions on or near the earth's surface. Since ω is directed toward the north, there will be a component of ω, ω_z, normal to the earth's surface in a local cartesian frame with its origin at the surface. This component will be outward in the Northern Hemisphere and inward in the

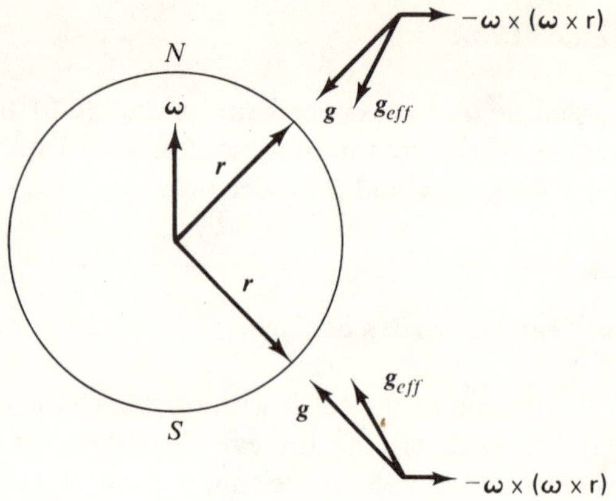

Fig. 6-2

Southern Hemisphere. If a particle is projected in the local tangent plane—that is, horizontally—then the Coriolis force $-2m\omega \times u$ will have a component in the plane of magnitude $2m\omega_z u$ directed to the right of the particle's initial motion in the Northern Hemisphere and to the left in the Southern Hemisphere. Since $\omega_z = \omega \sin \lambda$, where λ is the latitude, we see that this horizontal deflection force is maximum at the Poles and zero on the Equator. The Coriolis force is of major importance in determining the motion of the earth's atmosphere. If a local area of low pressure arises, the radial (in the horizontal plane) inward flow of air will be deflected by the Coriolis force such as to create a counterclockwise vortex in the Northern Hemisphere and a clockwise circulation in the Southern Hemisphere. This situation is depicted in Fig. 6-3. The Coriolis deflecting force will eventually establish an equilibrium with the pressure gradient force. This equilibrium circulation results in a generally eastward flow of air in the Temperate Zones and a westward flow

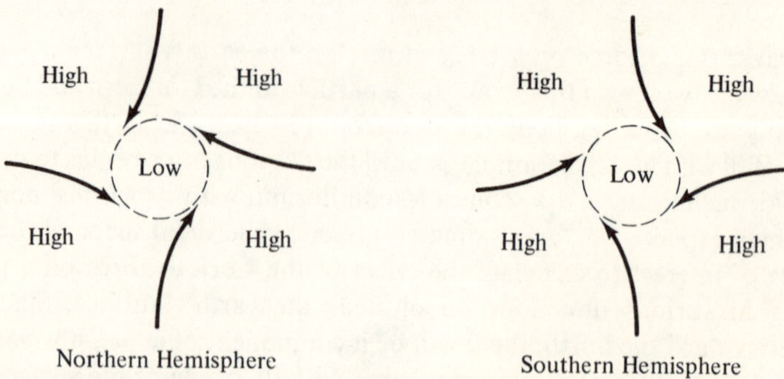

Fig. 6-3

(the trade winds) in the Tropics. The motion of water in whirlpools—for instance, the maelstroms found off the Norwegian coast—is similar in principle. In actuality, local perturbations dominate the Coriolis force, and whirlpools of both polarities are found in both hemispheres. In spite of the common misconception (which has often found its way into physics textbooks), the whirlpool formed when a bathtub drains, even long after filling, is governed by the direction of circulation established when it was filled and not by whether it is in London or Capetown.

Let us turn next to the horizontal deflection caused by the Coriolis force of a particle falling in the earth's gravitational field. Again we will use a local rectangular coordinate system with the x axis pointing east and the y axis pointing north. We will restrict ourselves to consider motion in the Northern Hemisphere. In terms of our coordinate system, the angular frequency vector is

$$\boldsymbol{\omega} = (0, \omega \cos \lambda, \omega \sin \lambda). \tag{6-11}$$

Although the Coriolis force will give rise to small velocity components in the horizontal plane, we can neglect them in comparison with u_z. Thus

$$\boldsymbol{u} \approx (0, 0, -gt). \tag{6-12}$$

We can now calculate the acceleration of the falling particle to be[2]

$$\boldsymbol{a} \approx (2\omega gt \cos \lambda, 0, -g). \tag{6-13}$$

The deflection of the particle eastward in time t is therefore

$$y(t) \approx \tfrac{1}{3}\omega gt^3 \cos \lambda. \tag{6-14}$$

Since the time taken for a particle to fall from a height h is approximately

$$t \approx \sqrt{\frac{2h}{g}}, \tag{6-15}$$

we see that a particle dropped from rest at a height h at latitude λ in the Northern Hemisphere will be deflected eastward by a distance (neglecting air resistance)

$$d \approx \tfrac{1}{3}\omega \cos \lambda \sqrt{\frac{8h^3}{g}}. \tag{6-16}$$

EXERCISES

6-2.1. Calculate the centrifugal accelerations caused by the earth's daily rotation on its axis and its yearly rotation about the sun and show that the ratio of the gravitational acceleration at the earth's surface to the spin and

---

[2] Here we have approximated g_{eff} by the constant acceleration of gravity g.

orbital centrifugal accelerations on a particle at the Equator are approximately 292 and 1650, respectively. Centrifugal acceleration due to the motion of the solar system about the galactic center is 20 million times smaller yet.

6-2.2. Using a local cartesian coordinate system with origin at the surface of the earth at latitude λ in the Northern Hemisphere wherein y is north, x is east, and z is up, show that the motion of a particle near the surface of the earth is described by

$$\ddot{x} = \omega(2\dot{y}\sin\lambda - 2\dot{z}\cos\lambda + x\omega\cos^2\lambda + y\omega\sin^2\lambda), \tag{6-17}$$

$$\ddot{y} = \omega\sin\lambda(-2\dot{x} + y\omega\sin\lambda - z\omega\cos\lambda), \tag{6-18}$$

$$\ddot{z} = -g + \omega\cos\lambda(2\dot{x} + z\omega\cos\lambda - y\omega\sin\lambda). \tag{6-19}$$

6-2.3. Using Eqs. (6-17) to (6-19), derive the simplest reasonable expression for the distance and direction between the launch point of a projectile aimed vertically and its point of impact with the earth. (*Hint:* Note that g is by far the largest term. The quantity $\omega\dot{z}$ or ωz is first order in ω. The quantities $\omega\dot{x}$, ωx and $\omega\dot{y}$, ωy are second and third order, respectively, in ω, involving as they do the Coriolis and centrifugal induced quantities x, $\dot{x}$ and y, $\dot{y}$. For this problem, omit all second- and third-order terms.)

6-2.4. Using the expressions for x, $\dot{x}$ and z, $\dot{z}$ obtained in the course of solving problem 6-2.3, retain the second-order terms in Eqs. (6-17) to (6-19) and again solve the problem of the horizontal deflection of a vertically launched projectile.

6-2.5. Derive an expression for the angle between g and g_{eff} at the surface of the earth for any latitude λ. Calculate the angle in seconds of arc for λ corresponding to the largest deviation.

6-2.6. A projectile is fired eastward from a point on the earth's surface at latitude λ in the Northern Hemisphere. The angle of inclination of the cannon with the horizontal is φ. Find the southerly deviation due to the rotation of the earth.

6-2.7. The origin of a local coordinate system located on the surface of the earth experiences an acceleration with respect to the "absolute" frame—that is, $\dot{\mathbf{U}}' \neq 0$. Estimate the magnitude of this term and show that its neglect for motion near the earth's surface is justified.

6-3 The Foucault Pendulum

In 1851 J. L. Foucault, a French physicist, established that the plane of oscillation of a pendulum is caused by the Coriolis force to precess about the vertical axis through the point of suspension. A pendulum is particularly useful as a sensor of "fictitious" forces, since it is detached from the rotary motion of its supporting framework and is only constrained to follow linear motions of the framework, which is, of course, at rest on the surface of the earth.

In Fig. 6-4 we show a pendulum bob under the action of the forces $m\mathbf{g}$ due to gravity and $\mathbf{T}$, the tension in the string. The equation of motion of the particle is therefore, neglecting higher-order terms in $\boldsymbol{\omega}$,

$$\ddot{\mathbf{r}} = \mathbf{g} + \frac{\mathbf{T}}{m} - 2\boldsymbol{\omega} \times \dot{\mathbf{r}} \tag{6-20}$$

or

$$\ddot{x} = 2\dot{y}\omega \sin \lambda - 2\dot{z}\omega \cos \lambda - \frac{T_x}{m},$$

$$\ddot{y} = -2\dot{x}\omega \sin \lambda - \frac{T_y}{m}, \tag{6-21}$$

$$\ddot{z} = 2\dot{x}\omega \cos \lambda + \frac{T_z}{m} - g.$$

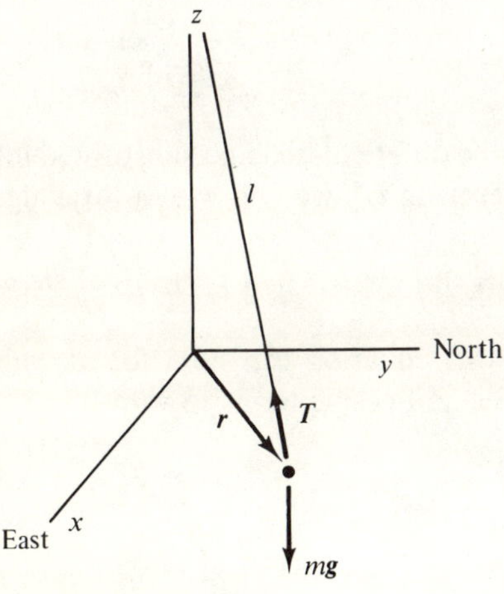

Fig. 6-4

This system of equations, including the Coriolis but not the centrifugal terms, can be simplified somewhat for our purposes. Clearly, the restriction of small displacement implies that $l \gg r$ and $\dot{z} \ll \dot{x}$ or $\dot{y}$. The components of the tension are

$$T \approx T_z = mg, \qquad T_x = \frac{Tx}{l}, \qquad T_y = \frac{Ty}{l}. \tag{6-22}$$

Making use of these consequences of small amplitude and writing $\omega_z = \omega \sin \lambda$, the equations of motion in the xy plane become

$$\ddot{x} - 2\omega_z \dot{y} + \frac{gx}{l} = 0, \tag{6-23}$$

$$\ddot{y} + 2\omega_z \dot{x} + \frac{gy}{l} = 0. \tag{6-24}$$

This set of coupled second-order equations can be solved mathematically by a straightforward approach. It is more satisfactory, however, to apply some simple physical reasoning. Clearly, the effect of decoupling oscillatory motion in a vertical plane from the local vertical component of the earth's rotation vector ω_z is, from the point of view of an observer on the earth, a precession of the plane of oscillation at a frequency $-\omega_z \hat{k}$—that is, clockwise in the Northern Hemisphere and counterclockwise in the Southern Hemisphere.

In order to demonstrate that this intuitive analysis is correct, let us introduce another cartesian coordinate system (x_1, y_1, z_1) having its origin and z_1 axis coincident with the origin and z axis of our laboratory frame but rotating clockwise from above at a frequency $\omega_z = \omega \sin \lambda$. The transformation from the x_1, y_1 to x, y is given by

$$x = x_1 \cos \omega_z t + y_1 \sin \omega_z t,$$
$$y = -x_1 \sin \omega_z t + y_1 \cos \omega_z t. \tag{6-25}$$

These equations can be differentiated and substituted into either Eq. (6-23) or (6-24). Neglecting terms in ω_z^2, we find with a little algebra

$$\left(\ddot{x}_1 + \frac{g}{l} x_1 \right) \cos \omega_z t + \left(\ddot{y}_1 + \frac{g}{l} y_1 \right) \sin \omega_z t = 0. \tag{6-26}$$

The only way that this equation can hold for all values of $\omega_z t$ is for the coefficients of the sine and cosine terms to vanish:

$$\ddot{x}_1 + \frac{g}{l} x_1 = 0, \tag{6-27}$$

$$\ddot{y}_1 + \frac{g}{l} y_1 = 0. \tag{6-28}$$

These equations are the familiar equations of simple harmonic oscillation in the $x_1 y_1$ plane. We see, therefore, that our intuitive picture of a simple pendulum precessing at a frequency $\omega \sin \lambda$ is confirmed.

EXERCISES

6-3.1. The bob of a pendulum is given an initial horizontal velocity such that the motion is circular. What is the effect of the Coriolis force on this motion?

6-3.2. Using Eqs. (6-23) and (6-24), define a complex quantity

$$z = x + iy$$

and show that the equation of the Foucault pendulum can be written in terms of z as

$$\ddot{z} + 2i\omega_z \dot{z} + \frac{g}{l} z = 0. \tag{6-29}$$

Solve this equation and show that the results agree with the conclusions that we reached intuitively.

6-4 Larmor Precession

The Coriolis force is peculiar in nature because of the fact that it always acts at right angles to the path of the particle. The magnetic force on a moving charged particle is

$$\boldsymbol{F} = q\boldsymbol{u} \times \boldsymbol{B}, \tag{6-30}$$

where $\boldsymbol{B}$ is the magnetic flux density. This force has the same form as the Coriolis force. Earlier we saw that the Coriolis force on a massive body can be made to vanish in a frame of reference rotating at a particular frequency with respect to the terrestrial (laboratory) frame. Does it not follow, as a result of the identical form of the equations, that the effect of a magnetic field on a charged particle can be made to vanish in some suitable rotating frame? The answer is yes.

In the laboratory frame, which we shall label with a prime, the equation of motion of a charged particle is

$$(\ddot{\boldsymbol{r}})' = \frac{q}{m}(\dot{\boldsymbol{r}})' \times \boldsymbol{B}. \tag{6-31}$$

In an effort to eliminate this force, we introduce a new coordinate system that has the same origin as the prime system and is in rotation about this origin with angular frequency $\boldsymbol{\omega}$. Since, by Eqs. (5-29) and (5-30),

$$(\dot{\boldsymbol{r}})' = \dot{\boldsymbol{r}} + \boldsymbol{\omega} \times \boldsymbol{r} \tag{6-32}$$

and

$$(\ddot{\boldsymbol{r}})' = \ddot{\boldsymbol{r}} + \boldsymbol{\omega} \times (\boldsymbol{\omega} \times \boldsymbol{r}) + 2\boldsymbol{\omega} \times \dot{\boldsymbol{r}} \tag{6-33}$$

for constant $\boldsymbol{\omega}$, we find for the equation of motion of the charge in the rotating system

$$\ddot{\boldsymbol{r}} = -\boldsymbol{\omega} \times (\boldsymbol{\omega} \times \boldsymbol{r}) + \frac{q}{m}(\boldsymbol{\omega} \times \boldsymbol{r}) \times \boldsymbol{B} + \dot{\boldsymbol{r}} \times \left(\frac{q\boldsymbol{B}}{m} + 2\boldsymbol{\omega}\right). \tag{6-34}$$

If we choose

$$\boldsymbol{\omega} = -\frac{q\boldsymbol{B}}{2m}, \tag{6-35}$$

the last term on the right-hand side of Eq. (6-34) vanishes and we are left with

$$\ddot{\boldsymbol{r}} = \left[\frac{q}{2m}\right]^2 \boldsymbol{B} \times (\boldsymbol{B} \times \boldsymbol{r}). \tag{6-36}$$

If the right-hand side of this equation is negligibly small, the magnetic effects can be said to be transformed away. Since the magnetic force that we have eliminated is linear in $\boldsymbol{B}$, whereas the magnetic term in Eq. (6-36) is quadratic in $\boldsymbol{B}$, it is possible in the limit of weak magnetic fields to neglect the latter. It is equivalent to requiring that the *Larmor frequency* given by Eq. (6-35) be much less than any frequency associated with the motion in the absence of the magnetic field.

To illustrate, consider a charged particle of mass m and charge q moving in a circle of radius r about an attractive center. An external magnetic field B lies in a direction making an angle θ with the normal to the plane of the orbit. The effect of the magnetic field is to give rise to a torque

$$\boldsymbol{N} = \boldsymbol{\mu}_m \times \boldsymbol{B} = \frac{q}{2}(\boldsymbol{r} \times \boldsymbol{u}) \times \boldsymbol{B}, \tag{6-37}$$

where $\boldsymbol{\mu}_m$ is the magnetic moment defined in Section 2-6. We shall set up the problem as shown in Fig. 6-5. The z' axis is parallel to $\boldsymbol{B}$. We assume on

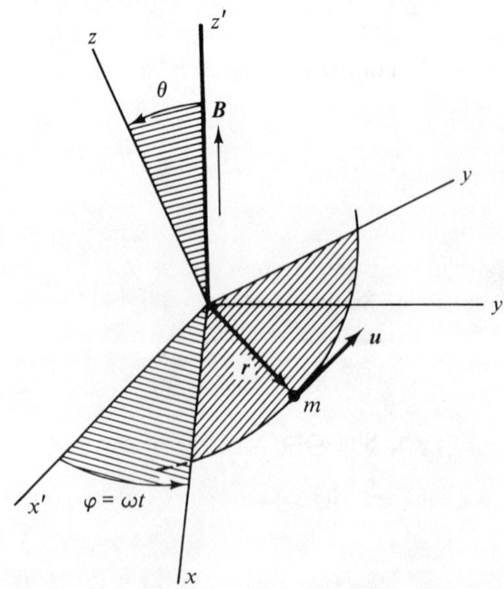

Fig. 6-5

the basis of experience that the magnetic moment is precessing at an angular frequency ω about $\boldsymbol{B}$, and so we take the axes x, y, z to rotate at angular frequency ω about the z axis. The orbit of the particle is in the xy plane, the x axis of which is constrained to remain in the $x'y'$ plane. The angular momentum arising from the orbital velocity is

$$\boldsymbol{L} = m\boldsymbol{r} \times \boldsymbol{u} = mr^2\Omega\hat{\boldsymbol{k}}. \tag{6-38}$$

The total angular momentum $\boldsymbol{J}$ is therefore

$$\begin{aligned}
\boldsymbol{J} &= \mathbf{I} \cdot \boldsymbol{\omega} + \boldsymbol{L} \\
&= I_x\omega_x\hat{\boldsymbol{i}} + I_y\omega_y\hat{\boldsymbol{j}} + (I_z\omega_z + mr^2\Omega)\hat{\boldsymbol{k}} \\
&= I_x\omega_x\hat{\boldsymbol{i}} + I_y\omega_y\hat{\boldsymbol{j}} + I_z(\omega_z + \Omega)\hat{\boldsymbol{k}}.
\end{aligned} \tag{6-39}$$

The derivative of $\boldsymbol{J}$ evaluated in the prime system is

$$\begin{aligned}
(\boldsymbol{J})' &= I_x\dot{\omega}_x\hat{\boldsymbol{i}} + I_y\dot{\omega}_y\hat{\boldsymbol{j}} + I_z(\dot{\omega}_z + \dot{\Omega})\hat{\boldsymbol{k}} \\
&\quad + I_x\omega_x\frac{d\hat{\boldsymbol{i}}}{dt} + I_y\omega_y\frac{d\hat{\boldsymbol{j}}}{dt} + I_z(\omega_z + \Omega)\frac{d\hat{\boldsymbol{k}}}{dt}.
\end{aligned} \tag{6-40}$$

Since $\dot{\hat{\boldsymbol{i}}} = \boldsymbol{\omega} \times \hat{\boldsymbol{i}}$ and so on, we find for the x component

$$(\dot{\boldsymbol{J}})'_x = I_x\dot{\omega}_x + (I_z - I_y)\omega_z\omega_y + I_z\omega_y\Omega. \tag{6-41}$$

The torque as given by Eq. (6-37) is in the negative x direction, and so we have

$$I_x\dot{\omega}_x + (I_z - I_y)\omega_z\omega_y + I_z\omega_y\Omega = -\tfrac{1}{2}qruB \sin\theta. \tag{6-42}$$

In order to simplify this equation, we will assume that Ω, the orbital angular frequency, is much larger than any of the components of ω, the precession frequency. In this spirit, we shall neglect the first two terms on the left. Thus we have

$$I_z\omega_y\Omega \approx -\frac{q}{2}ruB \sin\theta$$

or

$$\omega_y = -\frac{qruB \sin\theta}{2I_z\Omega}$$

$$= -\frac{qB}{2m}\sin\theta.$$

Hence

$$\omega = \frac{\omega_y}{\sin\theta} = -\frac{qB}{2m} \tag{6-43}$$

and again the Larmor frequency is found as the rate at which the magnetic moment vector precesses about the magnetic field vector or, equivalently, the precession rate of the coordinate frame in which magnetic effects are nil.

Larmor's theorem can be expanded to include a collection of charged particles, all with the same charge-to-mass ratio, acted on by mutual electrostatic forces and an external central force directed toward a common center. This generalization is possible because these forces depend only on the interparticle distances, and the distances of the particles from the attracting center of the external force which must coincide with the origin. These distances are the same in the primed or unprimed systems.

EXERCISES

6-4.1. Why should we not be surprised at the existence and, indeed, the form of the residual magnetic term

$$\left(\frac{q}{2m}\right)^2 B \times (B \times r)$$

seen by the particle in the unprimed frame?

6-4.2. It was shown in Section 2-6 that the motion of a charged particle in a plane normal to a magnetic field consists of circular orbits at a frequency twice the Larmor frequency. Why is Larmor's theorem inapplicable in this case?

ELASTIC BODIES

7-1 The Strain Tensor

In 1858 Helmholtz established the following theorem, which forms the basis for a kinematic description of elastic bodies.

The most general motion of a sufficiently small, deformable body can be described as the sum of a translation, a rotation, and an extension or contraction in three mutually orthogonal directions.

The proof of this theorem involves carrying out a Taylor expansion of the relative displacements of two neighboring points A and B in terms of their original coordinate separations. Let B be a point of the volume element under consideration with coordinates x, y, z in terms of a cartesian coordinate system, the origin of which, A, is within the element. If a general motion of the deformable body is induced, the points A and B will both experience a change of position, which we shall denote by the vectors $s_0(\xi_0, \eta_0, \zeta_0)$ and $s(\xi, \eta, \zeta)$, respectively, as indicated in Fig. 7-1. For a sufficiently small element, the leading terms of the Taylor expansion describe the displacement of B.

$$\xi = \xi_0 + \frac{\partial \xi}{\partial x} x + \frac{\partial \xi}{\partial y} y + \frac{\partial \xi}{\partial z} z + \cdots,$$

$$\eta = \eta_0 + \frac{\partial \eta}{\partial x} x + \frac{\partial \eta}{\partial y} y + \frac{\partial \eta}{\partial z} z + \cdots, \qquad (7\text{-}1)$$

$$\zeta = \zeta_0 + \frac{\partial \zeta}{\partial x} x + \frac{\partial \zeta}{\partial y} y + \frac{\partial \zeta}{\partial z} z + \cdots.$$

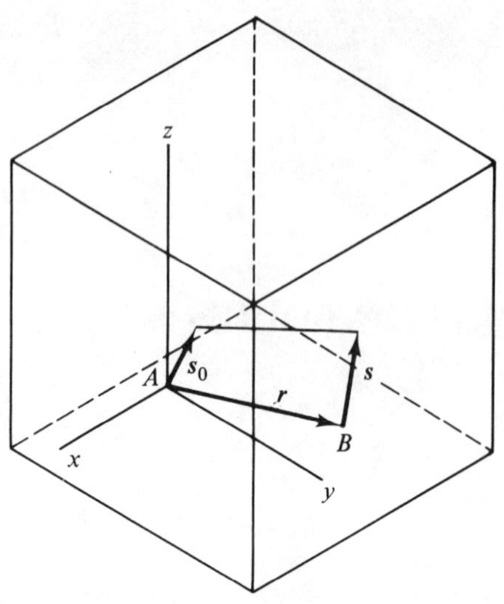

Fig. 7-1

Neglecting derivatives above the first and introducing the notation

$$\frac{\partial \xi}{\partial x} = a_{11}, \qquad \frac{\partial \xi}{\partial y} = a_{12}, \qquad \frac{\partial \xi}{\partial z} = a_{13},$$

$$\frac{\partial \eta}{\partial x} = a_{21}, \qquad \frac{\partial \eta}{\partial y} = a_{22}, \qquad \frac{\partial \eta}{\partial z} = a_{23}, \tag{7-2}$$

$$\frac{\partial \zeta}{\partial x} = a_{31}, \qquad \frac{\partial \zeta}{\partial y} = a_{32}, \qquad \frac{\partial \zeta}{\partial z} = a_{33},$$

Eq. (7-1) can be rewritten

$$\xi = \xi_0 + a_{11}x + a_{12}y + a_{13}z$$

$$= \xi_0 + \left[\left(\frac{a_{12} - a_{21}}{2} \right)y + \left(\frac{a_{13} - a_{31}}{2} \right)z \right]$$

$$+ \left[a_{11}x + \left(\frac{a_{12} + a_{21}}{2} \right)y + \left(\frac{a_{13} + a_{31}}{2} \right)z \right] = \xi_0 + \xi_1 + \xi_2, \tag{7-3}$$

$$\eta = \eta_0 + \left[\left(\frac{a_{21} - a_{12}}{2} \right)x + \left(\frac{a_{23} - a_{32}}{2} \right)z \right]$$

$$+ \left[\left(\frac{a_{21} + a_{12}}{2} \right)x + a_{22}y + \left(\frac{a_{23} + a_{32}}{2} \right)z \right] = \eta_0 + \eta_1 + \eta_2, \tag{7-4}$$

$$\zeta = \zeta_0 + \left[\left(\frac{a_{31} - a_{13}}{2}\right)x + \left(\frac{a_{32} - a_{23}}{2}\right)y\right]$$

$$+ \left[\left(\frac{a_{31} + a_{13}}{2}\right)x + \left(\frac{a_{32} + a_{23}}{2}\right)y + a_{33}z\right] = \zeta_0 + \zeta_1 + \zeta_2. \qquad (7\text{-}5)$$

Thus we have decomposed the motion into three terms

$$s = s_0 + s_1 + s_2. \qquad (7\text{-}6)$$

The displacement s_0 with components ξ_0, η_0, ζ_0 is the same for all points B of the volume element and is therefore recognizable as a translation. The term s_1 is a rotation. In order to show that this is so, we introduce the vector $\boldsymbol{\varphi}$ with components

$$\varphi_x = \frac{a_{32} - a_{23}}{2}, \qquad \varphi_y = \frac{a_{13} - a_{31}}{2}, \qquad \varphi_z = \frac{a_{21} - a_{12}}{2} \qquad (7\text{-}7)$$

and the position vector $\boldsymbol{r}$ directed from A to B. Then

$$s_1 = \boldsymbol{\varphi} \times \boldsymbol{r}, \qquad (7\text{-}8)$$

which is recognizable as an infinitesimal rotation. This motion induces no change in the length of the vector $\boldsymbol{r}$, a fact that is easily proven. The square of the distance AB is

$$|\boldsymbol{r}'|^2 = |\boldsymbol{r} + s_1|^2 = |\boldsymbol{r}|^2 + 2\boldsymbol{r} \cdot s_1 + |s_1|^2.$$

Since

$$\boldsymbol{r} \cdot s_1 = \boldsymbol{r} \cdot (\boldsymbol{\varphi} \times \boldsymbol{r}) = 0,$$

we find, neglecting second-order terms,

$$|\boldsymbol{r}'|^2 = |\boldsymbol{r}|^2 = r^2. \qquad (7\text{-}9)$$

Let us turn next to the third term, s_2. This displacement is a linear vector function of the position vector $\boldsymbol{r}$ with components ξ_2, η_2, ζ_2 given by

$$\xi_2 = \epsilon_{11}x + \epsilon_{12}y + \epsilon_{13}z,$$

$$\eta_2 = \epsilon_{21}x + \epsilon_{22}y + \epsilon_{23}z, \qquad (7\text{-}10)$$

$$\zeta_2 = \epsilon_{31}x + \epsilon_{32}y + \epsilon_{33}z,$$

where

$$\epsilon_{11} = a_{11} = \frac{\partial \xi}{\partial x}, \qquad \epsilon_{12} = \epsilon_{21} = \frac{a_{12} + a_{21}}{2} = \frac{1}{2}\left(\frac{\partial \xi}{\partial y} + \frac{\partial \eta}{\partial x}\right),$$

$$\epsilon_{22} = a_{22} = \frac{\partial \eta}{\partial y}, \qquad \epsilon_{23} = \epsilon_{32} = \frac{a_{23} + a_{32}}{2} = \frac{1}{2}\left(\frac{\partial \eta}{\partial z} + \frac{\partial \zeta}{\partial y}\right), \qquad (7\text{-}11)$$

$$\epsilon_{33} = a_{33} = \frac{\partial \zeta}{\partial z}, \qquad \epsilon_{31} = \epsilon_{13} = \frac{a_{31} + a_{13}}{2} = \frac{1}{2}\left(\frac{\partial \zeta}{\partial x} + \frac{\partial \xi}{\partial z}\right).$$

The ϵ_{ij} are the components of the *strain tensor*. This tensor is symmetric—that is, $\epsilon_{ij} = \epsilon_{ji}$. The coefficients of s_1, on the other hand, represent an antisymmetric tensor, $\varphi_{ij} = -\varphi_{ji}$ ($\varphi_{ii} = 0$), where

$$\varphi_x = \varphi_{23} = -\varphi_{32},$$

$$\varphi_y = \varphi_{13} = -\varphi_{31}, \tag{7-12}$$

$$\varphi_z = \varphi_{12} = -\varphi_{21}$$

[see Eq. (7-7)].

Consider the dot product

$$s_2 \cdot r = \epsilon_{11}x^2 + 2\epsilon_{12}xy + \epsilon_{22}y^2 + 2\epsilon_{23}yz$$

$$+ \epsilon_{22}z^2 + 2\epsilon_{13}xz = f(x, y, z). \tag{7-13}$$

Setting $f(x, y, z) = $ const, we obtain a surface of second order called the *strain quadric*. This surface is not necessarily an ellipsoid, as was the case with the inertia tensor. It may be any surface of second order—for example, a hyperboloid of one or two sheets, a pair of planes, and so forth. In order to ascertain the nature of the surface, we refer the strain quadric to principal axes. Introducing the corresponding cartesian coordinates X_1, X_2, X_3, Eq. (7-13) becomes

$$F(X_1, X_2, X_3) = \epsilon_1 X_1^2 + \epsilon_2 X_2^2 + \epsilon_3 X_3^2 = \text{const.} \tag{7-14}$$

The coefficients ϵ_1, ϵ_2, ϵ_3 are called *principal extensions* if positive or *principal contractions* if negative. In terms of Eq. (7-13), we can write Eq. (7-10) as

$$\xi_2 = \frac{1}{2}\frac{\partial f}{\partial x}, \qquad \eta_2 = \frac{1}{2}\frac{\partial f}{\partial y}, \qquad \zeta_2 = \frac{1}{2}\frac{\partial f}{\partial z}. \tag{7-15}$$

In principal coordinates, this relation becomes

$$\frac{1}{2}\frac{\partial F}{\partial X_i} = \epsilon_i X_i = \Xi_i, \tag{7-16}$$

where the principal components of displacement Ξ_i are the projections of the vector s_2 on the principal axes X_i.

Helmholtz's theorem has now been established with the derivation of Eq. (7-16), which indicates that the third partial motion consists of extensions or contractions in the mutually orthogonal directions of the principal axes of the strain quadric. According to Eq. (7-16), any point B of our volume element with coordinates X_i in the principal system goes over into a point B' with coordinates

$$X_i + \Xi_i = X_i(1 + \epsilon_i). \tag{7-17}$$

Now let us examine the cubical extension or *dilatation* Θ, which is defined as the specific change of volume

$$\Theta = \frac{\Delta\tau' - \Delta\tau}{\Delta\tau},$$ (7-18)

where $\Delta\tau$ is the original volume of the element and $\Delta\tau'$ is the expanded or contracted volume. For convenience, we take the volume element to be a rectangular solid with sides a_1, a_2, a_3 parallel to the principal axes. One corner is taken to coincide with the point A, the origin of the axes. Then $\Delta\tau = a_1 a_2 a_3$ and $\Delta\tau' = a_1' a_2' a_3'$, where, according to Eq. (7-17), $a_i' = a_i(1 + \epsilon_i)$. Thus

$$\Delta\tau' = \Delta\tau(1 + \epsilon_1)(1 + \epsilon_2)(1 + \epsilon_3)$$ (7-19)

and, by Eq. (7-18),

$$\Theta = (1 + \epsilon_1)(1 + \epsilon_2)(1 + \epsilon_3) - 1 = \epsilon_1 + \epsilon_2 + \epsilon_3$$ (7-20)

to first order in ϵ_i. This result is valid in any cartesian system—for example, the original system x, y, z. In order to establish this important assertion, we will prove that the sum of the diagonal elements T_{ii} of any tensor T, called the *trace* of T, is invariant with respect to the coordinate system to which the tensor is referred. Using Eqs. (5-90) and (1-14),

$$\mathrm{tr}\,\mathsf{T} = \sum_i T_{ii}$$

$$= \sum_i \sum_{j,l} \lambda_{ji} \lambda_{li} T'_{jl}$$

$$= \sum_{j,l} \left(\sum_i \lambda_{ji} \lambda_{li} \right) T'_{jl}$$

$$= \sum_{j,l} T'_{jl}\, \delta_{jl}$$

$$= \sum_j T'_{jj}.$$ (7-21)

Thus

$$\Theta = \mathrm{tr}\,\boldsymbol{\epsilon}.$$ (7-22)

Having established the significance of the *principal* extensions ϵ_i, we now wish to do the same for the components of the general strain tensor ϵ_{ij}. For the diagonal elements, say ϵ_{11}, the interpretation is the same as for ϵ_1. The quantity ϵ_{11} or ϵ_{xx} is simply the incremental change in length of a fiber lying along the x axis per unit of original length. We can write the specific change

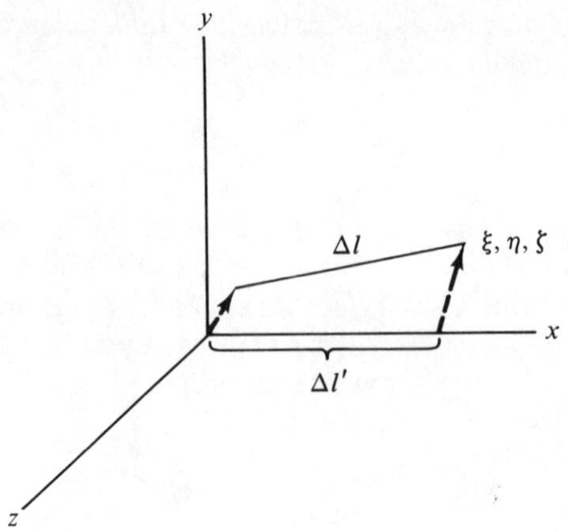

Fig. 7-2

in length as $(\Delta\ell' - \Delta\ell)/\Delta\ell$. The strained length $\Delta\ell'$ (see Fig. 7-2) can easily be evaluated. Since

$$\xi = \xi_0 + \Delta\ell\,\frac{\partial\xi}{\partial x},$$

$$\eta = \eta_0 + \Delta\ell\,\frac{\partial\eta}{\partial x}, \tag{7-23}$$

$$\zeta = \zeta_0 + \Delta\ell\,\frac{\partial\zeta}{\partial x},$$

we find

$$\Delta\ell' = \sqrt{\left(\Delta\ell + \Delta\ell\,\frac{\partial\xi}{\partial x}\right)^2 + \left(\Delta\ell\,\frac{\partial\eta}{\partial x}\right)^2 + \left(\Delta\ell\,\frac{\partial\zeta}{\partial x}\right)^2}$$

$$= \Delta\ell\sqrt{\left(1 + \frac{\partial\xi}{\partial x}\right)^2 + \left(\frac{\partial\eta}{\partial x}\right)^2 + \left(\frac{\partial\zeta}{\partial x}\right)^2}$$

$$= \Delta\ell\sqrt{1 + 2\frac{\partial\xi}{\partial x} + \left(\frac{\partial\xi}{\partial x}\right)^2 + \left(\frac{\partial\eta}{\partial x}\right)^2 + \left(\frac{\partial\zeta}{\partial x}\right)^2}$$

$$\approx \Delta\ell\sqrt{1 + 2\frac{\partial\xi}{\partial x}}$$

$$\approx \Delta\ell\left(1 + \frac{\partial\xi}{\partial x}\right). \tag{7-24}$$

Thus

$$\frac{\Delta\ell' - \Delta\ell}{\Delta\ell} = \frac{\partial\xi}{\partial x} = \epsilon_{11}. \tag{7-25}$$

Turning now to the off-diagonal elements, ϵ_{ij}, we consider an xy lamina as shown in Fig. 7-3. The rectangular lamina of dimension a, b is deformed in the process of straining into a parallelogram as shown. We have ignored displacements in the z direction, since, as was found for the x fiber, they produce only negligible higher-order contributions. Let us calculate the small angles γ_1 and γ_2 indicated in Fig. 7-3. For γ_1,

$$\gamma_1 \approx \tan \gamma_1 = \frac{\eta - \eta_0}{a + \xi - \xi_0}.$$

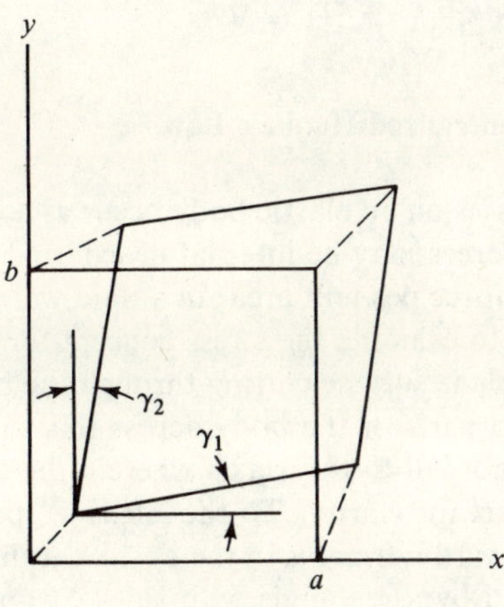

Fig. 7-3

Since

$$\eta = \eta_0 + \boldsymbol{r} \cdot \boldsymbol{\nabla}\eta = \eta_0 + a\frac{\partial \eta}{\partial x}$$

and
$$a \gg \xi - \xi_0,$$

we find

$$\gamma_1 = \frac{\eta_0 + a(\partial \eta/\partial x) - \eta_0}{a} = \frac{\partial \eta}{\partial x}, \qquad (7\text{-}26)$$

and similarly

$$\gamma_2 = \frac{\xi - \xi_0}{b + \eta} = \frac{\partial \xi}{\partial y}. \qquad (7\text{-}27)$$

Noting that

$$\epsilon_{12} = \frac{1}{2}\left(\frac{\partial \xi}{\partial y} + \frac{\partial \eta}{\partial x}\right),$$

we see that

$$\epsilon_{12} = \tfrac{1}{2}(\gamma_1 + \gamma_2) = \langle \gamma \rangle. \tag{7-28}$$

This is half the angular change in the originally right angle of the rectangular lamina *a, b* and is referred to as the *shearing strain*.

EXERCISE

7-1.1. Show that the dilatation Θ can also be written

$$\Theta = \nabla \cdot \mathbf{s}. \tag{7-29}$$

7-2 Stress and the Generalized Hooke's Law

Stresses on an elastic body occur as a result of forces acting on that body. Such forces may be internal or external in nature. We define *stress* as the reaction force per unit area; in a fluid we would use the concept of *pressure*. In order to examine the stress concept for the case of an elastic solid, we visualize a plane surface cutting through the body and examine the interaction of the two parts of the body across this surface. In Fig. 7-4 we show such a surface normal to the x axis wherein the two parts of the body have been moved apart for clarity. To the left is a "positive" x surface, so called because its outward normal is in the positive x direction. We take this surface to be bounded by a rectangle with sides Δy and Δz. The surface to the right is a negative x surface, since its outward normal is in the negative x direction.

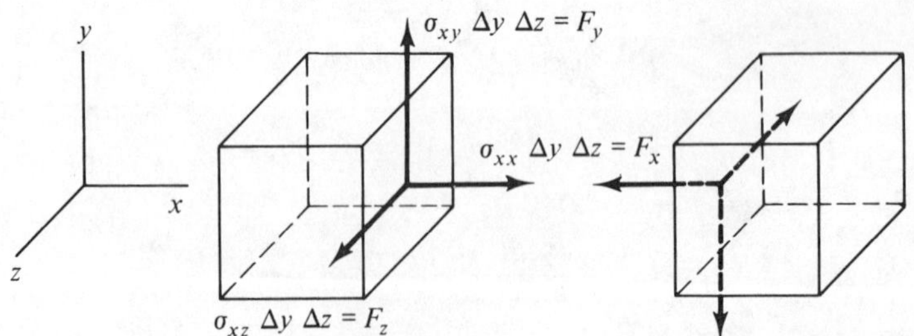

Fig. 7-4

The quantity σ_{xx} is a tensile stress, and the quantities σ_{xy} and σ_{xz} are shear stresses. In this notation the position of the surface is indicated by the first subscript (an x surface in the present case), and the direction of the stress is specified by the second subscript.

In order to complete our mathematical description of the stresses in a solid, we need to ascertain the symmetry properties of the array

$$\sigma = \begin{pmatrix} \sigma_{xx} & \sigma_{xy} & \sigma_{xz} \\ \sigma_{yx} & \sigma_{yy} & \sigma_{yz} \\ \sigma_{zx} & \sigma_{zy} & \sigma_{zz} \end{pmatrix} \tag{7-30}$$

and show that it is a tensor.

The symmetry can easily be established by requiring that the torque on a volume element due to the shear stresses vanishes. In Fig. 7-5 we show a

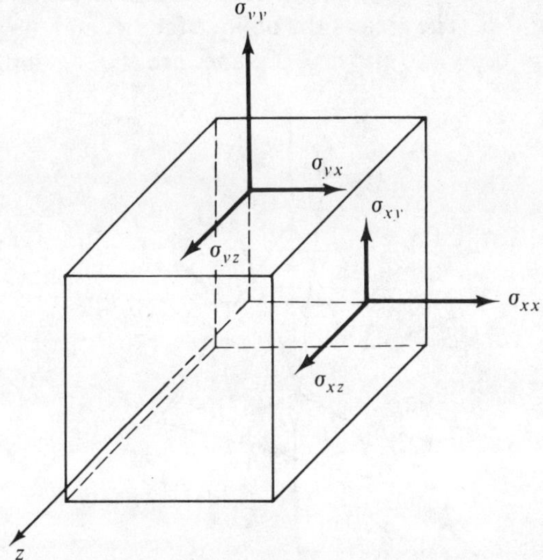

Fig. 7-5

volume element with sides Δx, Δy, Δz, and we wish to evaluate the torque about the z axis due to the forces on it. We see that the tensile stresses σ_{yy} and σ_{xx} do not contribute because the forces associated with them lie along lines of action that intersect the z axis. Consider the shear stresses σ_{xy} and σ_{yx}. The torque about the z axis that is due to these stresses is

$$N = (\sigma_{xy}\,\Delta y\,\Delta z)\left(\frac{\Delta x}{2}\right) - (\sigma_{yx}\,\Delta z\,\Delta x)\left(\frac{\Delta y}{2}\right)$$

$$= \tfrac{1}{2}(\sigma_{xy} - \sigma_{yx})\,\Delta\tau. \tag{7-31}$$

There are also shear stresses on the negative x and y surfaces. Let these stresses be σ'_{xy} and σ'_{yx}. Since these reactive stresses are equal to $-\sigma_{xy}$ and

$-\sigma_{yx}$, respectively, their contribution to the net torque is also given by $\frac{1}{2}(\sigma_{xy} - \sigma_{yx})\,\Delta\tau$, and the net torque, which must be zero for equilibrium, is

$$N = (\sigma_{xy} - \sigma_{yx})\,\Delta\tau = 0. \tag{7-32}$$

Since $\Delta\tau \neq 0$, then $\sigma_{xy} = \sigma_{yx}$, and the array given in Eq. (7-30) must (by simple extension) be symmetric about its diagonal—that is,

$$\sigma_{ij} = \sigma_{ji}. \tag{7-33}$$

Next, we wish to establish the tensor nature of σ. We do so by establishing its transformation properties. Consequently, we postulate a tetrahedronal volume, three faces of which coincide with the coordinate planes and the fourth of which is defined by an arbitrary normal direction $\hat{n}$. This situation is shown in Fig. 7-6. The area of the oblique face A_n and the areas of the faces lying in the $x = 0$, $y = 0$, and $z = 0$ planes are A_x, A_y, and A_z, respectively.

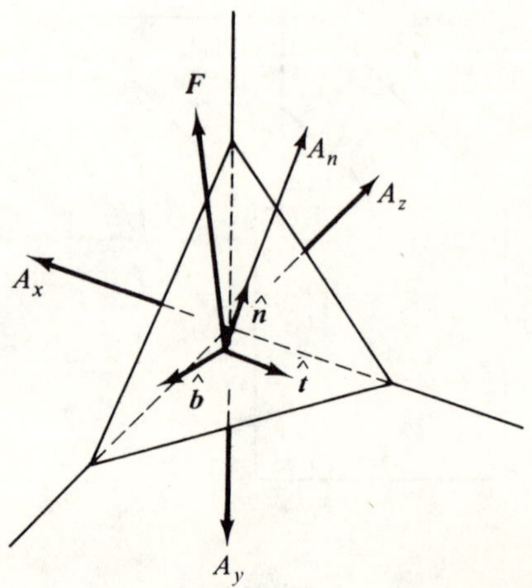

Fig. 7-6

Thus

$$A_x = A_n \cos(n, x) = A_n\alpha_1,$$
$$A_y = A_n \cos(n, y) = A_n\beta_1, \tag{7-34}$$
$$A_z = A_n \cos(n, z) = A_n\gamma_1.$$

The x, y, and z components of the force F (F not necessarily parallel to $\hat{n}$) applied to the oblique face are expressible as

$$F_x = \sigma_{nx}A_n = \sigma_{xx}A_x + \sigma_{yx}A_y + \sigma_{zx}A_z,$$
$$F_y = \sigma_{ny}A_n = \sigma_{xy}A_x + \sigma_{yy}A_y + \sigma_{zy}A_z, \tag{7-35}$$
$$F_z = \sigma_{nz}A_n = \sigma_{xz}A_x + \sigma_{yz}A_y + \sigma_{zz}A_z.$$

Using Eqs. (7-34) to eliminate A_x, A_y, and A_z, we find

$$\sigma_{nx} = \alpha_1 \sigma_{xx} + \beta_1 \sigma_{yx} + \gamma_1 \sigma_{zx},$$

$$\sigma_{ny} = \alpha_1 \sigma_{xy} + \beta_1 \sigma_{yy} + \gamma_1 \sigma_{zy}, \tag{7-36}$$

$$\sigma_{nz} = \alpha_1 \sigma_{xz} + \beta_1 \sigma_{yz} + \gamma_1 \sigma_{zz}.$$

Since

$$\boldsymbol{F} = (\sigma_{nx}\hat{\boldsymbol{i}} + \sigma_{ny}\hat{\boldsymbol{j}} + \sigma_{nz}\hat{\boldsymbol{k}})A_n,$$

then

$$\frac{\boldsymbol{F}}{A_n} \cdot \hat{\boldsymbol{n}} = (\sigma_{nx}\hat{\boldsymbol{i}} + \sigma_{ny}\hat{\boldsymbol{j}} + \sigma_{nz}\hat{\boldsymbol{k}}) \cdot \hat{\boldsymbol{n}}. \tag{7-37}$$

We note, however, that

$$\hat{\boldsymbol{i}} \cdot \hat{\boldsymbol{n}} = \alpha_1, \qquad \hat{\boldsymbol{j}} \cdot \hat{\boldsymbol{n}} = \beta_1, \qquad \hat{\boldsymbol{k}} \cdot \hat{\boldsymbol{n}} = \gamma_1, \tag{7-38}$$

and so

$$\frac{\boldsymbol{F}}{A_n} \cdot \hat{\boldsymbol{n}} = \sigma_{nn} = \sigma_{nx}\alpha_1 + \sigma_{ny}\beta_1 + \sigma_{nz}\gamma_1$$

$$= \sigma_{xx}\alpha_1^2 + \sigma_{yy}\beta_1^2 + \sigma_{zz}\gamma_1^2 + 2\sigma_{xy}\alpha_1\beta_1 + 2\sigma_{xz}\alpha_1\gamma_1 + 2\sigma_{yz}\beta_1\gamma_1. \tag{7-39}$$

This equation gives the normal stresses on the n face owing to the stresses on the x, y, and z faces.

If we then introduce a unit vector $\hat{\boldsymbol{t}}$ normal to $\hat{\boldsymbol{n}}$ and define

$$\alpha_2 = \cos(t, x), \qquad \beta_2 = \cos(t, y), \qquad \gamma_2 = \cos(t, z), \tag{7-40}$$

then the shear stress along $\hat{\boldsymbol{t}}$ can be found as

$$\sigma_{nt} = \sigma_{nx}\alpha_2 + \sigma_{ny}\beta_2 + \sigma_{nz}\gamma_2$$

$$= \sigma_{xx}\alpha_1\alpha_2 + \sigma_{yy}\beta_1\beta_2 + \sigma_{zz}\gamma_1\gamma_2$$

$$+ (\beta_1\alpha_2 + \beta_2\alpha_1)\sigma_{xy} + (\gamma_1\beta_2 + \gamma_2\beta_1)\sigma_{yz} + (\alpha_1\gamma_2 + \alpha_2\gamma_1)\sigma_{xz}. \tag{7-41}$$

If we go on to define a binormal unit vector $\hat{\boldsymbol{b}}$, mutually orthogonal to $\hat{\boldsymbol{n}}$ and $\hat{\boldsymbol{t}}$, with direction cosines α_3, β_3, γ_3, then an equation can be written for σ_{nb} that is identical with Eq. (7-41) except that the subscript 2 is replaced by 3.

Replacing the subscripts x, y, z and n, t, b with numerals and writing λ_{ik} for the nine direction cosines, Eqs. (7-39), (7-41), and the corresponding relation for σ_{nb} can be written

$$\sigma'_{ik} = \sum_{l,m} \lambda_{il}\lambda_{km}\sigma_{lm}, \tag{7-42}$$

which establishes the tensor nature of σ. The state of stress in the neighborhood of a point can therefore be visualized by means of a stress quadric.

We now wish to establish an equation of static equilibrium for elastic solids. If we consider a small rectangular volume element of sides $\Delta x \, \Delta y \, \Delta z$ located at x, y, z and subject it to a body force f per unit volume, the condition for static equilibrium that requires the vector sum of all forces to vanish may be written for the x component as

$$f_x \, \Delta x \, \Delta y \, \Delta z + \sigma_{xx}(x + \Delta x, y, z) \, \Delta y \, \Delta z - \sigma_{xx}(x, y, z) \, \Delta y \, \Delta z$$

$$+ \sigma_{yx}(y + \Delta y, x, z) \, \Delta x \, \Delta z - \sigma_{yx}(x, y, z) \, \Delta x \, \Delta z$$

$$+ \sigma_{zx}(x, y, z + \Delta z) \, \Delta x \, \Delta y - \sigma_{zx}(x, y, z) \, \Delta x \, \Delta y = 0 \qquad (7\text{-}43)$$

or by Taylor expansion

$$\left(f_x + \frac{\partial \sigma_{xx}}{\partial x} + \frac{\partial \sigma_{yx}}{\partial y} + \frac{\partial \sigma_{zx}}{\partial z} \right) \Delta \tau = 0. \qquad (7\text{-}44)$$

This expression, together with similar equations for the y and z components, can be lumped into a single tensor equation

$$\boldsymbol{f} + \boldsymbol{\nabla} \cdot \boldsymbol{\sigma} = 0. \qquad (7\text{-}45)$$

This amounts to three scalar equations of the form of Eq. (7-44) for the six independent components of the stress tensor. Obviously, additional relationships must be found in order to solve the problem of an elastic solid in a static state. What is wanted is the relationship between stress and strain.

If we suppose that the loads imposed on an elastic body are sufficiently small so that second and higher powers of stress and strain tensor components can be neglected, then the general form of the relations between stress and strain components becomes linear. Hooke, a contemporary of Newton, was the first to propose such a linear relationship, although in a scalar rather than a tensor formulation. We shall also assume that the elastic body is isotropic— that is, its physical properties are the same in all directions. A discussion of elastic materials outside the bounds of these two assumptions appears in the closing sections.

It will be convenient to select a volume element cut parallel to the principal axes of the stress quadric. The three pairs of faces are then subjected to the principal stresses σ_1, σ_2, σ_3, and shear stresses σ_{ij} do not occur. By the assumed isotropy, our initially rectangular volume element will remain rectangular under strain. Consequently, no angular changes arise and the shear strains ϵ_{ij} are therefore zero. As a result, we see that the principal axes of the stress quadric must coincide with the principal axes of the strain quadric for a lightly loaded isotropic solid.

The assumed linearity of the stress-strain relations implies for σ_1:

$$\sigma_1 = a\epsilon_1 + b\epsilon_2 + c\epsilon_3, \qquad (7\text{-}46)$$

where a, b, and c are constants. Since there is no preferred axis, two more equations must be obtainable simply by cyclic permutation of the subscripts:

$$\sigma_2 = a\epsilon_2 + b\epsilon_3 + c\epsilon_1, \tag{7-47}$$

and

$$\sigma_3 = a\epsilon_3 + b\epsilon_1 + c\epsilon_2. \tag{7-48}$$

Again, by virtue of isotropy, $b = c$; otherwise directions 2 and 3 would not be equivalent with regard to σ_1. So

$$\sigma_1 = a\epsilon_1 + b(\epsilon_2 + \epsilon_3)$$

or adding and subtracting $b\epsilon_1$

$$\sigma_1 = (a - b)\epsilon_1 + b(\epsilon_1 + \epsilon_2 + \epsilon_3)$$

$$= 2\mu\epsilon_1 + \lambda\Theta, \tag{7-49}$$

where μ and λ are called the *Lamé moduli* and Θ is the dilation, $\Theta = \operatorname{tr} \boldsymbol{\epsilon}$. Introducing this notation into Eqs. (7-47) and (7-48), we find

$$\sigma_i = 2\mu\epsilon_i + \lambda\Theta. \tag{7-50}$$

Our next task is obviously to derive a general form of Eq. (7-50)—that is, a relation that holds for a volume element of arbitrary orientation. Using the transformation relation

$$\sigma'_{ij} = \sum_{k,l} \lambda_{ik}\lambda_{jl}\sigma_{kl}, \tag{7-51}$$

where, in principal axes, σ_{kl} is given by

$$\sigma_{kl} = 2\mu\epsilon_k\,\delta_{kl} + \lambda\,\delta_{kl}\,\Theta, \tag{7-52}$$

we find

$$\sigma'_{ij} = \sum_{k,l} \lambda_{ik}\lambda_{jl}(2\mu\epsilon_k\,\delta_{kl} + \lambda\,\delta_{kl}\,\Theta)$$

$$= \sum_{k,l} \lambda_{ik}\lambda_{jl}2\mu\epsilon_k\,\delta_{kl} + \sum_{k,l} \lambda_{ik}\lambda_{jl}\lambda\,\delta_{kl}\,\Theta, \tag{7-53}$$

where the transformation matrices λ_{ij} should not be confused with the Lamé modulus λ. Invoking the orthogonality of the transformation matrices

$$\sum_{k,l} \lambda_{ik}\lambda_{jl} = \delta_{ij}\,\delta_{kl}, \tag{7-54}$$

we find

$$\sigma_{ij} = 2\mu\epsilon_{ij} + \lambda\,\delta_{ij}\,\Theta \tag{7-55}$$

as the general stress-strain relation for an arbitrary volume element.

Let us derive the inverse relations next. Taking the trace of σ, we find

$$\Sigma \equiv \mathrm{tr}\,\sigma = 2\mu\Theta + 3\lambda\Theta = (2\mu + 3\lambda)\Theta$$

or
$$\Theta = \frac{\Sigma}{2\mu + 3\lambda}. \tag{7-56}$$

Solving Eq. (7-55) for ϵ_{ij} and using Eq. (7-56), we have

$$\epsilon_{ij} = 2\mu'\sigma_{ij} + \lambda' \Sigma\,\delta_{ij}, \tag{7-57}$$

where
$$\mu' = \frac{1}{4\mu}, \qquad \lambda' = -\frac{\lambda}{2\mu(2\mu + 3\lambda)} \tag{7-58}$$

are known as the *constants of elasticity*.

The Lamé constants are most useful for general investigations, but their physical significance is not as obvious as *Young's modulus E* and *Poisson ratio v* common to engineering practice. For this reason, it is worthwhile to find the connection between these two formulations. Consequently, let us consider a vertical cylindrical rod of cross-sectional area A, fixed at the top end, with a force F acting downward on the bottom end. Obviously, the rod is free from shear stresses and is therefore in a principal axis system. Young's modulus E is defined by

$$\sigma = \frac{F}{A} = E\frac{\Delta l}{l} = E\epsilon, \tag{7-59}$$

where l is the unstrained length of the rod and Δl is the extension owing to the stress σ.

Poisson's ratio is defined as minus the ratio of the contraction normal to the tensile stress to the extension—that is,

$$v = -\frac{\epsilon'}{\epsilon}, \tag{7-60}$$

where ϵ' is the specific normal contraction along x or y and ϵ is the specific elongation along z. Thus

$$\epsilon' = -v\epsilon = -\frac{v\sigma}{E}. \tag{7-61}$$

Poisson not only introduced this ratio of transverse contraction to longitudinal extension, he also calculated its value on the basis of a somewhat oversimplified molecular theory and arrived at a value of 0.25. The modern theory of solid-state physics has shown that there is no universal value for v; it varies somewhat from one material to another.

Let us apply what we have gathered from the simple tension experiment to the general case. This we shall consider to be a superposition of three unidirectional stresses in the principal directions, the effect of each stress being the same as in the cylindrical rod problem. Because of the linearity of all relations, we can superimpose the stresses and associated strains. The extension along principal axis 1 is not determined solely by σ_1 but (if considered as a transverse direction) by the stresses in principal directions 2 and 3. So, using Eqs. (7-59) and (7-61), we find

$$\epsilon_1 = \frac{\sigma_1}{E} - \frac{\nu}{E}(\sigma_2 + \sigma_3),$$

$$\epsilon_2 = \frac{\sigma_2}{E} - \frac{\nu}{E}(\sigma_1 + \sigma_3), \tag{7-62}$$

$$\epsilon_3 = \frac{\sigma_3}{E} - \frac{\nu}{E}(\sigma_1 + \sigma_2).$$

Manipulating the first of Eqs. (7-62),

$$\epsilon_1 = \frac{\sigma_1}{E} + \frac{\nu\sigma_1}{E} - \frac{\nu\sigma_1}{E} - \frac{\nu}{E}(\sigma_2 + \sigma_3)$$

$$= \frac{\sigma_1}{E}(1 + \nu) - \frac{\nu}{E}(\sigma_1 + \sigma_2 + \sigma_3)$$

$$= \left(\frac{1 + \nu}{E}\right)\sigma_1 - \frac{\nu}{E}\Sigma$$

or, in general,

$$\epsilon_i = \left(\frac{1 + \nu}{E}\right)\sigma_i - \frac{\nu}{E}\Sigma, \tag{7-63}$$

where Σ is the trace of the stress tensor. Summing Eq. (7-63) over the index $i = 1, 2, 3$, we have

$$\text{tr } \boldsymbol{\epsilon} = \Theta = \left(\frac{1 + \nu}{E}\right)\Sigma - \left(\frac{3\nu}{E}\right)\Sigma = \left(\frac{1 - 2\nu}{E}\right)\Sigma$$

or

$$\Sigma = \left(\frac{E}{1 - 2\nu}\right)\Theta. \tag{7-64}$$

Thus we find for ϵ_i

$$\epsilon_i = \left(\frac{1 + \nu}{E}\right)\sigma_i - \frac{\nu}{E}\Sigma = \left(\frac{1 + \nu}{E}\right)\sigma_i - \frac{\nu}{E}\left(\frac{E}{1 - 2\nu}\right)\Theta$$

or

$$\sigma_i = \frac{E}{1 + \nu}\left[\epsilon_i + \left(\frac{\nu}{1 - 2\nu}\right)\Theta\right]. \tag{7-65}$$

Comparing this equation with Eq. (7-50), we obtain the relation between ν and E and the Lamé moduli:

$$2\mu = \frac{E}{1 + \nu}, \qquad \lambda = \frac{\nu E}{(1 + \nu)(1 - 2\nu)}. \tag{7-66}$$

It is interesting to note that in the expression for $\lambda(E, \nu)$ there is a singularity for $\nu = \frac{1}{2}$ that causes λ to blow up. Let us examine its significance. To do so, we postulate an equal pressure p from all directions and define a quantity K, the modulus of compressibility:

$$K = \frac{p}{\Theta}. \tag{7-67}$$

It follows from such a hydrostatic pressure that

$$\sigma_1 = \sigma_2 = \sigma_3 = -p \tag{7-68}$$

and

$$\Sigma = \operatorname{tr} \sigma = -3p. \tag{7-69}$$

Using Eq. (7-64), we find

$$K = \frac{p}{\Theta} = \frac{Ep}{(1 - 2\nu)\Sigma} = -\frac{E}{3(1 - 2\nu)}. \tag{7-70}$$

We see that as $\nu \to \frac{1}{2}$, $K \to \infty$, and so $\Theta \to 0$—that is, the material is totally incompressible. Thus $\nu = \frac{1}{2}$ represents the upper limit for ν; if $\nu > \frac{1}{2}$, the material would need to expand when a pressure is applied.

We are now in a position to write a complete set of equations for an elastic medium in the static state. It will be recalled that Eq. (7-45), which we rewrite here for convenience

$$f + \nabla \cdot \sigma = 0, \tag{7-71}$$

provided only three relations among the six independent components of the stress tensor. We write the x component of this equation

$$\frac{\partial \sigma_{xx}}{\partial x} + \frac{\partial \sigma_{xy}}{\partial y} + \frac{\partial \sigma_{xz}}{\partial z} + f_x = 0 \tag{7-72}$$

and substitute for σ_{ij}, using Eq. (7-55):

$$2\mu \frac{\partial \epsilon_{xx}}{\partial x} + \lambda \frac{\partial \Theta}{\partial x} + 2\mu \frac{\partial \epsilon_{xy}}{\partial y} + 2\mu \frac{\partial \epsilon_{xz}}{\partial z} + f_x = 0. \tag{7-73}$$

In terms of the displacements, this is

$$2\mu \frac{\partial^2 \xi}{\partial x^2} + \lambda \frac{\partial \Theta}{\partial x} + 2\mu \frac{\partial}{\partial y} \frac{1}{2} \left(\frac{\partial \eta}{\partial x} + \frac{\partial \xi}{\partial y} \right) + 2\mu \frac{\partial}{\partial z} \frac{1}{2} \left(\frac{\partial \zeta}{\partial x} + \frac{\partial \xi}{\partial z} \right) + f_x = 0$$

or

$$\mu \frac{\partial^2 \xi}{\partial x^2} + \mu \frac{\partial^2 \xi}{\partial y^2} + \mu \frac{\partial^2 \xi}{\partial z^2} + \mu \frac{\partial}{\partial x}\left(\frac{\partial \xi}{\partial x} + \frac{\partial \eta}{\partial y} + \frac{\partial \zeta}{\partial z}\right) + \lambda \frac{\partial \Theta}{\partial x} + f_x = 0.$$

Collecting terms and permuting to obtain the y and z components, we find for the equations of static equilibrium

$$\mu \, \nabla^2 \xi + (\mu + \lambda) \frac{\partial \Theta}{\partial x} + f_x = 0,$$

$$\mu \, \nabla^2 \eta + (\mu + \lambda) \frac{\partial \Theta}{\partial y} + f_y = 0, \qquad (7\text{-}74)$$

$$\mu \, \nabla^2 \zeta + (\mu + \lambda) \frac{\partial \Theta}{\partial z} + f_z = 0.$$

Here we have three interconnected (through Θ) equations for ξ, η, ζ, the displacement field. The solutions to these equations can then be used to find the components of ε and σ. Except for special cases, these equations are difficult to solve analytically.

The boundary conditions for a problem may be in terms of displacements or stresses (the derivatives of displacements) or some of each. When displacements are prescribed over the entire surface of the body, the boundary value problem is similar to that of potential theory but much more complicated. In order to specify the stresses, we must assume a system of external forces distributed over the surface of the body. The surface stresses are then balanced everywhere by the force per unit area f. If no external forces act on the boundary, the boundary conditions are

$$\sigma_{nx} = \sigma_{ny} = \sigma_{nz} = 0, \qquad (7\text{-}75)$$

as is the case for the sides of the bent beam to be considered in the next section.

The uniqueness of solution of Eq. (7-74) can be proven by assuming homogeneity of the material (which implies uniform temperature, since μ and λ are temperature dependent) and strictly elastic behavior. In practice, internal stresses are present because of imperfections in manufacture, and the strict uniqueness proofs are usually somewhat unrealistic.

As a final step, let us examine the stress-energy relationship. Consider the two x faces of the rectangular volume element $\Delta \tau = \Delta x \, \Delta y \, \Delta z$ as we impose an infinitesimal increment of displacement on the strain field:

$$\xi \rightarrow \xi + d\xi,$$

$$\eta \rightarrow \eta + d\eta, \qquad (7\text{-}76)$$

$$\zeta \rightarrow \zeta + d\zeta.$$

The work done by the stresses across the negative and positive x faces is, respectively,

$$dw_x = -(\sigma_{xx}\,d\xi + \sigma_{xy}\,d\eta + \sigma_{xz}\,d\zeta)\,\Delta y\,\Delta z \tag{7-77}$$

and
$$dw_{x+\Delta x} = (\sigma_{xx}\,d\xi + \sigma_{xy}\,d\eta + \sigma_{xz}\,d\zeta)\,\Delta y\,\Delta z$$

$$+ \frac{\partial}{\partial x}(\sigma_{xx}\,d\xi + \sigma_{xy}\,d\eta + \sigma_{xz}\,d\zeta)\,\Delta x\,\Delta y\,\Delta z. \tag{7-78}$$

Adding these two contributions to the work done by the external force f in the x direction, we find

$$\left[\frac{\partial}{\partial x}(\sigma_{xx}\,d\xi + \sigma_{xy}\,d\eta + \sigma_{xz}\,d\zeta) + f_x\,d\xi\right]d\tau.$$

Two analogous expressions are obtained for the work done in the displacement of the y and z surfaces. Adding all three contributions and carrying out the differentiations, we find, using Eq. (7-45) to eliminate f, that the work per unit volume is

$$dw = \sigma_{xx}\,d\frac{\partial\xi}{\partial x} + \sigma_{yy}\,d\frac{\partial\eta}{\partial y} + \sigma_{zz}\,d\frac{\partial\zeta}{\partial z}$$

$$+ \sigma_{xy}\,d\left(\frac{\partial\xi}{\partial y} + \frac{\partial\eta}{\partial x}\right) + \sigma_{yz}\,d\left(\frac{\partial\eta}{\partial z} + \frac{\partial\zeta}{\partial y}\right) + \sigma_{xz}\,d\left(\frac{\partial\zeta}{\partial x} + \frac{\partial\xi}{\partial z}\right) \tag{7-79}$$

or
$$dw = \sigma_{xx}\,d\epsilon_{xx} + \sigma_{yy}\,d\epsilon_{yy} + \sigma_{zz}\,d\epsilon_{zz}$$

$$+ 2\sigma_{xy}\,d\epsilon_{xy} + 2\sigma_{yz}\,d\epsilon_{yz} + 2\sigma_{xz}\,d\epsilon_{xz}$$

$$= \sum_{i,j}\sigma_{ij}\,d\epsilon_{ij}. \tag{7-80}$$

Eliminating σ_{ij} with the use of Eq. (7-55), we obtain

$$dw = \sum_{i,j}(2\mu\epsilon_{ij}\,d\epsilon_{ij} + \lambda\Theta\,\delta_{ij}\,d\epsilon_{ij})$$

$$= \mu\sum_{i,j}d(\epsilon_{ij}^2) + \lambda\,d\left(\frac{\Theta^2}{2}\right), \tag{7-81}$$

which is an exact differential. Thus the energy per unit volume w in the deformation is

$$w = \mu\sum_{i,j}\epsilon_{ij}^2 + \frac{\lambda}{2}\Theta^2 \tag{7-82}$$

if μ and λ remain constant—that is, if the process of deformation is an isothermal one. For an adiabatic or other nonisothermal deformation process, the temperature dependence of λ and μ must be taken into account; then dw is not an exact differential but depends on the path of the transition.

EXERCISES

7-2.1. Demonstrate explicitly for the case of a cylindrical rod of unstrained length l and radius a that if a tensile strain $\epsilon = \Delta l/l$ fails to change the volume, the Poisson ratio has the limiting value $\nu = \frac{1}{2}$.

7-2.2. Show that the equations of static equilibrium, Eqs. (7-74), can be written as a single vector equation

$$\mu\, \nabla \times \nabla \times s - (\lambda + 2\mu)\, \nabla(\nabla \cdot s) = f, \qquad (7\text{-}82)$$

where s is the displacement vector $s(\xi, \eta, \zeta)$.

7-2.3. Show that the constants of elasticity, μ' and λ', can be written in terms of Young's modulus and Poisson's ratio as

$$\lambda' = \frac{\nu}{E}, \qquad \mu' = \frac{1 + \nu}{2E}.$$

7-2.4. Rederive the energy relation in terms of stress rather than strain. Show that

$$w = \mu' \sum_{i,j} \sigma_{ij}^2 + \frac{\lambda'}{2}\Sigma^2,$$

where μ' and λ' are the constants of elasticity and Σ is the trace of the stress tensor. If the stress is applied as a hydrostatic pressure p, show that the energy per unit volume associated with the compression is given by

$$w = \frac{3p^2}{2(2\mu + 3\lambda)}.$$

7-3 The Bending of Beams

Bent and loaded beams are common elements in structures of all types. The theory for calculating the bending loads originated with Daniel Bernoulli over two hundred years ago.

In our analysis we will consider a straight beam subject to only slight bending. The cross section of the beam is arbitrary (for example, a channel, I-beam, or tubular beam); it is assumed, however, to be constant over its entire length. Let us initially ignore the weight of the beam. We shall consider the beam to be horizontal with one end fixed and with a vertical load applied to the other end. Bernoulli assumed that each plane cross section of the beam remains planar after bending. This assumption implies that the normal stresses transmitted by any cross section are linearly distributed as shown in Fig. 7-7(b). In this figure, which shows the beam before and after bending, we consider two neighboring sections originally separated by a

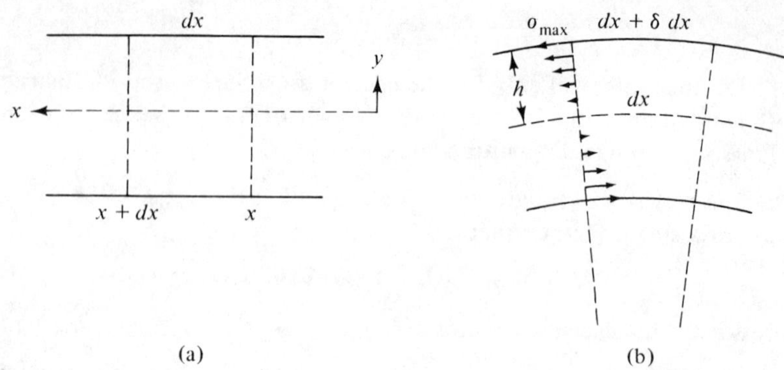

Fig. 7-7

distance dx. This separation distance is changed to $dx + \delta\, dx$ after bending. The extension of the fibers of the beam varies linearly from a maximum at the top of the beam to zero at some distance h from the top to a negative value below this layer—that is, a contraction. The strains and stresses are similarly linearly distributed

$$\sigma = \sigma_{\max}\frac{y}{h}. \tag{7-83}$$

In order to determine the maximum stress, we must examine the equilibrium of an isolated beam section of length x as shown in Fig. 7-8. Our main interest is in the equilibrium of torques taken about an axis normal to the page that intersects the cross section at x at its intersection with neutral layer O—O'. The torque due to the external force (or *bending moment* as it is usually called) about O is

$$N = Fx. \tag{7-84}$$

The torque due to the forces $\sigma\, da$, where da is the element of cross-sectional area, is found by using Eq. (7-83):

$$N = \int \sigma y\, da = \frac{\sigma_{\max}}{h}\int y^2\, da = \frac{\sigma_{\max}}{h} I, \tag{7-85}$$

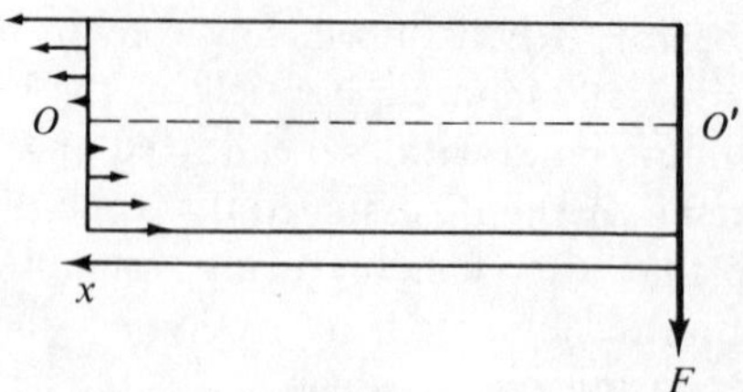

Fig. 7-8

where $I = \int y^2\, da$ is called the *moment of inertia of the cross section*. It is not quite the same as the moment of inertia considered in Chapter 5; there we were dealing with a mass distribution and here we are not. The dimensions of this cross-sectional moment of inertia are L^4 and not ML^2. Using Eqs. (7-83) and (7-85), the stress field is determined as

$$\sigma = \frac{F}{I}\,xy. \tag{7-86}$$

The maximum stress in the beam is therefore

$$\sigma_{\max} = \frac{F}{I}\,lh, \tag{7-87}$$

where l is the length of the beam and h is the vertical distance from the top of the beam to the unstrained plane 0—$0'$ ($y = 0$). This stress must be less than the maximum safe limit for the material

$$\sigma_{\max} < \sigma_s, \tag{7-88}$$

where, for example, $\sigma_s \approx 7 \times 10^4$ lb/in.^{2} for steel. For beam sections used in engineering, the values of I and I/h are tabulated in tables of moments of inertia and section moduli. For a circular cross section of radius a, $I = \pi a^4/4$; and for a rectangular cross section of width b and depth $2h$, $I = 2bh^3/3$. In both cases, we anticipate that the $y = 0$ plane passes through the centroid of the area.

Let us determine the shape of the neutral plane next. Referring to Fig. 7-9, we see that the triangles CDO and $OO'E$ are similar for small deflections.

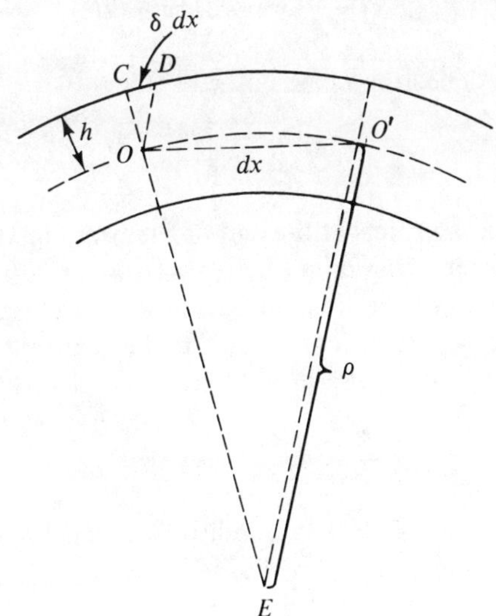

Fig. 7-9

So it follows that

$$\frac{dx}{\rho} = \frac{\delta\,dx}{h}.$$

(7-89)

Since

$$\sigma_{\max} = E\epsilon_{\max},$$

(7-90)

we see that

$$\frac{h}{\rho} = \frac{\delta\,dx}{dx} = \epsilon_{\max} = \frac{\sigma_{\max}}{E} = \frac{hFx}{IE};$$

(7-91)

thus

$$\frac{1}{\rho} = \frac{Fx}{IE} \approx \frac{d^2y}{dx^2}.$$

(7-92)

Integration of this equation yields

$$y = \frac{F}{EI}\left(\frac{x^3}{6} + Ax + B\right),$$

(7-93)

which is a cubic parabola. The integration constants A and B are evaluated by specifying the conditions at the point of support:

$$y\big|_{x=l} = \frac{dy}{dx}\bigg|_{x=l} = 0$$

(7-94)

from which we find $A = -l^2/2$ and $B = l^3/3$. Hence

$$y = \frac{F}{6EI}(x^3 - 3l^2x + 2l^3),$$

(7-95)

and the maximum deflection occurs at $x = 0$:

$$y_{\max} = \frac{Fl^3}{3EI}.$$

(7-96)

Suppose that a *couple* acts at the end of the beam instead of a single force as in Fig. 7-8. This situation might be realized by imposing two equal forces of opposite direction whose lines of action are in close proximity to each other. This couple N can be considered to be constant. The shape of the neutral plane is found from Eq. (7-92) with Fx replaced by N:

$$y = \frac{N}{EI}\left(\frac{x^2}{2} + Ax + B\right).$$

(7-97)

If the beam is fixed at $x = l$, then conditions (7-94) apply. We find for the integration constants $A = -l$, $B = l^2/2$. Hence

$$y = \frac{N}{2EI}(x^2 - 2lx + l^2)$$

(7-98)

and the deflection at $x = 0$ is

$$y_{\max} = \frac{Nl^2}{2EI}. \tag{7-99}$$

In this rather interesting case, the shape of the neutral plane can be described more accurately by $1/\rho = N/IE$, according to which it is a circular arc, since its radius of curvature ρ is constant.

A problem arises with the simple Bernoulli theory of beams because of the need to satisfy the equilibrium condition that the sum of all vertical force components vanish. This condition requires, in general, that a vertical shear force oppose the external loads at each cross section x. If we take the beam depicted in Fig. 7-8 as an example, shear stresses σ_{xy} must occur in addition to the normal stresses σ_{xx} such that for every cross section

$$\int \sigma_{xy} \, da = F. \tag{7-100}$$

This result, however, implies angular changes that cause the original cross section to become nonplanar. Thus Bernoulli's hypothesis of the permanent planarity of the cross sections and the assumption of linear variation of the normal stresses would seem to violate the fundamental principle of statics.

Before attempting to resolve this dilemma, let us calculate the required shear stresses. Equation (7-100) is not useful for this purpose; instead we shall use the x component of Eq. (7-45):

$$\frac{\partial \sigma_{xx}}{\partial x} + \frac{\partial \sigma_{xy}}{\partial y} + \frac{\partial \sigma_{xz}}{\partial z} + f_x = 0. \tag{7-101}$$

Since $f_x = 0$ and σ_{xz} also vanishes for a symmetrical cross section, we have

$$\frac{\partial \sigma_{xx}}{\partial x} + \frac{\partial \sigma_{xy}}{\partial y} = 0. \tag{7-102}$$

Despite the preceding critical remarks, Eq. (7-86) is a valid approximation, as we shall show; thus

$$\frac{\partial \sigma_{xx}}{\partial x} = \frac{Fy}{I} = -\frac{\partial \sigma_{xy}}{\partial y}, \tag{7-103}$$

and so

$$\sigma_{xy} = \frac{F}{2I}(A - y^2) = \frac{F}{2I}(h^2 - y^2), \tag{7-104}$$

where the integration constant A must equal h^2 such that $\sigma_{xy} = 0$ at $y = \pm h$. This result is necessary because the value of σ_{xy} at the upper and lower edges of the cross section determines the shear stresses acting within the face of the beam, which must be stress-free.

In 1855 St.-Venant resolved the apparent internal inconsistencies revealed above. In order to sort out the theory, it must be realized that the relations between stress and external force that we have used are insufficient to describe the elastic state completely. It is necessary to resort to the displacements ξ, η, ζ as found in Eqs. (7-74). Then we see that nowhere in the preceding arguments have the stresses σ_{yy}, σ_{xz}, σ_{yz} been mentioned. This omission occurs because they do not enter into the consideration of a bent beam and so have been set equal to zero. If we express them in terms of ξ, η, ζ, simple differential relations between these quantities result, through the use of which the fundamental static equations can be reduced to[1]

$$\left(\frac{\partial^2}{\partial x^2} = \frac{\partial^2}{\partial y^2} = \frac{\partial^2}{\partial z^2} = \frac{\partial^2}{\partial y\,\partial z}\right)\frac{\partial \xi}{\partial x} = 0. \tag{7-105}$$

From these relations we can conclude that the strain $\partial \xi/\partial x$, and consequently the stress σ_{xx}, is a linear function of the coordinates in the cross section. Thus the linear distribution law for σ_{xx} is a necessary consequence of the fundamental laws and does not depend on Bernoulli's ad hoc assumption.

Equation (7-105) further shows that the extension of the x fibers is a linear function of x. This fact, together with the relation

$$\sigma_{\max} = \frac{hM}{I}, \tag{7-106}$$

implies that the bending moment N is a linear function of x, which is true for one or more single loads or single moments but not for the continuously distributed load of the beam's weight. Whether the load is lumped or distributed can only make a difference in the neighborhood of the cross section in question, however, if the loads are statically equivalent. This statement can be shown to hold experimentally and goes under the name of *St.-Venant's principle.*

EXERCISES

7-3.1. A simple supported beam with a sectional moment of inertia I and length $l = a + b$ is loaded by a force F applied at a distance a from one end as shown in Fig. 7-10. The weight of the beam is negligible. Show that the shape of the neutral layer is given by

$$y_1 = \frac{-Fb}{6lIE}\,(x_1^3 + 2abx_1 - a^2x_1),$$

$$y_2 = \frac{-Fa}{6lIE}\,(x_2^3 + 2abx_2 - b^2x_2),$$

---

[1] For details of this calculation, see A. Foppl, *Vorl. ub. techn. Mechanik* (4th ed.). Leipzig, 1909, Vol. III, p. 73.

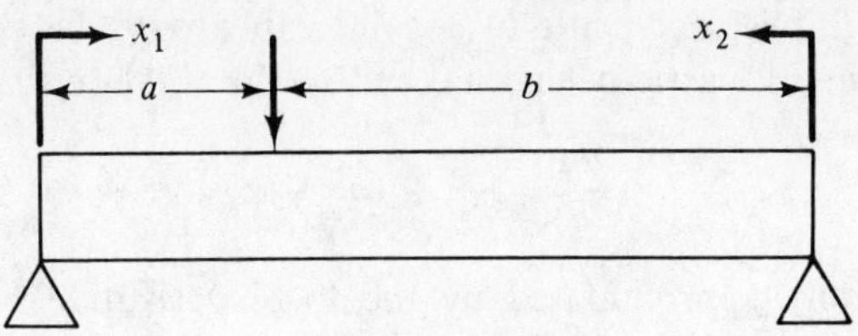

Fig. 7-10

where y_1 and y_2 refer to that part of the beam to the left and right of the point of application of the force, respectively, and x_1 and x_2 are measured from the ends inward.

7-3.2. Reconsider the simply supported beam for the case of no external loading. The beam is taken to be homogeneous with a weight per unit length β and an overall length l. Show that the equation of the neutral layer is given by

$$y = \frac{\beta}{24EI}(2lx^3 - x^4 - l^3x).$$

7-3.3. A homogeneous beam having a weight per unit length β is cantilevered as shown in Fig. 7-11. Show that the moment at any point in the beam is given by

$$N = \frac{\beta}{2}(l - x)^2$$

and
$$y = \frac{\beta}{24EI}(6l^2x^2 - 4lx^3 + x^4).$$

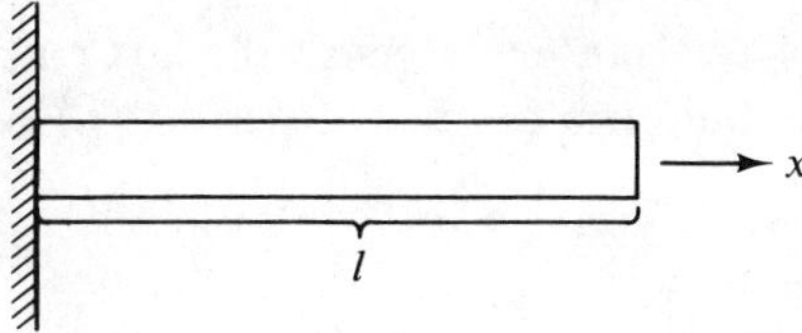

Fig. 7-11

7-4 Dynamics of Elastic Bodies

The dynamic equations of an elastic body follow directly from the static equilibrium condition by adding the inertia force to the

231

external force f. Displacements in a solid can always be considered small (within the range of elastic behavior), and so the total derivative

$$\frac{d\boldsymbol{u}}{dt} = \frac{\partial \boldsymbol{u}}{\partial t} + (\boldsymbol{u} \cdot \boldsymbol{\nabla})\boldsymbol{u} \tag{7-107}$$

can be adequately approximated by the local derivative $\partial \boldsymbol{u}/\partial t$. Using the static equilibrium condition in the form of Eq. (7-45), we thus obtain as the elastic equation of motion

$$\rho \frac{\partial^2 \boldsymbol{s}}{\partial t^2} = \boldsymbol{\nabla} \cdot \boldsymbol{\sigma} + \boldsymbol{f}, \tag{7-108}$$

where ρ is the mass density. For present purposes, however, it is preferable to use Eqs. (7-74) and write the equation of motion as

$$\rho \frac{\partial^2 \boldsymbol{s}}{\partial t^2} = (\mu + \lambda)\,\boldsymbol{\nabla}\Theta + \mu\,\boldsymbol{\nabla}^2 \boldsymbol{s} + \boldsymbol{f}, \tag{7-109}$$

where Θ is given in terms of s by Eq. (7-29):

$$\Theta = \boldsymbol{\nabla} \cdot \boldsymbol{s}. \tag{7-110}$$

Equation (7-109) holds only for cartesian coordinates owing to the vector Laplacian term; for other coordinate systems it must be rewritten by using Eq. (7-82).

We are only concerned here with free oscillations, and so we will set $f = 0$. This equation can be shown to describe waves of dilatation that are longitudinal and also torsion waves that are transversal in nature.

In order to obtain the longitudinal wave solutions, we take the divergence of Eq. (7-109). Using Eq. (7-110), we obtain

$$\rho \frac{\partial^2 \Theta}{\partial t^2} = (\mu + \lambda)\,\boldsymbol{\nabla}^2 \Theta + \mu\,\boldsymbol{\nabla} \cdot \boldsymbol{\nabla}^2 \boldsymbol{s}. \tag{7-111}$$

The divergence and Laplacian operators in the last term on the right-hand side commute, and so this term can be written

$$\mu\,\boldsymbol{\nabla} \cdot \boldsymbol{\nabla}^2 \boldsymbol{s} = \mu\,\boldsymbol{\nabla}^2(\boldsymbol{\nabla} \cdot \boldsymbol{s}) = \mu\,\boldsymbol{\nabla}^2 \Theta. \tag{7-112}$$

In this way, we obtain

$$\rho \frac{\partial^2 \Theta}{\partial t^2} = (2\mu + \lambda)\,\boldsymbol{\nabla}^2 \Theta, \tag{7-113}$$

which is recognizable as a wave equation (see Section 3-8). Since the standard form of such a simple wave equation is

$$\left(\boldsymbol{\nabla}^2 - \frac{1}{u^2}\frac{\partial^2}{\partial t^2}\right)\Theta = 0, \tag{7-114}$$

we find for the propagation velocity of the dilatation waves

$$u_d = \sqrt{\frac{2\mu + \lambda}{\rho}} = \frac{\omega}{k}, \tag{7-115}$$

where ω is the angular frequency of the waves and k is their wave number.

In order to extract the torsion wave solutions, we take the curl of Eq. (7-109):

$$\rho \frac{\partial^2}{\partial t^2} \nabla \times s = (\mu + \lambda) \nabla \times \nabla\Theta + \mu \nabla \times \nabla^2 s. \tag{7-116}$$

Noting from Eq. (1-98) that the curl of the gradient of any scalar vanishes and commuting the curl and Laplacian operators in the last term on the right-hand side, we find

$$\rho \frac{\partial^2}{\partial t^2} (\nabla \times s) = \mu \nabla^2 (\nabla \times s). \tag{7-117}$$

In order to show that we are indeed delving into torsion waves, we note that the only component of $s = s_0 + s_1 + s_2$ that has a nonzero curl is s_1, which we can write, using Eq. (7-8), as

$$s_1 = \varphi \times r. \tag{7-118}$$

Taking the curl of s_1 with the use of identity (1-95), we have

$$\nabla \times (\varphi \times r) = \varphi \nabla \cdot r - r \nabla \cdot \varphi + (r \cdot \nabla)\varphi - (\varphi \cdot \nabla)r. \tag{7-119}$$

Since we assume only small displacements, derivatives of φ are of higher order, which means that the middle two terms in Eq. (7-119) can be neglected. Using identity (1-101), we therefore find

$$\nabla \times s = 2\varphi \tag{7-120}$$

and Eq. (7-117) becomes

$$\rho \frac{\partial^2 \varphi}{\partial t^2} = \mu \nabla^2 \varphi. \tag{7-121}$$

Here φ represents the magnitude and axis of the torsion angle for the volume element under consideration. We see that the propagation velocity for these torsion waves is

$$u_t = \sqrt{\frac{\mu}{\rho}}. \tag{7-122}$$

For this reason, Lamé's modulus μ is sometimes referred to as the *torsion modulus*. Again, plane and spherical waves arise.

Since

$$\frac{u_d^2}{u_t^2} = 2 + \frac{\lambda}{\mu} = \frac{2 - 2\nu}{1 - 2\nu}, \tag{7-123}$$

we see that the dilatation waves are faster than the torsion waves. Moreover, since waves can generally be said to propagate at a speed proportional to the square root of the elastic resistance (or stiffness) divided by the inertial resistance, we see that the resistance to changes in volume is considerably larger than the resistance to changes in relative orientation.

The dilatation waves and torsion waves considered here are three-dimensional waves propagating in an unbounded elastic body. In addition, surface waves that occur only at the free surface of the body also exist. These waves can be shown to propagate at a speed given by

$$u_s = \sqrt{\frac{\mu}{\rho}} \left(1 - \frac{1}{24} \right), \tag{7-124}$$

slightly slower than the three-dimensional torsion waves.

In 1886, when the elastic surface waves were first treated, Lord Rayleigh voiced the opinion that they might play a major role in earthquakes. In the subsequent development of the science of seismology, largely due to Wiechert, such was found to be the case. As a result of the fact that the energy of spatial waves disperses in three dimensions, whereas surface waves are confined to two, the relative importance of the two-dimensional surface waves increases with increasing distance from the center of the disturbance in comparison with the spatial waves. In an earthquake detected at some point remote from its origin, the longitudinal waves form a weak precursor, followed by the transverse torsion waves, which are seen as the first part of the primary quake. Even though the propagation speed of these torsion waves is only about 4% slower than the surface waves, the time lag between their arrival and the arrival of the stronger surface waves is noticeable because of the longer path of the surface waves (arc versus chord).

Seismic waves not only give us information about the average elastic properties of the earth, they also provide our best source of knowledge concerning the terrestrial interior. It is a property of liquids that they will not support shear stresses. Hence the torsion waves (or S waves as they are known to seismologists) cannot propagate through liquid layers in the earth's core. Only the dilatation waves (P waves) are received, and they are refracted as shown in Fig. 7-12. By comparing the wave types, intensities, and arrival times at many seismic stations, conclusions regarding the terrestrial interior can be drawn.

EXERCISES

7-4.1. If P waves are found to travel at a speed of 12 km/sec and S waves at a speed of 7 km/sec in the earth, calculate the average values of the Lamé moduli. The density of the earth can be taken as 3 g/cm^3.

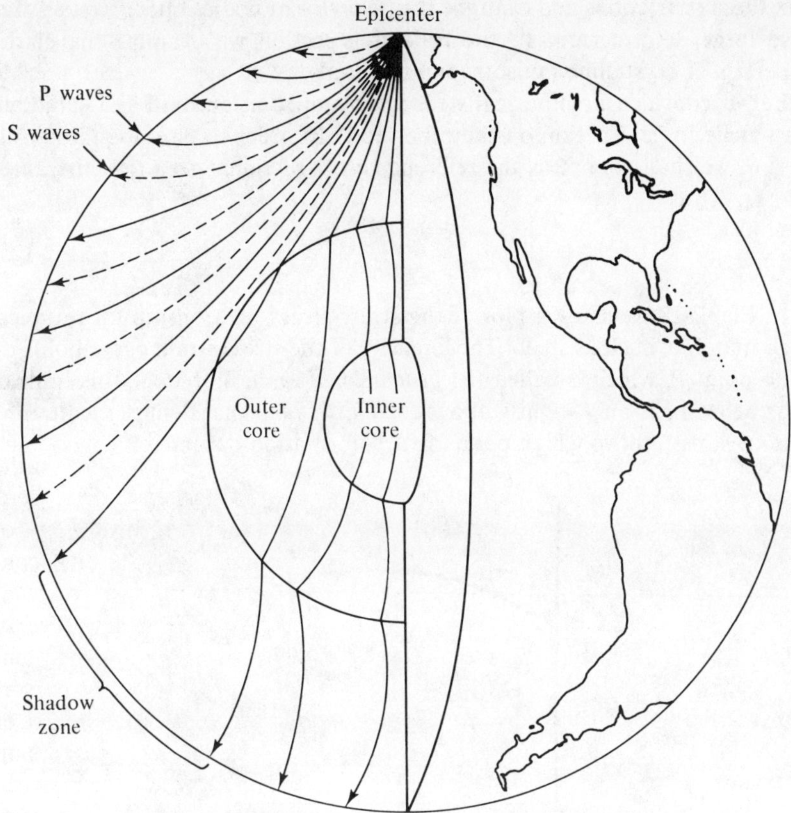

Fig. 7-12 The nature of the earth's interior as revealed by the propagation of elastic waves.

7-4.2. Two seismic stations are separated by 800 km, one being due south of the other. Both stations record an earthquake. The northern station records the arrival of the S wave 5 min and 4 sec after the P wave. The southern station records the arrival of the S wave exactly 5 min after the P wave. How far is the epicenter of the earthquake from each station? How long after the arrival of the S waves do the surface waves arrive at each station?

7-5 The Behavior of Materials Beyond Their Elastic Limits

In our discussion of the behavior of solid bodies under stress we have, up to this point, considered the bodies to be homogeneous and isotropic and have limited deformations to small values. In this section we

235

relax these restrictions and examine the behavior of bodies under stresses that cause large deformation. In the following section we examine the elastic properties of crystalline (anisotropic) materials.

Let us consider a cylindrical steel rod clamped at one end and subjected to a tensile force F acting on the other end. In order to describe the tensile stress σ, we shall choose as the reference cross-sectional area the unstrained area A_0. Thus

$$\sigma = \frac{F}{A_0}. \tag{7-125}$$

In Fig. 7-13 we show a plot of the stress-strain relationship for a typical tough material, such as steel. The linearity of the stress-strain curve holds up to the point P, which is called the *proportional limit*. For steel, this limit of linear behavior is only slightly above the *elastic limit* that defines the limiting value of stress, above which permanent deformation occurs.

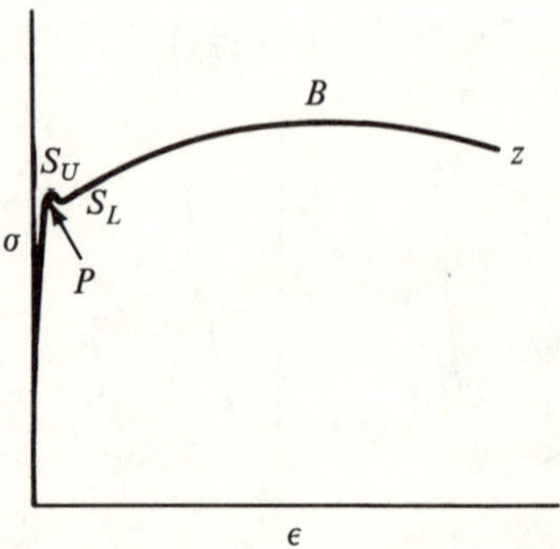

Fig. 7-13 The stress-strain curve for steel.

Beyond the proportional limit the slope of the stress-strain curve decreases and reaches a maximum S_U known as the plasticity limit or *upper yield point*. The curve then dips to the lower yield point S_L, beyond which the curve becomes more nearly parallel to the ϵ axis as plastic flow commences in earnest. In this range a small change in stress produces large changes in the strain. The curve rises to a maximum B corresponding to ultimate stress; fracture occurs, however, only when the point Z is reached.

This behavior is typical of such tough materials as steel. For a *brittle* material like cast iron there is almost no plastic range, and fracture occurs abruptly at the end of a very limited elastic range.

Moldable materials, such as clay or lead, exhibit almost no elastic properties; their behavior is plastic throughout.

In tests like those described for the steel rod it is assumed that the load is applied slowly, since the strain depends not only on the stress but also on the time duration of its application. Even in the elastic range the deformation associated with a particular load is reached only after a considerable time. This slow response of the material is known as *creep*. Various plastics, such as Teflon and nylon, are subject to large creep over long periods.

In analogy to the behavior of ferromagnetic materials, elastic materials experience a *hysteresis effect*. As in the magnetic case, the area enclosed by the hysteresis loop, $\oint \sigma \, d\epsilon$, is simply the energy input to deformation per cycle. If the load is not static or quasistatic but varies periodically, as is common in machine parts, fracture occurs at much lower values of stress than in the case of a static load. This effect is called *fatigue*. In Fig. 7-14 we have sketched a so-called Wöhler diagram of the average stress $\langle \sigma \rangle$ as a function of the number of load cycles until failure of the material, z. We see that this curve approaches a horizontal asymptote σ_s (the fatigue limit) as z increases.

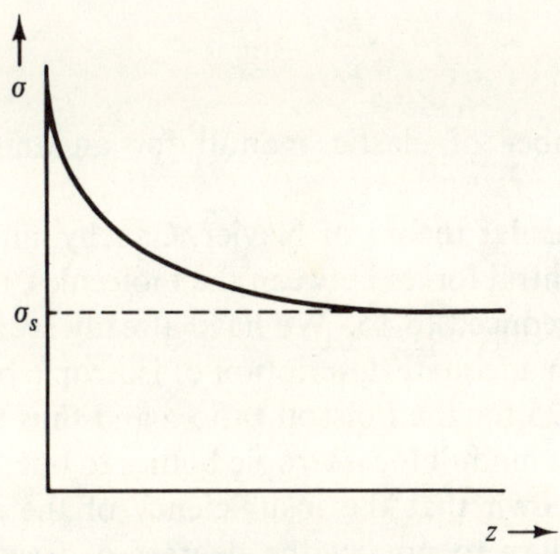

Fig. 7-14

7-6 Elastic Properties of Crystals

In concluding our discussion of elastic phenomena, let us examine the elastic properties of crystals. As before, we are concerned with deriving relationships between the stress and strain tensors. Since both tensors are of second order and symmetric, each having six independent components, the most general linear stress-strain relation must involve $6 \times 6 = 36$ coefficients.

For the present, it is convenient to replace the general double-index notation σ_{ij} and ϵ_{ij} by a single subscript notation:

$$\sigma_{11} = \sigma_1, \quad \sigma_{22} = \sigma_2, \quad \sigma_{33} = \sigma_3, \quad \sigma_{23} = \sigma_4, \quad \sigma_{31} = \sigma_5, \quad \sigma_{12} = \sigma_6 \quad (7\text{-}126)$$

and similarly for ϵ. In this notation the general stress-strain relation becomes

$$\sigma_i = \sum_{j=1}^{6} s_{ij}\epsilon_j, \qquad (7\text{-}127)$$

where the s_{ij} are the moduli of elasticity. The corresponding expression for the energy of deformation is

$$dw = \sum_{i=1}^{6} \sigma_i \, d\epsilon_i. \qquad (7\text{-}128)$$

These two equations imply that the s_{ij} array is symmetric—that is,

$$\frac{\partial^2 w}{\partial \epsilon_i \, \partial \epsilon_j} = s_{ij} = s_{ji} = \frac{\partial^2 w}{\partial \epsilon_j \, \partial \epsilon_i}, \qquad (7\text{-}129)$$

which is equivalent to requiring that dw be an exact differential and

$$\frac{\partial \sigma_i}{\partial \epsilon_j} = \frac{\partial \sigma_j}{\partial \epsilon_i}. \qquad (7\text{-}130)$$

The maximum number of elastic moduli for an anisotropic medium is therefore 21.

In the early molecular theory of Navier, Cauchy, and Poisson, based on the assumption of central forces between the molecules, the number of elastic moduli was further reduced to 15. We have already seen, however, that this theory fails to give an adequate description of isotropic bodies. It results in a universal value of 0.25 for the Poisson ratio ν and thus reduces the required number of two elastic moduli for isotropic bodies to one. The modern crystal lattice theory has shown that the insufficiency of the earlier central force theory lies in its failure to predict the degrees of freedom experienced by several superposed lattice structures capable of moving with respect to one another in elastic deformation.

We can denote the length of the crystal axes by a, b, c and the angles defining the direction of these axes by α, β, γ. In a *triclinic* crystal, which has the least symmetry of any, $a \neq b \neq c$ and $\alpha \neq \beta \neq \gamma$. The number of elastic moduli required to describe a triclinic system is the maximum, 21. In a *monoclinic* crystal $a \neq b \neq c$ and $\alpha = 90° \neq \beta \neq \gamma$. Thus one of the axes, say a, is an axis of symmetry for rotations through 180°. If we take this axis as the z axis of a cartesian coordinate system x, y, z and denote the elastic displacements as before by ξ, η, ζ, then a rotation of 180° about the z axis causes

$$
\begin{aligned}
x &\to -x \qquad & \xi &\to -\xi \\
y &\to -y \qquad & \eta &\to -\eta \\
z &\to z \qquad & \zeta &\to \zeta.
\end{aligned}
\qquad (7\text{-}131)
$$

The associated changes in the strains are, in terms of Eq. (7-126),

$$\epsilon_1 \rightarrow \epsilon_1 \qquad \epsilon_4 \rightarrow -\epsilon_4$$
$$\epsilon_2 \rightarrow \epsilon_2 \qquad \epsilon_5 \rightarrow -\epsilon_5 \qquad\qquad (7\text{-}132)$$
$$\epsilon_3 \rightarrow \epsilon_3 \qquad \epsilon_6 \rightarrow \epsilon_6.$$

Since the stresses σ_i behave in the same way as the strains, they must also obey the rules of Eqs. (7-132). We can now rewrite the stress-strain relations Eq. (7-127). Prior to the rotation,

$$\sigma_i = s_{i1}\epsilon_1 + s_{i2}\epsilon_2 + s_{i3}\epsilon_3 + s_{i4}\epsilon_4 + s_{i5}\epsilon_5 + s_{i6}\epsilon_6 \qquad (7\text{-}133)$$

and after the rotation

$$\sigma_i = s_{i1}\epsilon_1 + s_{i2}\epsilon_2 + s_{i3}\epsilon_3 - s_{i4}\epsilon_4 - s_{i5}\epsilon_5 + s_{i6}\epsilon_6 \qquad (7\text{-}134)$$

for $i = 1, 2, 3, 6$. For $i = 4, 5$, Eq. (7-134) becomes

$$-\sigma_i = s_{i1}\epsilon_1 + s_{i2}\epsilon_2 + s_{i3}\epsilon_3 - s_{i4}\epsilon_4 - s_{i5}\epsilon_5 + s_{i6}\epsilon_6. \qquad (7\text{-}135)$$

Since Eqs. (7-133) to (7-135) must be compatible—that is, the world must look the same with respect to 180° rotation about the symmetry axis—we find by subtracting Eq. (7-134) from Eq. (7-133):

$$s_{i4}\epsilon_4 + s_{i5}\epsilon_5 = 0, \qquad i = 1, 2, 3, 6, \qquad (7\text{-}136)$$

and adding Eqs. (7-133) and (7-135),

$$s_{i1}\epsilon_1 + s_{i2}\epsilon_2 + s_{i3}\epsilon_3 + s_{i6}\epsilon_6 = 0, \qquad i = 4, 5. \qquad (7\text{-}137)$$

From Eq. (7-136) we must conclude

$$s_{i4} = s_{i5}, \qquad i = 1, 2, 3, 6. \qquad (7\text{-}138)$$

The same conclusion follows from Eq. (7-137), owing to the symmetry of s_{ij}. Thus, for the monoclinic crystal, the number of elastic moduli required is reduced from 21 to 13:

$$s = \begin{pmatrix} s_{11} & s_{12} & s_{13} & 0 & 0 & s_{16} \\ s_{12} & s_{22} & s_{23} & 0 & 0 & s_{26} \\ s_{13} & s_{23} & s_{33} & 0 & 0 & s_{36} \\ 0 & 0 & 0 & s_{44} & s_{45} & 0 \\ 0 & 0 & 0 & s_{45} & s_{55} & 0 \\ s_{16} & s_{26} & s_{36} & 0 & 0 & s_{66} \end{pmatrix}. \qquad (7\text{-}139)$$

In an *orthorhombic* crystal, the a, b, and c axes are mutually orthogonal. So another axis of symmetry exists. The third symmetry axis that is also present is essentially a result of the other two and cannot be considered separately. By repeating the argument given above for a second symmetry axis, we find that the orthorhombic system has nine elastic constants:

$$
s = \begin{pmatrix}
s_{11} & s_{12} & s_{13} & 0 & 0 & 0 \\
s_{12} & s_{22} & s_{23} & 0 & 0 & 0 \\
s_{13} & s_{23} & s_{33} & 0 & 0 & 0 \\
0 & 0 & 0 & s_{44} & 0 & 0 \\
0 & 0 & 0 & 0 & s_{55} & 0 \\
0 & 0 & 0 & 0 & 0 & s_{66}
\end{pmatrix}.
\tag{7-140}
$$

In the case of the *cubic* crystal, $a = b = c$ and the axes are mutually orthogonal. Thus, in addition to the rotational invariance enjoyed by the orthorhombic system, the cubic system is also invariant to inversions. The number of elastic constants required to specify the elastic behavior of such a system is three:

$$
s = \begin{pmatrix}
s_{11} & s_{12} & s_{12} & 0 & 0 & 0 \\
s_{12} & s_{11} & s_{12} & 0 & 0 & 0 \\
s_{12} & s_{12} & s_{11} & 0 & 0 & 0 \\
0 & 0 & 0 & s_{44} & 0 & 0 \\
0 & 0 & 0 & 0 & s_{44} & 0 \\
0 & 0 & 0 & 0 & 0 & s_{44}
\end{pmatrix}.
\tag{7-141}
$$

This is still one more than is required to describe an isotropic body. As we know, only two constants, μ and λ, are required for the isotropic body. In terms of s_{ij},

$$
s_{11} = 2\mu + \lambda, \qquad s_{12} = \lambda, \qquad s_{44} = \mu,
\tag{7-142}
$$

and so, in addition to rotation and inversion invariance, the isotropic body also fulfills the condition

$$
s_{11} = s_{12} + 2s_{44},
\tag{7-143}
$$

which does not obtain in the case of the cubic crystal. From a physical point of view, this condition is associated with the fact that isotropic bodies have no uniform structure; they are polycrystals. Their elastic behavior is the result of averaging the elastic reaction of numerous single crystals oriented randomly with respect to one another.

EXERCISES

7-6.1. Show that the array of elastic moduli s_{ij} defined by Eq. (7-127) is symmetric.

7-6.2. Given Eq. (7-131), show that Eq. (7-132) follows.

7-6.3. Assume symmetry to 180° rotation about two mutually orthogonal axis and show that the number of elastic constants for an orthorhombic system is nine.

7-6.4. Derive Eq. (7-141) for cubic crystals.

ELEMENTS OF
STATISTICAL MECHANICS

In this chapter we begin a study of fluids by considering the fluid to consist of a very large number of discrete particles. The motions of these particles are subject to the laws of Newtonian mechanics; the number of particles, however, is such that we cannot hope to follow the motion of each and every mutually interacting particle, even using large-capacity computers. Since the dynamical state of the entire system is determined by the position and velocity of each of its constituent particles, we see that our knowledge of the dynamical state of any such system of numerous particles is always incomplete. The method applied, known as *statistical mechanics*, is formulated in such a way as to permit us to make statistically accurate predictions regarding the future variations of the macroscopic manifestations of a fluid system, starting from an incomplete knowledge of its initial state.

Most of the time we assume that the particles interact by percussive impact much as billiard balls do. This assumption suffices to describe gases of molecules. In the closing sections we consider *plasmas*—that is, fully ionized gases. Plasmas consist of electrically charged particles (ions and electrons) that interact primarily by means of the long-range coulomb force.

8-1 Concepts in Statistical Mechanics

We consider a system of N identical particles, each having mass m. The position of each particle at time t is given by the radius vector from the origin r_i; the velocity of each particle is given by u_i. In general, each particle experiences external forces that have as their resultant at the position of the ith particle the vector F_i. The internal force on the ith particle owing to the presence of the jth particle is F_{ij}. Both F_i and F_{ij} will be considered as depending only on the position of particle i and the relative positions of particles i and j.

Newton's second law of motion can be written for this system as two equations coupled by the velocity u_i:

$$\dot{r}_i = u_i,$$

$$\dot{u}_i = \frac{1}{m}\left(F_i + \sum_{\substack{j=1 \\ i \neq j}}^{N} F_{ij}\right). \tag{8-1}$$

We wish to rewrite the components of the vectors r_i and u_i as the coordinates of a point moving in a six- (or $6N$-) dimensional space known as *phase space*. It is useful to use the notation

$$x_{3i-2}, x_{3i-1}, x_{3i} = \text{components of } r_i,$$

$$x_{3(i+N)-2}, x_{3(i+N)-1}, x_{3(i+N)} = \text{components of } u_i,$$

$$v_{3i-2}, v_{3i-1}, v_{3i} = \text{components of } u_i,$$

$$v_{3(i+N)-2}, v_{3(i+N)-1}, v_{3(i+N)} = \text{components of } \frac{1}{m}\left(F_i + \sum_{\substack{j=1 \\ i \neq j}}^{N} F_{ij}\right),$$

where $i = 1, 2, \ldots, N$. In this notation Eqs. (8-1) can be written as a single equation

$$\dot{x}_i = v_i. \tag{8-2}$$

These equations constitute a set of N coupled equations. Given the initial conditions, Eq. (8-2) can (in principle at least) be integrated to yield

$$x_i = g_i(x_{01}, x_{02}, \ldots, x_{06N}, t), \tag{8-3}$$

where the index i runs from 1 to $6N$ and the subscript 0 signifies an initial value.

It is a fundamental property of our definition of the velocity and coordinate components in phase space, v_i and x_i, that

$$\frac{\partial v_i}{\partial x_i} = 0 \tag{8-4}$$

for all i. For $i \leq 3N$, $v_i = x_{3N+i}$, and so the result is evident in this range. For $i > 3N$, v_i is proportional to the forces F_i and F_{ij}, which have been assumed to be functions of the particle positions only. Thus v_i $(i > 3N)$ are functions only of x_i $(i \leq 3N)$ and Eq. (8-4) is proven. This result also holds for certain cases in which the external forces F_i are velocity dependent. For example, when charged particles move through a magnetic field, they experience a force

$$F = q(u \times B), \tag{8-5}$$

where q is the charge, u is the particle velocity, and B is the magnetic flux density. For this case, we see that for $i > 3N$

$$\frac{\partial v_i}{\partial x_i} \propto \frac{\partial F_i}{\partial u_i} = q \frac{\partial}{\partial u_i} (u \times B)_i = 0. \tag{8-6}$$

If we sum Eq. (8-4) over the index i, we define the six-dimensional divergence of the phase velocity vector v

$$\nabla \cdot v = \sum_{i=1}^{6} \frac{\partial v_i}{\partial x_i} = 0. \tag{8-7}$$

This feature of phase space has far-reaching consequences in statistical mechanics.

The six-dimensional phase space defined by $(x_1, x_2, x_3, x_4, x_5, x_6)$ is known as μ *space*. This is the phase space of an individual particle. The phase space for the system as a whole (an N-particle gas, say) has $6N$ degrees of freedom and is referred to as γ *space*. The volume in γ space occupied by an ensemble of points at time t can be written

$$V(t) = \int_{V_t} dx_1, dx_2, \ldots, dx_{6N}. \tag{8-8}$$

Equation (8-3) for x_i as a function of the initial coordinates of N particles and time can be used to write this volume in the form

$$V(t) = \int_{V_0} J \, dx_{01}, dx_{02}, \ldots, dx_{06N}, \tag{8-9}$$

where J is the Jacobian determinant of the variables x_i with respect to the variables x_{0i}.[1]

$$J = \frac{\partial(x_1, x_2, \ldots, x_{6N})}{\partial(x_{01}, x_{02}, \ldots, x_{06N})} = \begin{vmatrix} \dfrac{\partial x_1}{\partial x_{01}} & \dfrac{\partial x_1}{\partial x_{02}} & \cdots & \dfrac{\partial x_1}{\partial x_{06N}} \\[2mm] \dfrac{\partial x_2}{\partial x_{01}} & \dfrac{\partial x_2}{\partial x_{02}} & \cdots & \dfrac{\partial x_2}{\partial x_{06N}} \\[2mm] \vdots & \vdots & & \vdots \\[2mm] \dfrac{\partial x_{6N}}{\partial x_{01}} & \dfrac{\partial x_{6N}}{\partial x_{02}} & \cdots & \dfrac{\partial x_{6N}}{\partial x_{06N}} \end{vmatrix}. \tag{8-10}$$

---

[1] For a good discussion of Jacobian determinants, see F. E. Hildebrand, *Advanced Calculus for Applications*. Englewood Cliffs, N.J.: Prentice-Hall, Inc., 1962, pp. 340–348.

Since the initial coordinates x_{0i} are time independent, the establishment of the time independence of the Jacobian implies that the volume of the gas in phase space is a constant. Thus

$$\frac{\partial J}{\partial t} = \sum_{i=1}^{6N} \frac{\partial(x_1, \ldots, x_{i-1}, \partial x_i/\partial t, x_{i+1}, \ldots, x_{6N})}{\partial(x_{01}, x_{02}, \ldots, x_{06N})}. \tag{8-11}$$

Since $\partial x_i/\partial t = v_i$, the derivative of $\partial x_i/\partial t$ with respect to x_{0j} can be written

$$\frac{\partial v_i}{\partial x_{0j}} = \sum_{k=1}^{6N} \frac{\partial v_i}{\partial x_k} \frac{\partial x_k}{\partial x_{0j}} \tag{8-12}$$

and so

$$\frac{\partial J}{\partial t} = \sum_{k=1}^{6N} \frac{\partial v_i}{\partial x_k} \frac{\partial(x_1, \ldots, x_{i-1}, x_k, x_{i+1}, \ldots, x_{6N})}{\partial(x_{01}, x_{02}, \ldots, x_{06N})}. \tag{8-13}$$

The recast Jacobian determinant appearing on the right-hand side of Eq. (8-13) vanishes for $k \neq i$, since this condition causes the determinant to have two identical rows. For $k = i$, $\partial v_i/\partial x_i = 0$ as shown above. So we see that

$$\frac{\partial J}{\partial t} = J \nabla \cdot v = 0. \tag{8-14}$$

This property of the incompressibility of an ensemble of points in γ space was first stated by Liouville and is known as *Liouville's theorem*.

Let us turn to the probability of finding the point that represents the mechanical state of an N-particle system in γ space in some given region. Within the region in question we define a volume element

$$dr_1 \, dr_2 \ldots dr_N \, du_1 \, du_2 \ldots du_N,$$

which we shall abbreviate by $d\tau_N$. The probability that the point lies within a given volume is quite reasonably assumed to be proportional to the volume and independent of its shape. The probability that the point lies within the volume element $d\tau_N$ is denoted by

$$P = f_N(r_1, r_2, \ldots, r_N, u_1, u_2, \ldots, u_N, t) \, d\tau_N. \tag{8-15}$$

The function $f_N(r_1, r_2, \ldots, r_N, u_1, u_2, \ldots, u_N, t)$ is known as the *probability density* or N-particle *distribution function*.

Since the point representing the system must lie somewhere in γ space—that is, its constituent particles must all have position vectors and velocity vectors that lie within the range minus infinity to plus infinity—then

$$\int_{-\infty}^{\infty} f_N \, d\tau_N = 1. \tag{8-16}$$

The probability that a point is contained in the volume V_t at time t is the same as the probability that this same point is contained in the volume V_0 at the initial instant, since Eq. (8-3) defines a continuous mapping from volume V_0 to volume V_t. Thus $f_N \, d\tau_N$ is a constant of the motion. We have seen that the volume $d\tau_N$ is itself constant, and so it follows that the N-particle distribution function f_N is also constant in time. In other words,

$$\frac{df_N}{dt} = \frac{\partial f_N}{\partial t} + \sum_{i=1}^{N} \left[\boldsymbol{u}_i \cdot \frac{\partial f_N}{\partial \boldsymbol{r}_i} + \frac{1}{m}\left(\boldsymbol{F}_i + \sum_{\substack{j=1 \\ i \neq j}}^{N} \boldsymbol{F}_{ij} \right) \cdot \frac{\partial f_N}{\partial \boldsymbol{u}_i} \right] = 0, \qquad (8\text{-}17)$$

where $\partial f_N / \partial \boldsymbol{r}_i$ and $\partial f_N / \partial \boldsymbol{u}_i$ are vectors whose components are the partial derivatives of f_N with respect to the components of $\boldsymbol{r}_i$ and $\boldsymbol{u}_i$, respectively. The entire summation term is simply the convective part of the total derivative.

So far we have not effected any means for solving the N-particle gas. In order to know the detailed motion of the system, we would need $6N$ independent solutions of Eq. (8-17), known as the *Liouville equation*, corresponding to $6N$ independent first integrals of the motion of the form

$$f(\boldsymbol{r}_1, \boldsymbol{r}_2, \ldots, \boldsymbol{r}_N, \boldsymbol{u}_1, \boldsymbol{u}_2, \ldots, \boldsymbol{u}_N, t) = \text{const.} \qquad (8\text{-}18)$$

This is no easier than solving $6N$ coupled equations of motion, however.

What must be done is to *destroy* information in order to make accessible the answers that we want from the problem. We start with the N-particle distribution function f_N and integrate over a certain number of particles. For example,

$$f_1 = N \int f_N \, d\boldsymbol{r}_2 \, d\boldsymbol{r}_3 \ldots d\boldsymbol{r}_N \, d\boldsymbol{u}_2 \, d\boldsymbol{u}_3 \ldots d\boldsymbol{u}_N \qquad (8\text{-}19)$$

is called a one-particle distribution function; there are N of these functions corresponding to integrals of f_N over all particle positions and velocities except one. Since all particles are assumed to be identical, all $N f_1$'s are the same. The two-particle distribution function is defined by

$$f_{12} = N(N-1) \int f_N \, d\boldsymbol{r}_3 \, d\boldsymbol{r}_4 \ldots d\boldsymbol{r}_N \, d\boldsymbol{u}_3 \, d\boldsymbol{u}_4 \ldots d\boldsymbol{u}_N \qquad (8\text{-}20)$$

and in this way it is possible to go on to define distribution functions of higher order.

The Liouville equation can now be replaced by a sequence of equations, known as the *BBGKY hierarchy*, in terms of successive orders of distribution functions.[2] If we multiply Eq. (8-17) by $N \, d\boldsymbol{r}_2 \, d\boldsymbol{r}_3 \ldots d\boldsymbol{r}_N \, d\boldsymbol{u}_2 \, d\boldsymbol{u}_3 \ldots d\boldsymbol{u}_N$ and integrate over all space and velocities, we find, assuming that f_N vanishes for infinite velocities and at the wall of the gas container,

$$\frac{\partial f_1}{\partial t} + \boldsymbol{u}_1 \cdot \frac{\partial f_1}{\partial \boldsymbol{r}_1} + \frac{\boldsymbol{F}_1}{m} \cdot \frac{\partial f_1}{\partial \boldsymbol{u}_1} = -\int \frac{\boldsymbol{F}_{12}}{m} \cdot \frac{\partial f_{12}}{\partial \boldsymbol{u}_1} \, d\boldsymbol{r}_2 \, d\boldsymbol{u}_2. \qquad (8\text{-}21)$$

---

[2] Named for Bogoliubov, Born, Green, Kirkwood, and Yvon.

Similarly, if we multiply the Liouville equation by $N(N-1)\,dr_3\,dr_4\ldots$ $dr_N\,du_3\,du_4\ldots du_N$, we obtain an equation with two-particle distribution functions on the left and three-particle distribution functions on the right

$$\frac{\partial f_{12}}{\partial t} + u_1 \cdot \frac{\partial f_{12}}{\partial r_1} + u_2 \cdot \frac{\partial f_{12}}{\partial r_2} + \frac{F_1 + F_{12}}{m} \cdot \frac{\partial f_{12}}{\partial u_1} + \frac{F_2 + F_{21}}{m} \cdot \frac{\partial f_{12}}{\partial u_2}$$

$$= -\int \frac{F_{13}}{m} \cdot \frac{\partial f_{123}}{\partial u_1}\,dr_3\,du_3 - \int \frac{F_{23}}{m} \cdot \frac{\partial f_{123}}{\partial u_2}\,dr_3\,du_3 \qquad (8\text{-}22)$$

and so on. This hierarchy of equations is entirely equivalent to the Liouville equation. Obviously, the solution of this sequence of coupled equations is as difficult as the solution to the Liouville equation itself; the BBGKY hierarchy, however, lends itself to simplifying assumptions that make solution possible. Specifically, we shall reduce the hierarchy to a single equation involving f_1. The one-particle function f_1 contains much less information than the N-particle function, which contains *all* of the information; yet when N is quite large, f_1 is the best we can do.

The quantity

$$dr_1\,du_1 \int f_N\,dr_2 \ldots dr_N\,du_2 \ldots du_N = \frac{f_1}{N}\,dr_1\,du_1 \qquad (8\text{-}23)$$

represents the probability that the point corresponding to particle 1 lies within the volume element $dr_1\,du_1$ surrounding the point r_1, u_1 in six-dimensional phase space (μ space). Since the particles are indistinguishable, the probable number of particles in $dr_1\,du_1$ is the probability per particle given in Eq. (8-23) times the number of particles N—that is, $f_1\,dr_1\,du_1$. By integrating this expression over velocity, $dr_1 \int f_1\,du_1$, we obtain the probable number of particles in the spatial volume dr_1 surrounding the terminus of the vector r_1 in ordinary three-dimensional space. Thus

$$n = \int f_1\,du_1 \qquad (8\text{-}24)$$

is the probable number of particles per unit volume—that is, the number density.

EXERCISES

8-1.1. Using the x_i, v_i notation, show that the Liouville equation for a function $g(x_1, x_2, \ldots, x_{6N}, t)$ can be written

$$\frac{\partial g}{\partial t} + (v \cdot \nabla_N)g = 0,$$

where

$$v = \sum_{i=1}^{6N} v_i \hat{e}_i, \qquad \nabla_N = \sum_{i=1}^{6N} \frac{\partial}{\partial x_i} \hat{e}_i$$

and $\hat{e}_i$ is a unit vector directed along the x_i axis.

8-1.2. Show that all integrals of the form

$$\int u_i \cdot \frac{\partial f_N}{\partial r_i}\, dr_2\, dr_3 \ldots dr_N\, du_2\, du_3 \ldots du_N$$

and

$$\int F_i \cdot \frac{\partial f_N}{\partial u_i}\, dr_2\, dr_3 \ldots dr_N\, du_2\, du_3 \ldots du_N$$

vanish for $i \neq 1$ if F_i depends only on r_i.

8-1.3. Show that the result derived in problem 8-1.2 is still valid for $F_i = q(u \times B)_i$, where $B = B(r_i)$.

8-1.4. Show that integrals of the form

$$\int F_{ij} \cdot \frac{\partial f_N}{\partial u_i}\, dr_2\, dr_3 \ldots dr_N\, du_2\, du_3 \ldots du_N$$

vanish for $i \neq 1$ and that

$$\int \sum_{j=2}^{N} F_{ij} \cdot \frac{\partial f_N}{\partial u_1}\, dr_2 \ldots dr_N\, du_2 \ldots du_N = \frac{1}{N} \int F_{12} \cdot \frac{\partial f_{12}}{\partial u_1}\, dr_2\, du_2.$$

8-2 Basic Assumptions of Kinetic Theory

For gases at ordinary temperatures and pressures, we shall make several assumptions designed to render the BBGKY equations tractable.[3] Our first assumption is that only binary particle interactions are important and that the simultaneous interaction of three or more particles is rare and can be neglected. The consequence of this assumption so far as Eq. (8-22) is concerned is that, whenever $F_{12} \neq 0$, both F_{23} and F_{31} must vanish. Thus the right-hand side of Eq. (8-22), which represents triple correlations, can be neglected. One is then left with a Liouville equation for f_{12}:

$$\frac{\partial f_{12}}{\partial t} + u_1 \cdot \frac{\partial f_{12}}{\partial r_1} + u_2 \cdot \frac{\partial f_{12}}{\partial r_2} + \frac{F_1 + F_{12}}{m} \cdot \frac{\partial f_{12}}{\partial u_1} + \frac{F_2 + F_{21}}{m} \cdot \frac{\partial f_{12}}{\partial u_2} = 0. \quad (8\text{-}25)$$

If we replace r_2 by the variable $r = r_2 - r_1$, then Eq. (8-25) assumes the form

$$\frac{\partial f_{12}}{\partial t} + u_1 \cdot \frac{\partial f_{12}}{\partial r_1} + (u_2 - u_1) \cdot \frac{\partial f_{12}}{\partial r} + \frac{F_1 + F_{12}}{m} \cdot \frac{\partial f_{12}}{\partial u_1}$$

$$+ \frac{F_2 + F_{21}}{m} \cdot \frac{\partial f_{12}}{\partial u_2} = 0. \quad (8\text{-}26)$$

---

[3] By ordinary temperatures and densities we mean that the gas should not have any appreciable degree of ionization or be so dense that quantum mechanical effects need be taken into account.

Our next assumption is that the range of the interaction force is finite—that is, F_{12} vanishes for $|r| > r_0$. We shall denote by s_0 the sphere centered at r_1 with radius r_0. Now let us multiply Eq. (8-26) by $dr\,du_2$ and integrate over all u_2 and over all r interior to s_0; we obtain

$$\left(\frac{\partial}{\partial t} + u_1 \cdot \frac{\partial}{\partial r_1} + \frac{F_1}{m} \cdot \frac{\partial}{\partial u_1}\right) \int_{s_0} \int_{u_2} f_{12}\, dr\, du_2$$

$$= -\int_{s_0} \int_{u_2} \frac{F_{12}}{m} \cdot \frac{\partial f_{12}}{\partial u_1}\, dr\, du_2 - \int_{s_0} \int_{u_2} (u_2 - u_1) \cdot \frac{\partial f_{12}}{\partial r}\, dr\, du_2. \quad (8\text{-}27)$$

The first integral term on the right-hand side is equal to the right-hand side of Eq. (8-21); we shall label this integral $\mathscr{I}$. Regarding the left-hand side of Eq. (8-27), let us examine the integral $\int_{s_0} f_{12}\, dr\, du_2$, which represents the probable number of particle pairs, such that one particle lies in the unit volume surrounding r_1, u_1, whereas the other particle is within the interaction sphere s_0 and moves at arbitrary speed u_2. If the radius r_0 of the sphere s_0 is assumed to be sufficiently small and f_{12} and its derivatives are sufficiently regular, then the left-hand side of Eq. (8-27) must be negligible in comparison with each term on the right-hand side. So

$$\mathscr{I} = -\int_{s_0} \int_{u_2} \frac{F_{12}}{m} \cdot \frac{\partial f_{12}}{\partial u_1}\, dr\, du_2 = \int_{s_0} \int_{u_2} (u_2 - u_1) \cdot \frac{\partial f_{12}}{\partial r}\, dr\, du_2. \quad (8\text{-}28)$$

The spatial part of this integral can be rewritten, using Eq. (1-129) as a surface integral over s_0:

$$\mathscr{I} = \oint_{s_0} \int_{u_2} f_{12}(u_2 - u_1) \cdot d\sigma\, du_2; \quad (8\text{-}29)$$

then Eq. (8-21) can be written

$$\frac{\partial f_1}{\partial t} + u_1 \cdot \frac{\partial f_1}{\partial r_1} + \frac{F_1}{m} \cdot \frac{\partial f_1}{\partial u_1} = \oint_{s_0} \int_{u_2} f_{12}(u_2 - u_1) \cdot d\sigma\, du_2. \quad (8\text{-}30)$$

We wish to express the interaction operator $\mathscr{I} = (\partial f_1/\partial t)_{\text{coll}}$ solely in terms of the one-particle distribution function f_1. Consequently, we assume that the particles are distributed randomly outside the sphere s_0. This assumption implies that the probable number of particle pairs, such that one member of the pair is located in a unit volume surrounding the point r_1, u_1 and the other member is located in a unit volume surrounding r_2, u_2, is equal to the product of the probable number of particles in the first volume and the probable number in the second volume—that is,

$$f_{12}(r_1, r_2, u_1, u_2, t) = f_1(r_1, u_1, t)f_2(r_2, u_2, t) \quad (8\text{-}31)$$

for $|r_2 - r_1| > r_0$. This is called the *hypothesis of molecular chaos*. With this assumption, the interaction integral becomes

$$\mathscr{I} = \oint_{s_0} \int_{u_2} (u_2 - u_1) \cdot f_1 f_2 \, d\sigma \, du_2. \qquad (8\text{-}32)$$

In order to carry out the integration over the sphere, we divide the sphere into two hemispheres about the equatorial plane perpendicular to the vector $(u_2 - u_1)$. We label the hemisphere on which $(u_2 - u_1) \cdot d\sigma > 0$ by s_0^+ and the hemisphere on which $(u_2 - u_1) \cdot d\sigma < 0$ by s_0^-. The sense of this division is that those particles having positions and velocities corresponding to points on s_0^+ have already experienced interaction within the sphere; those particles corresponding to points on s_0^- are just about to experience interaction. We shall denote f_{12} on s_0^+ by $f_1^+ f_2^+$ and f_{12} on s_0^- by $f_1^- f_2^-$. As integration parameters over the sphere, we shall select polar coordinates ρ and ϕ of the projection of the point r_2 onto the plane dividing s_0^+ from s_0^-. The angle ϕ can, for example, be measured from the intersection of this plane with the fixed plane passing through r_1. The range of interaction on the equatorial plane is the disk $\mathscr{D}$ ($0 < \rho < r_0, 0 \leq \phi < 2\pi$) covered twice, since two points on the sphere map into one on the disk. For one of these points

$$(u_2 - u_1) \cdot d\sigma = |u_2 - u_1| \rho \, d\rho \, d\phi$$

and for the other

$$(u_2 - u_1) \cdot d\sigma = -|u_2 - u_1| \rho \, d\rho \, d\phi.$$

In this way, the collision integral becomes

$$\mathscr{I} = \int_{\mathscr{D}} \int_{u_2} (f_1^+ f_2^+ - f_1^- f_2^-) |u_2 - u_1| \rho \, d\rho \, d\phi \, du_2, \qquad (8\text{-}33)$$

where the integral covers velocity space and the disk.

Our final assumption, made in order to evaluate Eq. (8-33), is that the distribution functions vary only slightly as the result of a binary encounter. The assumption of the smallness of r_0 (that is, $r_2 \approx r_1$) necessary in order to write Eq. (8-28) and the very short time duration of the encounter enable us to write

$$f_1^+ f_2^+ = f_1(r_1, u_1', t') f_2(r_2^+, u_2', t')$$

$$\approx f_1(r_1, u_2', t) f_2(r_1, u_2', t) = f_1' f_2' \qquad (8\text{-}34)$$

and

$$f_1^- f_2^- = f_1(r_1, u_1, t) f_2(r_2^-, u_2, t)$$

$$\approx f_1(r_1, u_1, t) f_2(r_1, u_2, t) = f_1 f_2, \qquad (8\text{-}35)$$

where the prime superscript refers to values after the encounter (the positive side of the sphere). Using these definitions, we write

$$\mathscr{I} = \int_{\mathscr{D}} \int_{u_2} (f_1' f_2' - f_1 f_2)|u_2 - u_1|\rho \, d\rho \, d\phi \, du_2. \qquad (8\text{-}36)$$

This integral is therefore a function of the same variables as f_1 itself—r_1, u_1, and t.

EXERCISE

8-2.1. Explain in detail, in regard to Eq. (8-27), why

$$\left(\frac{\partial}{\partial t} + u_1 \cdot \frac{\partial}{\partial r_1} + (u_2 - u_1) \cdot \frac{\partial}{\partial r} + \frac{F_1 + F_{12}}{m} \cdot \frac{\partial}{\partial u_1}\right) \int_{s_0} \int_{u_2} f_{12} \, dr \, du_2$$

$$\approx \left(\frac{F_{12}}{m} \cdot \frac{\partial}{\partial u_1} + (u_2 - u_1) \cdot \frac{\partial}{\partial r}\right) \int_{s_0} \int_{u_2} f_{12} \, dr \, du_2$$

for $r_0 \to 0$. Why does the term

$$\int_{s_0} \int_{u_2} \frac{F_2 + F_{21}}{m} \cdot \frac{\partial f_{12}}{\partial u_2} \, dr \, du_2$$

drop out in going from Eq. (8-26) to (8-27)?

8-3 The Boltzmann Equation and the Maxwell–Boltzmann Distribution

Combining Eqs. (8-29), (8-30), and (8-36), we can now write

$$\frac{\partial f_1}{\partial t} + u_1 \cdot \frac{\partial f_1}{\partial r_1} + \frac{F_1}{m} \cdot \frac{\partial f_1}{\partial u_1} = \int_{\mathscr{D}} \int_{u_2} (f_1' f_2' - f_1 f_2)|u_2 - u_1|\rho \, d\rho \, d\phi \, du_2. \qquad (8\text{-}37)$$

At this point we must destroy yet more information by an averaging process. We define the *velocity distribution f* as the time average of f_1 over a time period τ that is long compared to the duration of a binary encounter but short in terms of the time scale of change for macroscopic quantities:

$$f = \langle f_1(r_1, u_1, t)\rangle = \frac{1}{\tau}\int_{-\tau/2}^{\tau/2} f_1(r_1, u_1, t + \tau) \, d\tau. \qquad (8\text{-}38)$$

The time average of the product of two one-particle functions $\langle f_1 f_2 \rangle$ is

$$\langle f_1 f_2 \rangle = \frac{1}{\tau}\int_{\tau/2}^{\tau/2} f_1(r_1, u_1, t + \tau)f_2(r_2, u_2, t + \tau) \, d\tau. \qquad (8\text{-}39)$$

For very small τ, a Taylor expansion of f_1 and f_2 about the instant t may be substituted for the integrand and integrated term by term. It is left as an exercise for the reader to show that to second order in τ,

$$\langle f_1 f_2 \rangle = f_1(t)f_2(t) + \frac{\tau^2}{24}\left(f_1 \frac{\partial^2 f_2}{\partial t^2} + 2 \frac{\partial f_1}{\partial t}\frac{\partial f_2}{\partial t} + f_2 \frac{\partial^2 f_1}{\partial t_2} \right) + \cdots . \quad (8\text{-}40)$$

If we now consider in the same way the product of two time-averaged functions, $\langle f_1 \rangle \langle f_2 \rangle$, we find

$$\langle f_1 \rangle \langle f_2 \rangle = f_1(t)f_2(t) + \frac{\tau^2}{24}\left(f_1 \frac{\partial^2 f_2}{\partial t^2} + f_2 \frac{\partial^2 f_1}{\partial t^2} \right) + \cdots . \quad (8\text{-}41)$$

The difference between the mean of the product and the product of means is therefore second order in τ:

$$\langle f_1 f_2 \rangle - \langle f_1 \rangle \langle f_2 \rangle = \frac{\tau^2}{12}\frac{\partial f_1}{\partial t}\frac{\partial f_2}{\partial t} + \text{higher-order terms} \quad (8\text{-}42)$$

and we shall consider this difference to be negligible.

Having established these properties of the time averages, we time average Eq. (8-37) and find

$$\frac{\partial f}{\partial t} + \boldsymbol{u}\cdot\frac{\partial f}{\partial \boldsymbol{r}} + \frac{\boldsymbol{F}}{m}\cdot\frac{\partial f}{\partial \boldsymbol{u}} = \int (f'f'_* - ff_*)|\boldsymbol{u}_* - \boldsymbol{u}|\rho\,d\rho\,d\phi\,d\boldsymbol{u}_*, \quad (8\text{-}43)$$

where we have written

$$\langle f_1 \rangle = f, \qquad \langle f_2 \rangle = f_* .$$

This equation is called the *Boltzmann equation*. The approximations that have been made in deriving the Boltzmann equation from the BBGKY equations are

1. *Only binary interactions take place.*
2. *The range of the interaction force is very short.*
3. *Outside this range, velocities are distributed randomly.*
4. *The distribution functions vary only slightly in an encounter.*

These approximations are quite valid for spherical particles undergoing percussive collisions. If the interaction force varies as r^{-s}, then r_0 is infinite and the second assumption breaks down. The van der Waals force between molecules varies as r^{-5}; for $s \geq 5$, long-range collisions do not make a significant contribution to the collision integral, and so the Boltzmann equation remains valid.

The solution of the Boltzmann equation is, in general, quite difficult. Maxwell was able to find a very simple solution in 1859 that we now know as the *Maxwell–Boltzmann velocity distribution*. We shall present what was essentially Maxwell's original derivation of this distribution function. We first inquire what fraction of the total number density n of molecules has x components of velocity between some arbitrary value u_x and some slightly larger value $u_x + du_x$. We shall represent this fraction as dn_{u_x}/n; it will depend on the location in velocity space, and so we can write

$$dn_{u_x} = ng(u_x)\,du_x, \tag{8-44}$$

where $g(u_x)$ is an arbitrary function of u_x that is to be determined. In the absence of streaming, all directions in velocity space must be equivalent. Thus we must also have

$$dn_{u_y} = ng(u_y)\,du_y, \tag{8-45}$$

and
$$dn_{u_z} = ng(u_z)\,du_z. \tag{8-46}$$

We next wish to know what fraction of the molecules with x components of velocity between u_x and $u_x + du_x$ has at the same time y components of velocity between u_y and $u_y + du_y$. Although the subgroup dn_{u_x} of molecules is a small fraction of n, the total number density of molecules, it must still constitute a large number of molecules. It amounts to the same thing to say that du_x must be large enough so that dn_{u_x} is large, still keeping du_x small compared to the total velocity range. This assumption is one of the weaker points in Maxwell's theory and is best justified by the fact that the distribution agrees well with experiment.

We shall write the fractional number density of molecules having simultaneously x components of velocity in the range u_x to $u_x + du_x$ and y components between u_y and $u_y + du_y$ as $d^2n_{u_xu_y}$. We can then write the fractional number of u_x-component molecules with y components of velocity between u_y and $u_y + du_y$ as $d^2n_{u_xu_y}/dn_{u_x}$.

The fraction of the total number density with y components in the range u_y to $u_y + du_y$ is given by Eq. (8-45). Since we assume n to be so large that these fractions are the same, we find with the use of Eqs. (8-44) and (8-45)

$$d^2n_{u_xu_y} = ng(u_x)g(u_y)\,du_x\,du_y. \tag{8-47}$$

By extension of the foregoing reasoning we can write for the number of molecules simultaneously having their x components of velocity between u_x and $u_x + du_x$, y components between u_y and $u_y + du_y$, and z components between u_z and $u_z + du_z$

$$d^3n_{u_xu_yu_z} = ng(u_x)g(u_y)g(u_z)\,du_x\,du_y\,du_z. \tag{8-48}$$

Since Eq. (8-48) describes the number of representative points in the volume element $du_x\,du_y\,du_z$, we can define the density of points in phase space f as

$$f = \frac{d^3n_{u_xu_yu_z}}{du_x\,du_y\,du_z} = ng(u_x)g(u_y)g(u_z). \tag{8-49}$$

Since the velocity distribution is isotropic, the phase space density or distribution function f must be the same in any other volume element at the same radial distance $u = \sqrt{u_x^2 + u_y^2 + y_z^2}$ from the origin.

Consider, next, a second volume element near the first one in which the distribution function differs from that in the first. The change in f due to changes in u_x, u_y, and u_z is

$$df = \frac{\partial f}{\partial u_x}\, du_x + \frac{\partial f}{\partial u_y}\, du_y + \frac{\partial f}{\partial u_z}\, du_z. \tag{8-50}$$

Since the functions $g(u_x)$, $g(u_y)$, $g(u_z)$ are functions, respectively, of u_x, u_y, and u_z only, we find

$$\frac{\partial f}{\partial u_x} = ng'(u_x)g(u_y)g(u_z),$$

$$\frac{\partial f}{\partial u_y} = ng(u_x)g'(u_y)g(u_z), \tag{8-51}$$

$$\frac{\partial f}{\partial u_z} = ng(u_x)g(u_y)g'(u_z),$$

where the prime superscript indicates differentiation with respect to the argument.

Let us consider the special case in which the second volume element lies at the same distance from the origin as the first and $df = 0$. Setting Eq. (8-50) equal to zero and using Eqs. (8-51), we find

$$\frac{g'(u_x)}{g(u_x)}\, du_x + \frac{g'(u_y)}{g(u_y)}\, du_y + \frac{g'(u_z)}{g(u_z)}\, du_z = 0 \tag{8-52}$$

with the auxiliary condition obtained by differentiating $u_x^2 + u_y^2 + u_z^2 = u^2 = \text{const}$,

$$u_x\, du_x + u_y\, du_y + u_z\, du_z = 0. \tag{8-53}$$

If du_x, du_y, du_z were independent, which, because of Eq. (8-53), they are not, then the coefficients of each of these differentials in Eq. (8-52) would need to equal zero. Since this is not the case, we must use a method devised by Lagrange and known as the method of *undetermined multipliers*. Applying this technique, we can combine Eqs. (8-52) and (8-53) to obtain an equation in which du_x, du_y, and du_z *are* independent. We multiply Eq. (8-53) by a constant λ, the value of which will subsequently be determined, and add the resulting equation to Eq. (8-52). Thus we obtain

$$\left[\frac{g'(u_x)}{g(u_x)} + \lambda u_x\right] du_x + \left[\frac{g'(u_y)}{g(u_y)} + \lambda u_y\right] du_y + \left[\frac{g'(u_z)}{g(u_z)} + \lambda u_z\right] du_z = 0. \tag{8-54}$$

We then assign a value to λ such that

$$\frac{g'(u_x)}{g(u_x)} + \lambda u_x = 0, \tag{8-55}$$

which causes Eq. (8-54) to reduce to

$$\left[\frac{g'(u_y)}{g(u_y)} + \lambda u_y\right] du_y + \left[\frac{g'(u_z)}{g(u_z)} + \lambda u_z\right] du_z = 0. \tag{8-56}$$

Any two of the differentials du_x, du_y, du_z can be considered as independent, and we shall select du_y and du_z. Then, according to Eq. (8-56), the coefficients must vanish:

$$\frac{g'(u_y)}{g(u_y)} + \lambda u_y = 0, \tag{8-57}$$

$$\frac{g'(u_z)}{g(u_z)} + \lambda u_z = 0, \tag{8-58}$$

and so the result of using the Lagrange multiplier technique to incorporate the auxiliary condition Eq. (8-53) is to render du_x, du_y, du_z independent. We can now solve any of Eqs. (8-55), (8-57), or (8-58) to find the form of the function g. Using Eq. (8-55), we obtain

$$g(u_x) = \alpha \exp\left(-\beta^2 u_x^2\right), \tag{8-59}$$

where α is the integration constant and $\beta^2 = \lambda/2$. Since $g(u_y)$ and $g(u_z)$ obey the same equation as $g(u_x)$, naturally

$$g(u_y) = \alpha \exp\left(-\beta^2 u_y^2\right), \tag{8-60}$$

and $$g(u_z) = \alpha \exp\left(-\beta^2 u_z^2\right). \tag{8-61}$$

These results can then be substituted into Eq. (8-48) in order to evaluate d^3n. We find

$$d^3n = n\alpha^3 \exp\left(-\beta^2 u^2\right) du_x \, du_y \, du_z, \tag{8-62}$$

where $u^2 = u_x^2 + u_y^2 + u_z^2$. In terms of f, the distribution function

$$f = \frac{d^3n}{du_x \, du_y \, du_z} = n\alpha^3 \exp\left(-\beta^2 u^2\right). \tag{8-63}$$

In order to describe the number density of molecules having *speeds* within a certain range, say u to $u + du$, including all directions of space, we recall that f is constant in any thin spherical shell of radius u. The volume of the shell in question is $4\pi u^2 \, du$, and so

$$dn_u = 4\pi u^2 \, du \cdot f = 4\pi n\alpha^3 u^2 \exp\left(-\beta^2 u^2\right) du. \tag{8-64}$$

The ratio dn_u/du is the Maxwell–Boltzmann distribution function of molecular speeds. Unlike the velocity distribution function f, it does not represent the density in phase space but rather the number density of molecules in a speed range du. This distribution is plotted in Fig. 8-1. Because of the factor u^2, which is absent in the velocity distribution function, dn_u/du is zero for $u = 0$ and rises to a maximum, beyond which it falls off monotonically with increasing u.

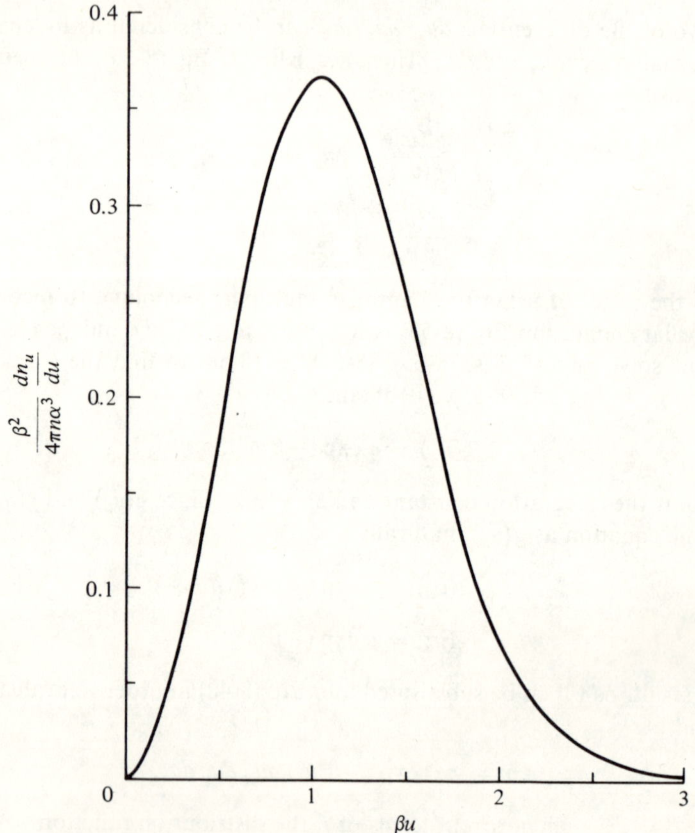

Fig. 8-1 The Maxwell–Boltzmann distribution of molecular speeds.

The next step is to consider the evaluation of the constants α and β. The amplitude constant can be found by integrating dn_u over all values of u from zero to infinity and setting the result equal to the total number density of molecules n.

$$n = \int dn_u = 4\pi n\alpha^3 \int_0^\infty u^2 \exp(-\beta^2 u^2)\, du. \tag{8-65}$$

The definite integral can be found in any good integral table; its value is $\sqrt{\pi}/4\beta^3$. Thus

$$\alpha^3 = \left(\frac{\beta}{\sqrt{\pi}}\right)^3 \tag{8-66}$$

and the speed distribution function can be expressed wholly in terms of β:

$$dn_u = \frac{4n\beta^3}{\sqrt{\pi}}\, u^2 \exp\left(-\beta^2 u^2\right) du. \tag{8-67}$$

In order to evaluate β, we shall calculate the root-mean-square speed, defined by

$$u_{\text{rms}} = \sqrt{\langle u^2 \rangle} = \left(\frac{1}{n}\int u^2\, dn_u\right)^{1/2} = \left[\frac{4\beta^3}{\sqrt{\pi}}\int_0^\infty u^4 \exp\left(-\beta^2 u^2\right) du\right]^{1/2}. \tag{8-68}$$

The value of the definite integral in Eq. (8-68) is $3\sqrt{\pi}/8\beta^5$, and so we find

$$\beta = \sqrt{\frac{3}{2}}\,\frac{1}{u_{\text{rms}}}. \tag{8-69}$$

It can be shown that the mean translational kinetic energy of a molecule is associated with the absolute temperature of the gas sample by the relation

$$\tfrac{1}{2}m\langle u^2\rangle = \tfrac{3}{2}kT, \tag{8-70}$$

where k is Boltzmann's constant ($1.3803 \cdot 10^{-23}$ J/molecule-deg), m is the molecular mass, and T is the absolute temperature in degrees Kelvin.[4] Combining Eqs. (8-69) and (8-70), we therefore find for β

$$\beta = \sqrt{\frac{m}{2kT}} \tag{8-71}$$

and the Maxwell–Boltzmann distribution of molecular speeds becomes

$$dn_u = \frac{4n}{\sqrt{\pi}}\left(\frac{m}{2kT}\right)^{3/2} u^2 \exp\left(\frac{-mu^2}{2kT}\right) du. \tag{8-72}$$

In terms of velocities, the distribution is

$$d^3n_{u_x u_y u_z} = n\left(\frac{m}{2\pi kT}\right)^{3/2} \exp\left(\frac{-mu^2}{2kT}\right) du_x\, du_y\, du_z. \tag{8-73}$$

Having derived this distribution in a roundabout fashion, it remains to show that it satisfies the Boltzmann equation, Eq. (8-43). First, we note that

---

[4] For the proof of this statement, see any good text on kinetic theory—for instance, Sir James Jeans, *An Introduction to the Kinetic Theory of Gases*. Cambridge: Cambridge University Press, 1962, pp. 25–28.

$f = d^3 n_{u_x u_y u_z}/du_x\, du_y\, du_z$ is independent of time and position. If the external force F vanishes, then clearly the left-hand side of the Boltzmann equation

$$\frac{\partial f}{\partial t} + \boldsymbol{u} \cdot \frac{\partial f}{\partial \boldsymbol{r}} + \frac{\boldsymbol{F}}{m} \cdot \frac{\partial f}{\partial \boldsymbol{u}} = 0. \tag{8-74}$$

The collision integral can also be shown to vanish, since the argument of the exponential function $(-m/2kT)(u^2 + u_*^2)$ possesses the same value before and after an elastic collision. Hence $f'f'_* = ff_* = 0$.

EXERCISES

8-3.1. Show that, to second order in τ, the time average of the product $f_1 f_2$ is given by Eq. (8-40).

8-3.2. What is the fractional number of molecules having a speed higher than some value u_0 in a Maxwellian gas?

8-3.3. Calculate the average speed of the molecules in a Maxwellian gas in terms of kT/m.

8-3.4. Derive an expression for the most probable speed u_m of a molecule in a Maxwellian gas. This is the speed corresponding to the peak of the distribution curve.

8-3.5. Calculate u_m, $\langle u \rangle$, and u_{rms} for oxygen molecules at a temperature of 300°K (room temperature).

8-3.6. The mean free path λ is the most probable distance traveled by a molecule between two successive collisions. For spherical molecules of radius ρ, show that the mean free path is given by

$$\lambda = \frac{1}{n\sigma},$$

where $\sigma = 4\pi\rho^2$ and n is the number density.

8-4 The Transport Equations

Using the velocity distribution function $f(\boldsymbol{r}, \boldsymbol{u}, t)$ and integrating over velocity space, we can calculate various macroscopic quantities that are used in fluid mechanics to formulate a continuum theory. We introduced one of them in Eq. (8-24), the number density

$$n(\boldsymbol{r}, t) = \int f(\boldsymbol{r}, \boldsymbol{u}, t)\, d\boldsymbol{u}. \tag{8-75}$$

In general, given a quantity $\phi(r, u, t)$ associated with a particle whose position and velocity are r and u at time t, the average value of ϕ for all particles near r at the same instant t is given by

$$\langle \phi(r, t) \rangle = \frac{1}{n} \int \phi(r, u, t) f(r, u, t) \, du. \tag{8-76}$$

For example, the average velocity $u_0(r, t)$ is

$$u_0(r, t) = \frac{1}{n} \int u f(r, u, t) \, du. \tag{8-77}$$

In practice, *only* the average values defined in this way are measurable.

The *random velocity* v is simply the difference between u and u_0, $v = u - u_0$; the mean value of this random velocity is identically zero. The kinetic energy of a particle is $\frac{1}{2}mu^2 = \frac{1}{2}m(u_0 + v)^2$. The sum of the kinetic energies of all the particles is the total energy of the fluid. We can therefore write the total energy per unit volume as

$$\int \tfrac{1}{2}m(u + v)^2 f \, du = \tfrac{1}{2}\rho u_0^2 + \int \tfrac{1}{2}mv^2 f \, du, \tag{8-78}$$

where ρ is the mass density of the fluid, $\rho = nm$. The first term on the right is the average kinetic energy density associated with streaming; the second term is the kinetic energy density of random motion. We refer to this latter term as the internal energy density of the fluid. It is just this internal energy that we had in mind when we wrote Eq. (8-70), which we can now rewrite as

$$\frac{1}{n} \int \tfrac{1}{2}mv^2 f \, du = \tfrac{3}{2}kT(r, t), \tag{8-79}$$

where $T(r, t)$ is defined as the absolute temperature of the gas.

For every r and t we can associate with $\phi(r, u, t)$ and a unit vector $\hat{e}$ a quantity known as the *flux* of ϕ in the $\hat{e}$ direction. In order to describe this flux, we denote by $d\sigma$ an element of area normal to $\hat{e}$ and situated in the neighborhood of r at time t. We shall assume that this area element moves with a velocity u'. The particles with velocities in the range u to $u + du$ that cross $d\sigma$ during the time interval t to $t + dt$ are those that lie within a cylinder of base $d\sigma$ and length $(u - u') \, dt$ at time t. The number of particles interior to such a cylinder is simply

$$\hat{e} \cdot (u - u') \, dt \, d\sigma \, f(u) \, du.$$

The total amount of the quantity ϕ associated with these particles is

$$\hat{e} \cdot (u - u') \phi(u) f(u) \, du \, dt \, d\sigma.$$

Dividing by $dt \, d\sigma$, we thus obtain an expression for the amount of ϕ transported in the direction $\hat{e}$ at the particular location, per unit time and area, by particles with velocities in the range u to $u + du$. Integrating over velocity space therefore yields the total flux of ϕ in the $\hat{e}$ direction,

$$\int \hat{e} \cdot (u - u')\phi(u)f(u)\,du = n\langle\phi(u - u')\rangle \cdot \hat{e}. \tag{8-80}$$

Consider the simplest case where $\phi = 1$. We find $n\langle u - u'\rangle \cdot \hat{e}$ for the number of particles crossing a unit surface in unit time. For $\phi = \int \frac{1}{2}mv^2 f\,du$, the internal energy density, and u', the velocity of the surface across which we calculate the flux equal to the mean velocity of the fluid u_0, we obtain

$$q = \tfrac{1}{2}\rho\langle v^2 v\rangle, \tag{8-81}$$

where q is called the *heat flux vector*. The quantity of heat crossing a surface having a unit normal $\hat{e}$, per unit area per unit time, is simply $q \cdot \hat{e}$.

Let us examine one more case of particular importance, $\boldsymbol{\phi} = mu$, the particle momentum. For this case, we shall consider a gas without streaming ($u_0 = 0$) in which the molecules rebound elastically from the walls of their container, which is at rest with respect to the observer ($u' = 0$). The momentum flux can therefore be written

$$\boldsymbol{p}_e = \rho\langle uu\rangle \cdot \hat{e} \tag{8-82}$$

or, since $u = v$, the random velocity

$$\boldsymbol{p}_e = \rho\langle vv\rangle \cdot \hat{e}. \tag{8-83}$$

The product vv is called a *dyad* and amounts to a second-rank tensor with components given by

$$vv = \begin{pmatrix} v_x v_x & v_x v_y & v_x v_z \\ v_y v_x & v_y v_y & v_y v_z \\ v_z v_x & v_z v_y & v_z v_z \end{pmatrix}. \tag{8-84}$$

We therefore define the *pressure tensor* p as

$$\mathsf{p} = \rho\langle vv\rangle. \tag{8-85}$$

The pressure exerted on the wall is then equal to the dot product of the pressure tensor with the unit normal

$$\boldsymbol{p}_e = \mathsf{p} \cdot \hat{e}. \tag{8-86}$$

In order to calculate q and p, we must know the velocity distribution. If we consider the Maxwell–Boltzmann distribution, which we can write in the form

$$f = n\left(\frac{m}{2\pi kT}\right)^{3/2} \exp\left(\frac{-mv^2}{2kT}\right), \tag{8-87}$$

then the heat flux vector is given by

$$q = \frac{\rho}{2}\left(\frac{m}{2\pi kT}\right)^{3/2}\int_{-\infty}^{\infty} v^2 \boldsymbol{v}\exp\left(\frac{-mv^2}{2kT}\right)dv. \qquad (8\text{-}88)$$

This integral vanishes, since it involves an even function of the components of v. This fact implies that Maxwellian fluids do not support heat propagation.

In the case of the pressure tensor for a Maxwellian fluid, we find that the off-diagonal elements of the tensor vanish, since they also involve integrals of odd functions of the velocity components. The diagonal terms are all equal, owing to the isotropy of the Maxwellian distribution. If we denote these diagonal elements by the scalar p, then

$$\mathsf{p} = p\mathsf{E}, \qquad (8\text{-}89)$$

where E is the unit tensor and

$$p = \frac{\rho}{3}\left(\frac{m}{2\pi kT}\right)^{3/2}\int\!\!\!\int\!\!\!\int_{-\infty}^{\infty} v^2\exp\left(\frac{-mv^2}{2kT}\right)dv$$

$$= \frac{8}{3\sqrt{\pi}}nkT\int_{0}^{\infty} x^4 e^{-x^2}\,dx = nkT. \qquad (8\text{-}90)$$

Having explored the averaging process in some detail, we are now prepared to use the Boltzmann equation to derive equations in terms of the macroscopic variables. This step is done by multiplying the Boltzmann equation by certain functions $\phi(\boldsymbol{r}, \boldsymbol{u}, t)$ and integrating over velocity space. Taking the left-hand side of the Boltzmann equation term by term, we find

$$\int \phi\frac{\partial f}{\partial t}\,d\boldsymbol{u} = \frac{\partial}{\partial t}\int \phi f\,d\boldsymbol{u} - \int f\frac{\partial \phi}{\partial t}\,d\boldsymbol{u}$$

$$= \frac{\partial}{\partial t}(n\langle\phi\rangle) - n\frac{\partial}{\partial t}\langle\phi\rangle; \qquad (8\text{-}91)$$

considering the x component of the second term

$$\int \phi u_x\frac{\partial f}{\partial x}\,du_x = \frac{\partial}{\partial x}\int \phi u_x f\,du_x - \int f\frac{\partial}{\partial x}(\phi u_x)\,du_x$$

$$= \frac{\partial}{\partial x}(n\langle\phi u_x\rangle) - n\left\langle u_x\left(\frac{\partial\phi}{\partial x}\right)\right\rangle,$$

then

$$\int \phi\left(\boldsymbol{u}\cdot\frac{\partial f}{\partial \boldsymbol{r}}\right)d\boldsymbol{u} = \boldsymbol{\nabla}\cdot(n\langle\phi\boldsymbol{u}\rangle) - n\langle(\boldsymbol{u}\cdot\boldsymbol{\nabla})\phi\rangle; \qquad (8\text{-}92)$$

for calculating the third term

$$\int \phi F_x \frac{\partial f}{\partial u_x}\, du = \int du_y\, du_z \left(\phi F_x f \Big|_{u_x=-\infty}^{n_x=+\infty} - \int f \frac{\partial}{\partial u_x} (\phi F_x)\, du \right)$$

$$= -n \left\langle \frac{\partial(\phi F_x)}{\partial u_x} \right\rangle,$$

since f must vanish for infinite velocities. For forces with components F_i independent of the corresponding velocity components u_i,

$$\int \phi \frac{\boldsymbol{F}}{m} \cdot \frac{\partial f}{\partial \boldsymbol{u}}\, d\boldsymbol{u} = -n \left\langle \frac{\boldsymbol{F}}{m} \cdot \frac{\partial \phi}{\partial \boldsymbol{u}} \right\rangle. \tag{8-93}$$

Thus if the Boltzmann equation is abbreviated as

$$\mathscr{D}f = \left(\frac{\partial f}{\partial t} \right)_{\text{coll}}$$

then

$$\int \phi \mathscr{D}f\, d\boldsymbol{u} = \frac{\partial}{\partial t}\, (n\langle \phi \rangle) + \boldsymbol{\nabla} \cdot (n\langle \phi \boldsymbol{u} \rangle)$$

$$= -n \left[\frac{\partial}{\partial t} \langle \phi \rangle + \langle (\boldsymbol{u} \cdot \boldsymbol{\nabla})\phi \rangle + \left\langle \frac{\boldsymbol{F}}{m} \cdot \frac{\partial \phi}{\partial \boldsymbol{u}} \right\rangle \right]$$

$$= n\, \Delta\langle \phi \rangle, \tag{8-94}$$

where $n\, \Delta\langle \phi \rangle$ is given by

$$\int \left(\frac{\partial f}{\partial t} \right)_{\text{coll}} \phi\, d\boldsymbol{u} = n\, \Delta\langle \phi \rangle. \tag{8-95}$$

At this point it becomes convenient to introduce the random velocity v as a variable in place of $\boldsymbol{u}$. Doing so leads to the following replacement formulas.

$$\frac{\partial \phi}{\partial \boldsymbol{u}} \rightarrow \frac{\partial \phi}{\partial \boldsymbol{v}},$$

$$\frac{\partial \phi}{\partial t} \rightarrow \frac{\partial \phi}{\partial t} + \frac{\partial \phi}{\partial \boldsymbol{u}} \cdot \frac{\partial \boldsymbol{u}_0}{\partial t} = \frac{\partial \phi}{\partial t} - \frac{\partial \phi}{\partial \boldsymbol{v}} \cdot \frac{\partial \boldsymbol{u}_0}{\partial t},$$

$$\frac{\partial \phi}{\partial x} \rightarrow \frac{\partial \phi}{\partial x} + \frac{\partial \phi}{\partial \boldsymbol{u}_0} \cdot \frac{\partial \boldsymbol{u}_0}{\partial x} = \frac{\partial \phi}{\partial x} - \frac{\partial \phi}{\partial \boldsymbol{v}} \cdot \frac{\partial \boldsymbol{u}_0}{\partial x}, \tag{8-96}$$

$$u_x \frac{\partial \phi}{\partial x} \rightarrow u_x \frac{\partial \phi}{\partial x} - u_x \frac{\partial \phi}{\partial \boldsymbol{v}} \cdot \frac{\partial \boldsymbol{u}_0}{\partial x},$$

$$\boldsymbol{u} \cdot \frac{\partial \phi}{\partial \boldsymbol{r}} \rightarrow \boldsymbol{u} \cdot \frac{\partial \phi}{\partial \boldsymbol{r}} - \frac{\partial \phi}{\partial \boldsymbol{v}} \cdot (\boldsymbol{u} \cdot \boldsymbol{\nabla})\boldsymbol{u}_0.$$

Using Eqs. (8-96) to transform Eq. (8-94), we find

$$\int \phi \mathscr{D} f \, du = \frac{\partial}{\partial t}(n\langle\phi\rangle) + \boldsymbol{\nabla} \cdot [n\langle\phi(v + u_0)\rangle]$$

$$- n\left[\frac{\partial\langle\phi\rangle}{\partial t} - \left\langle\frac{\partial\phi}{\partial v} \cdot \frac{\partial u_0}{\partial t}\right\rangle + (v + u_0) \cdot \boldsymbol{\nabla}\phi\right.$$

$$\left. - \left\langle\frac{\partial\phi}{\partial v} \cdot [(u_0 + v) \cdot \boldsymbol{\nabla}]u_0\right\rangle + \left\langle\frac{F}{m} \cdot \frac{\partial\phi}{\partial v}\right\rangle\right]. \qquad (8\text{-}97)$$

Writing

$$\frac{d}{dt} = \frac{\partial}{\partial t} + u_0 \cdot \boldsymbol{\nabla},$$

the total or *substantial* derivative, and rearranging Eq. (8-97) somewhat, we find

$$\int \phi \mathscr{D} f \, du = \frac{d}{dt}(n\langle\phi\rangle) + n\langle\phi\rangle \, \boldsymbol{\nabla} \cdot u_0 + \boldsymbol{\nabla} \cdot n\langle\phi v\rangle$$

$$- n\left[\frac{d}{dt}\langle\phi\rangle + \langle v \cdot \boldsymbol{\nabla}\phi\rangle + \left(\frac{F}{m} - \frac{du_0}{dt}\right) \cdot \left\langle\frac{\partial\phi}{\partial v}\right\rangle\right.$$

$$\left. - \frac{\partial\phi}{\partial v} \cdot (v \cdot \boldsymbol{\nabla})u_0\right] = n\,\Delta\langle\phi\rangle. \qquad (8\text{-}98)$$

In choosing the function ϕ, we are attracted to functions for which $n\,\Delta\langle\phi\rangle = 0$—that is, functions that are collision invariant. Since $v = u - u_0$, we see that

$$\phi^{(0)} = m,$$

$$\boldsymbol{\phi}^{(1)} = mv,$$

$$\phi^{(2)} = mv^2,$$

satisfy this requirement. Using these three functions in Eq. (8-98), we find the three *transport equations*

$$\frac{d\rho}{dt} + \rho \, \boldsymbol{\nabla} \cdot u_0 = 0, \qquad (8\text{-}99)$$

$$\rho \frac{du_0}{dt} + \boldsymbol{\nabla} \cdot \mathrm{p} = nF, \qquad (8\text{-}100)$$

$$\frac{3}{2}\frac{d}{dt}(nkT) + \frac{3}{2}nkT \, \boldsymbol{\nabla} \cdot u_0 + \boldsymbol{\nabla} \cdot q + (\mathrm{p} \cdot \boldsymbol{\nabla})u_0 = 0. \qquad (8\text{-}101)$$

If the fluid in question is locally Maxwellian, then the second and third transport equations are simplified to

$$\rho \frac{d\boldsymbol{u}_0}{dt} + \nabla p = n\boldsymbol{F} \tag{8-102}$$

and

$$\frac{d}{dt}[n(kT)^{-3/2}] = 0. \tag{8-103}$$

EXERCISES

8-4.1. Show that a dyad such as Eq. (8-84) transforms as a tensor.

8-4.2. Carry out the steps indicated in Eq. (8-90). In particular, show that the triple integral in the top line can be transformed into a single integral.

8-4.3. Show that the quantity mv^2v is not conserved in collisions.

8-4.4. Carry out the derivation of Eqs. (8-99) to (8-101), using Eq. (8-98).

8-4.5. Show that Eqs. (8-102) and (8-103) follow from Eqs. (8-100) and (8-101), respectively, if the fluid in question is Maxwellian.

8-5 An Introduction to Plasmas

Plasma physics, the study of gases at temperatures high enough for there to be appreciable ionization, has become an important area of research for two reasons. First, it became evident to astrophysicists in the early 1940s that the plasma condition of matter[5] is the overwhelmingly dominant form of matter in the universe. This fact alone would be reason enough to include a discussion of plasma in an intermediate mechanics text-book. In addition, the worldwide effort to realize controlled nuclear fusion reactions as a cheaply and abundantly fueled energy source has involved the effort to confine a plasma magnetically, at temperatures of several hundred million degrees Kelvin, for times long enough for useful energy to be extracted from the fusion reactions. Because of scaling relations, to be discussed in the next chapter, cosmical plasmas and laboratory fusion-reactor plasmas occupy much the same parameter range. It is this stimulation of interest in both fields that has moved plasma physics from unfashionable obscurity to a major item of research.

The rich variety of phenomena in plasma physics arises from the simple fact that charged particles interact with one another through long-range $1/r^2$

---

[5] One hesitates to say "state," since phase transition is not involved; however, the analogy of solid, liquid, gas, and plasma with the old Greek elements earth, water, air, and fire is tempting.

forces. Since the total number of charges increases with r^3, it is possible for the electric and magnetic fields of a gas of charged particles to superpose in a coherent manner. This simple fact gives plasmas collective modes of behavior not found in ordinary gases, where the particles interact by short-range forces only.

Strictly speaking, the term *plasma* is only applied to a collection of charged particles sufficiently dense such that space-charge effects induce coherent behavior. A quantitative criterion may be obtained by considering two plane-parallel conducting surfaces separated by a distance $2x$ and each held at the same electrical potential V_0. Imagine the space between the plates to be occupied by a uniform number density n of particles carrying an identical charge q each. Using Gauss' theorem for the electric field[6]

$$\oiint \boldsymbol{E} \cdot d\boldsymbol{\sigma} = \frac{1}{\epsilon_0} \iiint nq \, d\tau \tag{8-104}$$

and

$$\boldsymbol{E} = -\nabla V, \tag{8-105}$$

we can find for the potential at the midplane

$$V_m = V_0 + \frac{nqx^2}{2\epsilon_0}. \tag{8-106}$$

Thus the energy required to move a particle from the surface to the midplane is

$$\mathscr{E} = q(V_m - V_0) = \frac{nq^2x^2}{2\epsilon_0}. \tag{8-107}$$

If we suppose that the particles have a random motion characterized by an absolute temperature T, then the collective (space-charge) effects will dominate over the random thermal motion if

$$\frac{nq^2x^2}{2\epsilon_0} > \tfrac{1}{2}kT,$$

that is,

$$x > \sqrt{\frac{\epsilon_0 kT}{nq^2}} \equiv \lambda_{\mathrm{D}}. \tag{8-108}$$

The quantity λ_{D}, first derived by Peter Debye in his theory of electrolytes and known as the *debye length*, must be less than the dimensions of the sample if it is to be a plasma, as distinct from a collection of charged particles too small for collective effects.

---

[6] The constant in Coulomb's law is equal to unity for gaussian CGS units and is $1/(4\pi\epsilon_0)$ for rationalized MKS, where $\epsilon_0 = 8.86 \cdot 10^{-12}$ farad/m is called the *vacuum permittivity*.

Although our basic definition of plasma includes systems with a substantial net charge density, such cases are exceptional and cannot exist in static equilibrium. All commonly encountered plasmas are mixtures of positively and negatively charged particles, ions and electrons, in which almost complete neutralization over distances of the order of a debye length must obtain. In order to illustrate this point, imagine that there is an excess density of electrons over ions Δn in some locality. This situation gives rise to a space-charge density $e(\Delta n)$, where e is the charge on the electron ($e = 1.6 \cdot 10^{-19}$ C in MKS). The electric field occurring as a result of the charge separation is

$$E = 1.8 \cdot 10^{-8} x\, \Delta n, \qquad \frac{V}{m}. \tag{8-109}$$

If the basic number density of the plasma is $5 \cdot 10^{22}/\text{m}^3$, one thousandth atmospheric density, and we allow $\Delta n/n$ of one part in 10^6 over a distance of one centimeter, then $x\, \Delta n = 5 \cdot 10^{14}$ and the space-charge field is 90,000 V per centimeter!

This example illustrates the principle of *quasineutrality*. Plasmas satisfy this principle to a high degree, since seemingly insignificant violations produce enormous restoring forces. A more physical picture emerges if we inquire into the form of the potential distribution about a particle of charge q embedded in a plasma in thermal equilibrium. The potential is found to be

$$V = \frac{q}{4\pi\epsilon_0 r} \exp\left(\frac{-r}{\lambda_D}\right) \tag{8-110}$$

if the mean ionic charge number is one (hydrogenic plasma). Thus a shielding cloud described by the exponential function with a radius of order λ_D forms around each particle in the plasma. Consequently, in spite of the long range of the coulomb force, charged particles in a plasma do not interact with each other *individually* at distances much greater than a debye length.

The preceding arguments are only valid if there are many charged particles within a *debye sphere*—that is,

$$\frac{4\pi}{3} n\lambda_D^3 \gg 1 \tag{8-111}$$

in order that our description be statistically meaningful. Fortunately, Eq. (8-111) is well satisfied for most cases of interest. This is equivalent to requiring that the kinetic energy kT be much larger than the potential energy, which must hold true so that recombination does not destroy the ionization.

If we now introduce two more lengths—the distance of closest approach in a head-on collision (the distance at which the potential energy equals the kinetic energy), $p_c = q^2/4\pi\epsilon_0 kT$, and the simplest mean free path for large-angle Rutherford scattering, $\lambda_c = (4\pi p_c^2 n)^{-1}$—we find

$$\lambda_D^2 = \lambda_c p_c. \tag{8-112}$$

This relation, together with Eq. (8-111), suggests that plasmas are typically described by

$$p_c \ll n^{-1/3} \ll \lambda_D \ll \lambda_c. \tag{8-113}$$

So the mean free path for large-angle binary encounters is much longer than the debye length and in laboratory plasmas is often much longer than the dimensions of the plasma.

Since collective effects dominate plasmas, they constitute a much more complex many-body problem than ordinary gases.

EXERCISES

8-5.1. Consider the region surrounding an ion in a plasma. Electrons in thermodynamic equilibrium in the potential well of the ion have a number density given by

$$n_e = n_0 \exp\left(\frac{qV}{kT}\right),$$

where V is the local potential and $n_0 = n_i$ is the average number density. Using this relation and Eqs. (8-104) and (8-105), together with the assumption of singly ionized ions, show that if $qV \ll kT$, then the equation for the potential about an ion is

$$\nabla^2 V = \frac{V}{\lambda_D^2},$$

which has as its solution Eq. (8-110).

8-5.2. Show that the assumption $qV \ll kT$ is equivalent to assuming that the number of particles in a debye sphere is large.

8-6 The Statistical Mechanics of Plasmas

If λ_c is the order of or greater than the dimensions of the plasma, as is true for all high-temperature laboratory plasmas, then the plasma particles all interact with each other simultaneously via the microscopic electromagnetic field. In this sense, the plasma can be regarded as *collisionless* and its one-particle distribution functions (f^+ for ions, f^- for electrons) can be described by a Liouville-type equation

$$\frac{\partial f^{\pm}}{\partial t} + \boldsymbol{u} \cdot \frac{\partial f^{\pm}}{\partial \boldsymbol{r}} \pm \frac{Ze}{m}(\boldsymbol{E} + \boldsymbol{u} \times \boldsymbol{B}) \cdot \frac{\partial f^{\pm}}{\partial \boldsymbol{u}} = 0 \tag{8-114}$$

known as the *Vlasov equation.* Here $Z = 1$ for electrons (the minus sign) and equals the ionic charge number q/e for the ions. This equation, together with Maxwell's equations

$$\nabla \times \boldsymbol{B} = \boldsymbol{J} + \frac{1}{c^2}\dot{\boldsymbol{E}},$$

$$\nabla \times \boldsymbol{E} = -\dot{\boldsymbol{B}},$$

$$\nabla \cdot \boldsymbol{E} = \frac{\rho_c}{\epsilon_0}, \qquad (8\text{-}115)$$

$$\nabla \cdot \boldsymbol{B} = 0,$$

where $\mu_0 = 4\pi \cdot 10^{-7}\ \text{H/m} = 1/\epsilon_0 c^2$, and

$$\rho_c = Ze \int f^+\, d\boldsymbol{u} - e \int f^-\, d\boldsymbol{u} \qquad (8\text{-}116)$$

and
$$\boldsymbol{J} = Ze \int \boldsymbol{u} f^+\, d\boldsymbol{u} - e \int \boldsymbol{u} f^-\, d\boldsymbol{u} \qquad (8\text{-}117)$$

form a self-consistent set of equations. Knowing $\boldsymbol{E}(\boldsymbol{r}, t)$ and $\boldsymbol{B}(\boldsymbol{r}, t)$, the Vlasov equation (8-114) yields f^+ and f^-. These functions yield, through Eqs. (8-116) and (8-117), the charge densities and currents, which, in turn, act as the sources of the electric and magnetic fields according to Maxwell's equations (8-115). This theory is useful, however, only for plasmas in which λ_D is much less than the dimensions of the plasma and λ_c is much greater. We would like to know how to formulate a more general theory that would include the effect of collisions—that is, close individual encounters.

It is useful to ask: How do we specify an *ideal* collisionless plasma? In answer, we chop each plasma particle very fine and distribute the pieces evenly. This process of "liquification" in no way changes any macroscopic quantities, since q/m, qn, λ_D, and so on are all invariant. The effect of reducing the graininess is that the microscopic $\boldsymbol{E}$ and $\boldsymbol{B}$ fields will tend in detail to the macroscopic (measurable) fields and collisionless theory will be exact. The acceleration of a test particle by the macroscopic fields remains invariant under liquification, whereas the acceleration of the test particle by a single particle (collision) goes as the charge on the particle q. So by going to the limit $q \to 0$, $m \to 0$, holding q/m and qn constant, one approaches the perfect collisionless plasma. The degree of liquification is described by the dimensionless parameter $g = (n\lambda_\mathrm{D}^3)^{-1}$. If $g \to 0$, we are at the collisionless limit, and small but finite values of g should yield collisional correction terms.

If we go back to the BBGKY hierarchy of equations, which are exact for fluids of all kinds, we find that by going to the limit $g \to 0$, the one-particle distribution functions obey the Vlasov equation. More significantly, a series expansion in g can be carried out and the correction to f from the first-order term in g can be obtained. The resulting equation for f has the usual derivative operator on the left, $\mathscr{D}f$, and new terms on the right that we identify as the expression for $(\partial f/\partial t)_{\text{coll}}$ in the small-g approximation. These terms are known as the *Fokker–Planck terms*, and the equation is the Fokker–Planck equation. Having stated how the Fokker–Planck equation arises, we shall derive it in a somewhat more direct way.

The ratio of large-angle to small-angle collisions goes approximately as $(p_c/\lambda_{\mathrm{D}})^2$, which is very much smaller than one. We therefore conclude that the overwhelming majority of particle encounters in a plasma result in small deflections. If a particle undergoes two or more collisions simultaneously with particles within the debye sphere, then the deflection angle is approximately the sum of the individual deflections.

Let $P(u, \Delta u)$ be the probability that a particle changes its velocity from u to $u + \Delta u$ in a time interval Δt as a result of multiple small-angle encounters. If only collisions change the spatially uniform distribution function, then its value at time t is

$$f(u, t) = \int f(u - \Delta u, t - \Delta t)P(u - \Delta u, \Delta u)\, d\,\Delta u. \qquad (8\text{-}118)$$

Since Δu is small if Δt is, we can expand Eq. (8-118):

$$f(u, t) = \int \left\{ f(u, t)P(u, \Delta u) - \Delta t\, \frac{\partial f}{\partial t} P(u, \Delta u) \right.$$

$$- \Delta u\, \frac{\partial}{\partial u}\, [f(u, t)P(u, \Delta u)] + \tfrac{1}{2}\, \Delta u_i\, \Delta u_j\, \frac{\partial^2}{\partial u_i\, \partial u_j}$$

$$\left. \times\, [f(u, t)P(u, \Delta u)] + \cdots \right\} d\,\Delta u. \qquad (8\text{-}119)$$

Noting that the probability of some transition is unity

$$\int P(u, \Delta u)\, d\,\Delta u = 1 \qquad (8\text{-}120)$$

and defining the average velocity change per unit time

$$\langle \Delta u \rangle = \frac{1}{\Delta t} \int f(u, t)P(u, \Delta u)\, \Delta u\, d\,\Delta u, \qquad (8\text{-}121)$$

and

$$\langle \Delta u_i\, \Delta u_j \rangle = \frac{1}{\Delta t} \int f(u, t)P(u, \Delta u)\, \Delta u_i\, \Delta u_j\, d\,\Delta u \qquad (8\text{-}122)$$

and neglecting higher-order terms, we find for the change of f per unit time owing to multiple small-angle collisions

$$\left(\frac{\partial f}{\partial t}\right)_{\text{coll}} = -\frac{\partial}{\partial u_i}\left[\langle \Delta u_i\rangle f(\mathbf{u}, t)\right] + \frac{1}{2}\frac{\partial^2}{\partial u_i\,\partial u_j}\left[\langle \Delta u_i\,\Delta u_j\rangle f(\mathbf{u}, t)\right]. \quad (8\text{-}123)$$

These are the Fokker–Planck collision terms.

As a means of gaining some physical insight into these collision terms, we can rewrite Eq. (8-123) as a continuity equation

$$\left(\frac{\partial f}{\partial t}\right)_{\text{coll}} + \nabla_u \cdot \boldsymbol{\gamma} = 0, \quad (8\text{-}124)$$

where ∇_u is the nabla operator in velocity space and the flow vector $\boldsymbol{\gamma}$ is given by

$$\boldsymbol{\gamma} = \langle \Delta\mathbf{u}\rangle f - \tfrac{1}{2}\nabla_u \cdot \left[\langle \Delta\mathbf{u}\,\Delta\mathbf{u}\rangle f\right]. \quad (8\text{-}125)$$

The physical significance of $\langle \Delta\mathbf{u}\rangle$ and $\langle \Delta\mathbf{u}\,\Delta\mathbf{u}\rangle$ can be illustrated by examining the behavior of a stream of test particles injected with velocity $\mathbf{u}$ into the plasma. As time passes, the test particles will be slowed by collisions (on the average) and will undergo a diffusion in velocity space. The average change in velocity per unit time is $\langle \Delta\mathbf{u}\rangle$, and the rate of diffusion of the stream in velocity space is characterized by $\langle \Delta\mathbf{u}\,\Delta\mathbf{u}\rangle$.

In order to apply Eq. (8-123) to plasmas, the quantities $\langle \Delta\mathbf{u}\rangle$ and $\langle \Delta\mathbf{u}\,\Delta\mathbf{u}\rangle$ must be evaluated for coulomb collisions. The results of this calculation are

$$\langle \Delta\mathbf{u}\rangle = \Gamma\,\frac{\partial h}{\partial \mathbf{u}} \quad (8\text{-}126)$$

and

$$\langle \Delta u_i\,\Delta u_j\rangle = \Gamma\,\frac{\partial^2 g}{\partial u_i\,\partial u_j}, \quad (8\text{-}127)$$

where

$$\Gamma = \frac{(q^2 q_f^2)\,\ln\,(\lambda_{\mathrm{D}}/p_c)}{4\pi\epsilon_0^2 m^2} \quad (8\text{-}128)$$

and

$$h(\mathbf{u}) = \frac{m}{\mu}\int \frac{f(\mathbf{u}_f)}{|\mathbf{u} - \mathbf{u}_f|}\,d\mathbf{u}_f, \quad (8\text{-}129)$$

$$g(\mathbf{u}) = \int f(\mathbf{u}_f)|\mathbf{u} - \mathbf{u}_f|\,d\mathbf{u}_f. \quad (8\text{-}130)$$

Here μ refers to the reduced mass. The Fokker–Planck equation for plasmas can therefore be written

$$\frac{1}{\Gamma}\,\mathscr{D}f = -\frac{\partial}{\partial u_i}\left[f\,\frac{\partial h}{\partial u_i}\right] + \frac{1}{2}\frac{\partial^2}{\partial u_i\,\partial u_j}\left[f\,\frac{\partial^2 g}{\partial u_i\,\partial u_j}\right]. \quad (8\text{-}131)$$

It can be shown, using this equation, that a non-Maxwellian plasma, if left to its own devices, will relax to thermal equilibrium. First, the electrons will thermalize among themselves in a time proportional to $T^{3/2}$. In a time longer by roughly a factor $(m_{ion}/m_{elec})^{1/2}$ the ion component thermalizes, reaching, in general, a temperature different from that of the electrons. Subsequently a common equilibrium is reached via a relatively slow ion-electron energy exchange process.

EXERCISE

8-6.1. Prove that the debye length λ_D is invariant to the liquification process in which $q \to 0$, $m \to 0$, whereas q/m and qn are held constant.

REFERENCES

CABANNES, H., *General Mechanics*. Waltham, Mass.: Blaisdell Publishing Co., 1968, Part 2, Chapter 1.

CHAPMAN, S., and T. G. COWLING, *The Mathematical Theory of Non-Uniform Gases*. Cambridge: Cambridge University Press, 1960.

FRANK-KAMENETSKII, D. A., *Plasma, the Fourth State of Matter*. New York: Plenum Press, 1972.

JEANS, SIR JAMES, *The Dynamical Theory of Gases* (4th ed.). New York: Dover Publications, Inc., 1954.

JEANS, SIR JAMES, *An Introduction to the Kinetic Theory of Gases*. Cambridge: Cambridge University Press, 1962.

SCHMIDT, G., *Physics of High Temperature Plasmas*. New York: Academic Press, 1966, Chapters 3 and 9.

VLASOV, A. A., *Many-Particle Theory and Its Application to Plasma*. New York: Gordon and Breach Science Publishers, Inc., 1961.

ELEMENTS OF
FLUID MECHANICS

We continue our study of fluids by exploiting the continuum equations (8-99) to (8-101), derived by taking moments of the Boltzmann equation. We also consider electrically conductive fluids under the action of electromagnetic forces, a relatively recent field of study known as *magnetohydrodynamics*.

Physicists of the author's generation were largely unexposed to fluid mechanical equations in their formal education because, one supposes, of their basic nonlinearity. The present trend, however, is to include such studies in the physics curriculum—and rightly so, since a wide range of highly interesting phenomena is encompassed.

9-1 Static Solutions

The equation of motion, known as *Euler's equation*, appropriate to a fluid in thermodynamic equilibrium is Eq. (8-102). This equation can be written

$$\rho \frac{d\boldsymbol{u}}{dt} + \nabla p = \boldsymbol{f}, \tag{9-1}$$

where $\boldsymbol{f} = n\boldsymbol{F}$ is the external force per unit volume and we have dropped the subscript 0 on the velocity for convenience. In the case of a static fluid, the velocity is identically zero and the static equilibrium relation is

$$\nabla p = \boldsymbol{f}. \tag{9-2}$$

In considering the static equilibrium of a solid, we required a second equilibrium relation on the torques. This is not necessary for fluids, however. The normal pressure cannot induce a torque about any axis; such a torque would require shear stresses p_{ij} and they are all zero. If the external force field f is continuous, then the lines of force must be parallel in any infinitesimal region (the gravity field, for example) and no torques can arise from this source. So Eq. (9-2) is not only a *necessary* condition for static equilibrium, it is also *sufficient*.

An immediate and interesting result of Eq. (9-2) is that only external forces expressible as the gradient of some scalar potential can produce equilibrium. Thus

$$f = -\nabla\phi, \tag{9-3}$$

where the potential ϕ must be single valued within the space occupied by the fluid. Under this assumption the pressure distribution $p(r)$ can be calculated in the form

$$\nabla(p + \phi) = 0 \rightarrow p + \phi = \text{const.} \tag{9-4}$$

The integration constant on the right-hand side of this equation can be determined by the boundary conditions.

As a common example, let us consider a fluid in static equilibrium in the gravitational field. We shall take the surface of the fluid to be located at $z = 0$ and consider the positive z axis to point downward. In this case,

$$f_z = \rho g \rightarrow \phi_g = -\rho g z. \tag{9-5}$$

If we take the pressure at the surface of the fluid (liquid, actually, since we are considering an incompressible fluid) to be equal to the atmospheric pressure p_0, then

$$p = p_0 - \phi_g = p_0 + \rho g z; \tag{9-6}$$

if we wish p to represent the overpressure relative to the atmosphere (gage pressure), then

$$p = \rho g z. \tag{9-7}$$

Let us examine a stationary solution—that is, a solution in which u is nonzero but constant in time. A liquid in a rotating drum rotates in equilibrium with a constant angular frequency about the vertical axis of the drum. The centrifugal force per unit volume is $f = f_r \hat{r}$, where

$$f_r = \rho r \omega^2. \tag{9-8}$$

The corresponding potential is

$$\phi_c = -\tfrac{1}{2}\rho r^2 \omega^2. \tag{9-9}$$

The total potential is therefore

$$\phi = \phi_g + \phi_c = -(\rho g z + \tfrac{1}{2}\rho r^2 \omega^2) \tag{9-10}$$

and the pressure distribution in the liquid is

$$p = \rho g z + \tfrac{1}{2}\rho r^2 \omega^2 + \text{const.} \tag{9-11}$$

If we take the liquid level on the axis of the drum ($r = 0$) to be z_0 and if p is the gage pressure, then

$$0 = \rho g z_0 + \text{const},$$

and
$$p = \rho g(z - z_0) + \tfrac{1}{2}\rho r^2 \omega^2. \tag{9-12}$$

The free surface characterized by $p = 0$ is then described by

$$z_0 - z = \frac{r^2 \omega^2}{2g}, \tag{9-13}$$

a paraboloid of revolution where, since z is positive downward, $z_0 - z$ is the ordinate of the parabolic meridian curve counted positive upward. Denoting by h the difference of the liquid levels at the circumference and the center and by u the circumferential speed, h takes on the familiar form of the velocity head

$$h = \frac{\omega^2}{2g}. \tag{9-14}$$

Consider, next, the more involved problem of the static state of a compressible fluid (gas). In this case, the density depends on the pressure, and we must write an equation of state, which we shall take to be

$$\frac{d}{dt}\left(\frac{p}{\rho^n}\right) = 0 \tag{9-15}$$

or
$$\frac{p}{p_0} = \left(\frac{\rho}{\rho_0}\right)^n. \tag{9-16}$$

This is called a *polytropic equation*; n is the polytropic exponent, and p_0, ρ_0 are constants referring to a normal state. If $n = 1$, the temperature is constant throughout the fluid (the *isothermal state*) as a consequence of the *perfect gas law*

$$\frac{p}{\rho} = \frac{kT}{m}, \tag{9-17}$$

where m is the mass of a gas molecule. In the *adiabatic state* (convective equilibrium such as results from a vigorous mixing process without heat exchange), n is equal to the ratio c_p/c_v of the specific heats at constant pressure and constant volume and is given by $(f + 2)/f$, where f is the number of degrees of freedom of the gas molecule. For a diatomic molecule, such as N_2 or O_2, $f = 5$ (three degrees of translation and two of rotation), and so $n = 1.4$.

For compressible fluids, the force per unit volume f is no longer a useful concept; instead we shall refer the external force to the unit of mass and denote it by $\mathscr{F}$. Thus

$$f = \rho\mathscr{F}; \tag{9-18}$$

in the case of gravity, we therefore find

$$\mathscr{F}_z = g. \tag{9-19}$$

The advantage of this new notation becomes apparent on writing the equilibrium relation, which now reads

$$\frac{\nabla p}{\rho} = \mathscr{F}. \tag{9-20}$$

The right-hand side is a known quantity and the left-hand side can be expressed in terms of either p or ρ through the equation of state. Using Eq. (9-16), Eq. (9-20) assumes the form

$$\nabla\mathscr{P} = \mathscr{F}, \tag{9-21}$$

$$\mathscr{P} = \frac{p_0^{1/n}}{\rho_0}\frac{p^{1-(1/n)}}{1-(1/n)}. \tag{9-22}$$

where

Equation (9-21) also holds for any nonpolytropic ρ, p relationship if we define the quantity $\mathscr{P}$ by

$$\mathscr{P} = \int_A^B \frac{dp}{\rho}, \tag{9-23}$$

where A is a fixed point and B is the variable field point having coordinates x, y, z.

We see that, as in the incompressible case, equilibrium is only possible when the external force—here the force per unit mass—is conservative—that is, $\mathscr{F} = -\nabla\varphi$. Thus

$$\nabla(\mathscr{P} + \varphi) = 0 \rightarrow \mathscr{P} + \varphi = \text{const}. \tag{9-24}$$

For a uniform gravitational field, $\varphi = gz$, where z is positive upward. In this case, using Eq. (9-22), the pressure as a function of height is found to be

$$\frac{p^{1/n}}{\rho_0}p^{1-(1/n)} = \frac{n-1}{n}(\text{const} - gz). \tag{9-25}$$

If we establish our reference pressure p_0 at $z = 0$, the integration constant is found from

$$\frac{p_0}{\rho_0} = \text{const}\left(\frac{n-1}{n}\right), \tag{9-26}$$

and Eq. (9–25) can then be manipulated into the form

$$p = p_0 \left[1 - \left(\frac{n-1}{n} \right) \frac{\rho_0}{p_0} gz \right]^{n/(n-1)}.$$

(9-27)

This equation expresses the pressure distribution in a polytropic atmosphere. The height of the atmosphere h is determined by setting $p = 0$, $z = h$:

$$h = \frac{n}{n-1} \frac{p_0}{\rho_0 g}.$$

(9-28)

By introducing h into E. (9-27) and using Eq. (9-16), the pressure and density distributions can be written

$$p = p_0 \left(1 - \frac{z}{h} \right)^{n/(n-1)},$$

(9-29)

$$\rho = \rho_0 \left(1 - \frac{z}{h} \right)^{1/(n-1)}.$$

(9-30)

Using these two equations to calculate p/ρ, the temperature distribution of the polytropic atmosphere can be determined through Eq. (9-17):

$$kT = m \frac{p_0}{\rho_0} \left(1 - \frac{z}{h} \right).$$

(9-31)

According to this formula, the temperature of the atmosphere is a linear function of altitude, decreasing from its sea level value at $z = 0$ to zero at $z = h$.

The assumption of a constant temperature for the terrestrial atmosphere results in a calculated height of infinity, as opposed to 28 km for $n = 1.4$, values appropriate for air in the adiabatic case. Neither assumption constitutes a realistic model. In the lowest level of the atmosphere (the troposphere) the temperature decreases with altitude. In the stratosphere the temperature remains constant over a considerable interval of height and then increases to several million degrees as the atmosphere fades into the interplanetary plasma known as the solar wind. For small-altitude intervals, the assumption of constant temperature is allowable, however. The pressure and density distribution in this case can be derived directly by using Eqs. (9-20) and (9-17). We find

$$\frac{1}{p} \frac{dp}{dz} = -\frac{mg}{kT},$$

(9-32)

which yields on integration

$$p = p_0 \exp \left(\frac{-mgz}{kT} \right).$$

(9-33)

This is known as the *barometric equation*. Since $\rho = \rho_0 p/p_0$, the same formula applies to the density. This equation can also be derived directly from Eq. (9-27) by taking the limit as $n \to 1$ and using

$$e^x = \lim_{k \to \infty} \left(1 + \frac{x}{k} \right)^k. \tag{9-34}$$

Let us turn to electrically conductive fluids in the presence of electromagnetic fields. These fluids are governed by the same set of equations as ordinary fluids except that the force per unit volume f is dominated by the electromagnetic force

$$f = J \times B, \tag{9-35}$$

which serves to couple the equation of motion (9-1) to the Maxwell equations (8-115). In static equilibrium, the velocity u and the time derivatives vanish and the resulting magnetohydrostatic equations are

$$\nabla p = J \times B, \tag{9-36}$$

$$\nabla \times B = \mu_0 J, \tag{9-37}$$

$$\nabla \cdot B = 0. \tag{9-38}$$

We can eliminate the current density J between Eqs. (9-36) and reduce our static equations to the equivalent set

$$\frac{1}{\mu_0} (\nabla \times B) \times B = \nabla p, \tag{9-39}$$

$$\nabla \cdot B = 0.$$

Two general classes of solution of Eqs. (9-39) are of interest: those in which $\nabla p \neq 0$, which we shall call *pressure-balanced* solutions, and those for which $\nabla p = 0$, which shall be referred to as *force-free* solutions.

In the first case, it is useful to use Eq. (1-93) to rewrite Eq. (9-39) in the form

$$\nabla\left(p + \frac{B^2}{2\mu_0} \right) = \frac{1}{\mu_0} (B \cdot \nabla)B. \tag{9-40}$$

If the magnetic field lines are straight and parallel, then Eq. (9-40) reduces to

$$\nabla\left(p + \frac{B^2}{2\mu_0} \right) = 0 \to p + \frac{B^2}{2\mu_0} = \frac{B_0^2}{2\mu_0}, \tag{9-41}$$

where the integration constant $B_0^2/2\mu_0$ is the *magnetic pressure* in the fluid-free region. We see from this equation that the magnetic field is reduced in the region occupied by the fluid (diamagnetic effect). If the vacuum magnetic field B_0 is known and the distribution of fluid pressure is given, then the magnetic field distribution in the fluid can be calculated by using the pressure-balance equation (9-41).

Equations (9-36) to (9-39) have some interesting results. From Eq. (9-36) we see that

$$J \cdot \nabla p = 0 \tag{9-42}$$

and

$$B \cdot \nabla p = 0. \tag{9-43}$$

Thus the $p = $ const surfaces are both "magnetic surfaces" and "current surfaces"—that is, the vectors in question lie entirely within the $p = $ const surfaces and do not cross them. From Eq. (9-37), using identity (1-99), we see that

$$\nabla \cdot J = 0. \tag{9-44}$$

Since both B and J are divergence-free, the magnetic and the current lines must either extend to infinity or be closed (or ergodic). So the $p = $ const magnetic and current surfaces are either tubes of infinite length or assume the form of toroids.

In the case of force-free static equilibrium, the electromagnetic force $J \times B$ is zero even though J and B are both finite. In this case, Eqs. (9-39) reduce to

$$(\nabla \times B) \times B = 0,$$
$$\nabla \cdot B = 0, \tag{9-45}$$

which implies that B is a Beltrami vector field given by

$$\nabla \times B = \alpha(r)B, \tag{9-46}$$

where $\alpha(r)$ is a scalar function of position. Taking the divergence of Eq. (9-46) and using Eq. (9-38) and identities (1-91) and (1-99), we have

$$B \cdot \nabla \alpha = 0. \tag{9-47}$$

Consequently, the $\alpha = $ const surfaces are also magnetic surfaces.

Ohm's law relating current density to electric field strength in a static magnetofluid is

$$J = \sigma E = \frac{1}{\mu_0} (\nabla \times B). \tag{9-48}$$

In the case of a force-free magnetic field, it reduces to

$$E = \frac{\alpha(r)}{\sigma \mu_0} B. \tag{9-49}$$

Taking the curl of this equation and using the Faraday induction relation

$$\nabla \times E = -\frac{\partial B}{\partial t} \tag{9-50}$$

to eliminate E, we find

$$\sigma\mu_0 \frac{\partial B}{\partial t} = -\nabla \times (\alpha B)$$

$$= B \times \nabla\alpha - \alpha \nabla \times B$$

$$= B \times \nabla\alpha - \alpha^2 B. \tag{9-51}$$

If the magnetic field is to remain force-free as the currents decay due to Ohmic heating, we must have

$$B = B_0 f(t), \tag{9-52}$$

where B_0 is the field at time $t = 0$ and $f(t)$ is a function of time only. Substitution of Eq. (9-52) into Eqs. (9-46), (9-47), and (9-51) implies that α is independent of t and

$$B_0 \cdot \nabla\alpha = B_0 \times \nabla\alpha = 0; \tag{9-53}$$

hence α is a constant in space as well. Thus Eq. (9-51) reduces to

$$\sigma\mu_0 \frac{\partial B}{\partial t} = -\alpha^2 B, \tag{9-54}$$

which has as its solution

$$B = B_0 \exp\left(-\frac{\alpha^2 t}{\sigma\mu_0}\right). \tag{9-55}$$

Force-free vector fields with constant α are known as *Trkalian fields*.

If the electrical conductivity of the fluid is considered to be infinite, α is not necessarily constant. However, it can be shown that Trkalian magnetic fields represent the states of lowest magnetic energy that a closed system may attain. Furthermore, in a system in which magnetic forces initially dominate and in which there is some mechanism to dissipate fluid motions (which need not be zero in a force-free state), Trkalian fields appear to be the natural end configuration.

EXERCISES

9-1.1. Consider a conducting fluid (mercury, say) in a shallow cylindrical container with an insulated bottom and conducting walls as shown in Fig. 9-1.

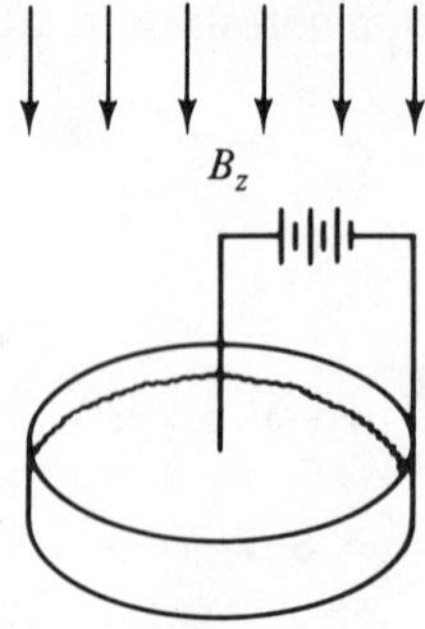

Fig. 9-1

A potential is applied between a vertical wire on the symmetry axis and the wall, giving rise to a current density

$$J_r(r) = \frac{I}{2\pi rh},$$

where I is the total current and h is the depth of the fluid. A uniform magnetic field B_z is applied as shown. Derive a potential for the electromagnetic force $f = J \times B$ and show that static equilibrium cannot exist for such a potential.

9-1.2. At what depth in water is a gage pressure of one atmosphere attained?

9-1.3. Find the pressure as a function of altitude above the surface of the earth for a polytropic atmosphere, considering the surface to be spherical rather than planar. Show that under this condition the pressure in an isothermal atmosphere does not vanish at any finite distance from the earth or, surprisingly, at infinity.

9-1.4. Reconsider the centrifugal problem discussed in this section for the case of a compressible isothermal fluid. Show that we again obtain paraboloids for the $p = $ const surfaces but that no free surface exists.

9-1.5. Show that the following two relations follow from the basic equations of magnetohydrostatics:

$$(B \cdot \nabla)J = (J \cdot \nabla)B$$

and
$$\nabla \cdot (B \times \nabla p) = 0.$$

9-1.6. Show that a Trkalian vector obeys the vector Helmholtz equation

$$\nabla^2 B + \alpha^2 B = 0.$$

Solve this equation to find the simplest Trkalian field that depends on x only.

9-1.7. Show that α as defined by Eq. (9-46) can be expressed as $\alpha = \hat{\imath} \cdot \nabla \times \hat{\imath}$, where $\hat{\imath}$ is a unit vector tangent to the magnetic field vector B.

280

9-2 First Integrals of the Euler Equation

Our first task in this section is to recast the Euler equation

$$\rho\frac{d\boldsymbol{u}}{dt} + \nabla p = \boldsymbol{f} \tag{9-56}$$

in an invariant form for any curvilinear coordinates. Using the expanded form of the substantial derivative

$$\frac{d\boldsymbol{u}}{dt} = \frac{\partial\boldsymbol{u}}{\partial t} + (\boldsymbol{u}\cdot\nabla)\boldsymbol{u} \tag{9-57}$$

to separate the partial and convective derivatives, it can be shown (see problem 9-2.1) that the convective term is misleading if we try to write it in noncartesian coordinates. This dilemma is solved by rewriting identity (1-93) for $\boldsymbol{P} = \boldsymbol{Q} = \boldsymbol{u}$:

$$(\boldsymbol{u}\cdot\nabla)\boldsymbol{u} = \nabla\frac{u^2}{2} - \boldsymbol{u}\times(\nabla\times\boldsymbol{u}). \tag{9-58}$$

Using Eqs. (9-57), (9-58), and the incompressibility condition $\nabla\cdot\boldsymbol{u} = 0$, the Euler equation can be rewritten in the invariant form

$$\rho\left[\frac{\partial\boldsymbol{u}}{\partial t} - \boldsymbol{u}\times(\nabla\times\boldsymbol{u})\right] + \nabla\left(\rho\frac{u^2}{2} + p\right) = \boldsymbol{f}. \tag{9-59}$$

The second term in square brackets is sometimes referred to as the *Magnus force*. It is the force that furnishes most of the lift on an aerofoil. The term $\rho u^2/2$ is known as the *stagnation pressure*.

We notice immediately the nonlinearity of the Euler equation written in this form, a factor due to the convective terms. It is this nonlinearity that makes the integration incomparably more difficult than, for example, the seemingly more complicated Maxwell equations of electromagnetics. For nonlinear equations, the principle of superposition of solutions by which more general integrals of linear equations can be constructed does not apply.

Only in the case of *irrotational flow*—that is,

$$\nabla\times\boldsymbol{u} = 0, \tag{9-60}$$

can a *first integral* (or constant of the motion) be found immediately.

Initially, we shall impose a further restriction—namely, steady flow

$$\frac{\partial\boldsymbol{u}}{\partial t} = 0, \tag{9-61}$$

and further assume that the external force is conservative

$$\boldsymbol{f} = -\nabla\phi. \tag{9-62}$$

281

With these restrictions, Eq. (959) reduces to

$$\nabla\left(\rho\,\frac{u^2}{2} + p + \phi\right) = 0, \tag{9-63}$$

which can immediately be integrated to yield Bernoulli's famous equation

$$\rho\,\frac{u^2}{2} + p + \phi = \text{const}. \tag{9-64}$$

Bernoulli's equation is probably the most important theorem in elementary fluid mechanics. The first term is the kinetic energy per unit volume; the third term is the potential energy per unit volume of the external force. The pressure p takes on the role of the potential energy of internal forces and accounts for the dynamic interaction between neighboring incompressible fluid elements.

Suppose that we take the force of gravity as the external force and rewrite Eq. (9-64) as

$$\frac{u^2}{2} + \frac{p}{\rho} + gz = \text{const} \tag{9-65}$$

and look at several simple examples. Imagine a tank filled with liquid, having an orifice at the level of the bottom as shown in Fig. 9-2. The free surface is

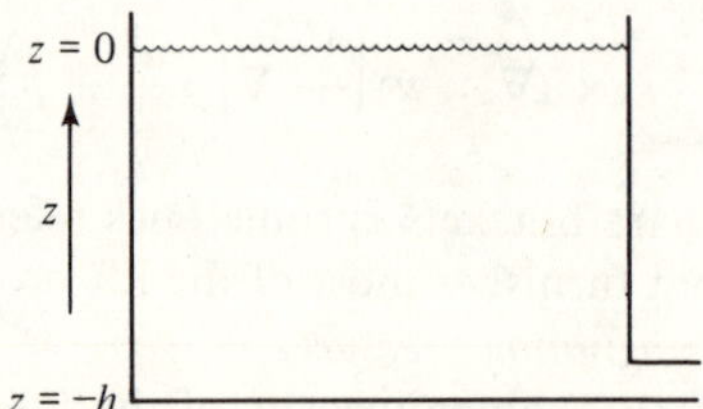

Fig. 9-2

located at $z = 0$ and the orifice at $z = -h$. As long as the orifice is closed, $u = 0$ everywhere; the gage pressure at $z = 0$ is $p = 0$. Thus the constant in Eq. (9-65) is zero. We have seen from our study of static equilibrium that the pressure is $p = \rho gh$ at the bottom. When the orifice is opened, we can assume that at the surface u remains approximately zero, and so const $= 0$ as before. Since p is now also zero at the orifice, we find from Eq. (9-65) for $z = -h$

$$u = \sqrt{2gh} \tag{9-66}$$

as in the case of a freely falling body. This is Torricelli's theorem.

Another example is the steady flow through a horizontal tube of variable cross section. Since the volume flow through each cross section is constant for an incompressible fluid, u must increase with decreasing cross section and vice versa. According to Eq. (9-65), the pressure must change conversely.

This situation is illustrated in Fig. 9-3, where $A_1 > A_2$ represents the respective cross-sectional areas. Since continuity of mass requires

$$A_1 u_1 = A_2 u_2, \tag{9-67}$$

we see from Bernoulli's equation that the change in pressure from one cross section to another is

$$p_1 - p_2 = \frac{\rho}{2}(u_2^2 - u_1^2). \tag{9-68}$$

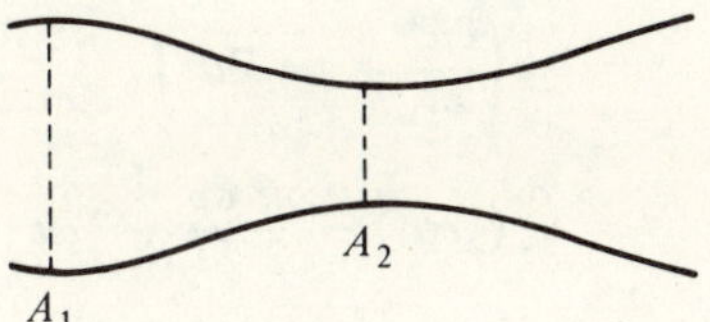

Fig. 9-3

Let us generalize the Bernoulli equation for a nonsteady flow $\partial u/\partial t \neq 0$, still excluding rotational flows and requiring the external force to be conservative. As we saw in Chapter 1, Eq. (9-60) is the necessary and sufficient condition for the vector u to be conservative. So we can write

$$u = -\nabla\psi, \tag{9-69}$$

where ψ is the velocity potential. Since $\nabla \cdot u = 0$, ψ satisfies Laplace's equation

$$\nabla^2\psi = 0. \tag{9-70}$$

The Euler equation now becomes

$$\nabla\left(-\rho\frac{\partial\psi}{\partial t} + \rho\frac{u^2}{2} + p + \phi\right) = 0, \tag{9-71}$$

which is integrated to

$$-\frac{\partial\psi}{\partial t} + \tfrac{1}{2}(\nabla\psi)^2 + \frac{1}{\rho}(p + \phi) = \text{const}, \tag{9-72}$$

the *generalized Bernoulli equation*. Note that here the integration constant, independent of x, y, z, may, in general, still depend on the time.

In the even more general case where $\nabla \times u \neq 0$, integration is still possible if we integrate in the direction of a stream line or, in the nonsteady case, along the vector field u. In this way, the integral over $u \times (\nabla \times u)$ vanishes.

In conclusion, suppose that we examine the hydromagnetic equations without restrictions other than the assumption of infinite conductivity. We take the scalar product of u with the Euler equation

$$\rho u \cdot \frac{du}{dt} = -u \cdot \nabla p + \frac{1}{\mu_0} u \cdot (\nabla \times B) \times B \tag{9-73}$$

and manipulate it term by term. The left-hand term becomes

$$\rho u \cdot \frac{du}{dt} = \rho u \cdot \left[\frac{\partial u}{\partial t} + (u \cdot \nabla)u \right]$$

$$= \tfrac{1}{2}\rho \left(\frac{\partial u^2}{\partial t} + u \cdot \nabla u^2 \right)$$

$$= \frac{\partial}{\partial t} (\tfrac{1}{2}\rho u^2) - \frac{u^2}{2} \frac{\partial \rho}{\partial t} + \tfrac{1}{2}\rho u \cdot \nabla u^2$$

$$= \frac{\partial}{\partial t} (\tfrac{1}{2}\rho u^2) + \frac{u^2}{2} \nabla \cdot \rho u + \tfrac{1}{2}\rho u \cdot \nabla u^2$$

$$= \frac{\partial}{\partial t} (\tfrac{1}{2}\rho u^2) + \nabla \cdot \left(\frac{u^2}{2} \rho u \right), \tag{9-74}$$

where the equation of continuity in the form

$$\frac{\partial \rho}{\partial t} + \nabla \cdot (\rho u) = 0 \tag{9-75}$$

and identity (1-93) have been used. In order to transform the first term on the right-hand side of Eq. (9-73), we write the continuity equation and the polytropic relation as

$$\frac{d\rho}{dt} = -\rho \nabla \cdot u \tag{9-76}$$

and
$$\frac{dp}{dt} \rho^{-n} - n\rho^{-(n+1)}p \frac{d\rho}{dt} = 0. \tag{9-77}$$

Combining these two equations, we have

$$\frac{\partial p}{\partial t} + u \cdot \nabla p + np \nabla \cdot u = 0 \tag{9-78}$$

or
$$\frac{\partial p}{\partial t} + (1 - n)u \cdot \nabla p + n \nabla \cdot (pu) = 0 \tag{9-79}$$

from which we obtain

$$u \cdot \nabla p = \frac{1}{n-1} \frac{\partial p}{\partial t} + \frac{n}{n-1} \nabla \cdot (pu). \tag{9-80}$$

The last term on the right-hand side of Eq. (9-73) can be rewritten

$$\frac{1}{\mu_0} \mathbf{u} \cdot (\nabla \times \mathbf{B}) \times \mathbf{B} = -\frac{1}{\mu_0} (\mathbf{u} \times \mathbf{B}) \cdot (\nabla \times \mathbf{B})$$

$$= \frac{1}{\mu_0} \mathbf{E} \cdot \nabla \times \mathbf{B}$$

$$= \frac{1}{\mu_0} \mathbf{B} \cdot \nabla \times \mathbf{E} - \frac{1}{\mu_0} \nabla \cdot (\mathbf{E} \times \mathbf{B})$$

$$= -\frac{\partial}{\partial t} \frac{B^2}{2\mu_0} - \nabla \cdot (\mathbf{E} \times \mathbf{H}), \qquad (9\text{-}81)$$

where use has been made of Eq. (1-94), the second of Eqs. (8-115) (the Faraday induction law), the condition equation for infinite conductivity

$$\mathbf{E} + \mathbf{u} \times \mathbf{B} = 0, \qquad (9\text{-}82)$$

and the constitutive equation

$$\mathbf{B} = \mu_0 \mathbf{H}. \qquad (9\text{-}83)$$

Summing the contribution from Eqs. (9-74), (9-80), and (9-81), we obtain

$$\frac{\partial}{\partial t} \left(\tfrac{1}{2}\rho u^2 + \frac{p}{n-1} + \frac{B^2}{2\mu_0} \right) + \nabla \cdot \left[\frac{u^2}{2} \rho \mathbf{u} + \frac{n}{n-1} p\mathbf{u} + \mathbf{E} \times \mathbf{H} \right] = 0. \quad (9\text{-}84)$$

We now integrate this equation over all space. The divergence term yields a surface integral by virtue of the Gauss divergence theorem. Since this surface is located at infinity, its contribution vanishes and we obtain the energy integral

$$\iiint \left(\tfrac{1}{2}\rho u^2 + \frac{p}{n-1} + \frac{B^2}{2\mu_0} \right) d\tau = \text{const.} \qquad (9\text{-}85)$$

The first of these integrals is the kinetic energy of the fluid, the second is the thermal free energy, and the last is the magnetic field energy.

EXERCISES

9-2.1. Show that there is a difference between the specification of the convective term given in Eq. (9-58) and the result obtained by taking a "simple-minded" interpretation of the operator $(\mathbf{u} \cdot \nabla)$. Take cylindrical coordinates as an example, where one might be tempted to write the convective operator as

$$(\mathbf{u} \cdot \nabla) \equiv u_\rho \frac{\partial}{\partial \rho} + \frac{u_\theta}{\rho} \frac{\partial}{\partial \theta} + u_z \frac{\partial}{\partial z}.$$

Show that the discrepancy occurs in the ρ and θ components but not in the z component (which is cartesian). Relate the missing terms to the centripetal and Coriolis accelerations.

9-2.2. The upper surface of water in a cylindrical vessel is at a height H above a horizontal plane. At what depth h can a small hole drilled into the vessel wall produce an emerging horizontal water stream that strikes the plane at a maximum distance from the vessel? What is this maximum distance?

9-2.3. Air streams horizontally past an airplane wing of area A. The weight of the airplane is W. What must the air velocity differential between the top and bottom of the wing be in order that the airplane fly? (The assumption of incompressible flow characteristics and the applicability of Bernoulli's equation for air is valid for speeds up to an appreciable fraction of the speed of sound.)

9-2.4. Consider a stream of fluid of density ρ and speed u_1 passing *abruptly* from a cylindrical tube of cross-sectional area A_1 into a larger cylindrical tube of cross-sectional area A_2. As a steady state is reached, the jet from the small tube will mix with the surrounding fluid and will then flow on almost uniformly with an average speed u_2 as shown in Fig. 9-4. Without reference to the

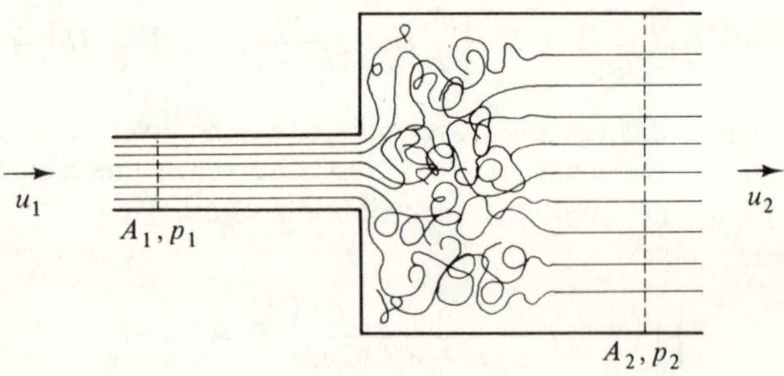

Fig. 9-4

details of the mixing process, use momentum ideas to show that the pressure increase due to the mixing is approximately

$$p_2 - p_1 = \rho u_2(u_1 - u_2)$$

instead of

$$p_2 - p_1 = \tfrac{1}{2}\rho(u_1^2 - u_2^2),$$

which would be realized for a gradual change of cross section. Explain the loss of pressure $\tfrac{1}{2}\rho(u_1 - u_2)^2$ due to the abrupt change in area.

9-2.5. The dispersion relation for gravity waves in deep water can be found by using the generalized Bernoulli equation (9-72). The wave is taken to propagate in the x direction, and the y coordinate is normal to the unperturbed air/water interface. The velocity potential for water particles associated with the wave can be assumed to be

$$\psi = Ae^{-ky} \exp\left[i(kx - \omega t)\right].$$

The equation of the surface profile is

$$y = a \exp\left[i(kx - \omega t)\right].$$

In order that these assumptions be consistent, the y velocity of the free surface, as calculated from each of these relations, must agree. Using this condition and Eq. (9-72) for the free surface ($p = 0$) and ignoring the quadratic term, show that the dispersion relation for waves in the ocean is $\omega^2 = gk$, and so the propagation speed is $\omega/k = U = \sqrt{g\lambda/2\pi}$, where $\lambda = 2\pi/k$ is the wavelength.

9-3 Viscous Flows

We now wish to consider the steady flow of a viscous fluid. We shall assume the velocity of the fluid to be everywhere parallel to the x axis, varying in some manner with y:

$$u = [u_x(y), 0, 0]. \tag{9-86}$$

The reactions due to viscous friction that are transferred through a surface normal to the y axis are, according to Newton, who did the pioneering work, proportional to the velocity gradient. The factor of proportionality μ (not to be confused with the Lamé modulus) depends on the nature of the fluid. We shall refer the reactive forces to unit area and term them viscous pressures. Newton's assumption then reads

$$p_{yx} = -\mu \frac{\partial u_x}{\partial y}. \tag{9-87}$$

In Fig. 9-5 we indicate a surface $s - s'$, which is a positive y surface relative to the fluid beneath it. The fluid above $s - s'$, flowing faster, tries to accelerate the fluid beneath it and in doing so experiences a retardation. This factor explains the minus sign in Eq. (9-87). Let us look at the situation in terms of the strain tensor ϵ. In a fluid, the time rate of change of the displacement rather than the displacement itself is the quantity of interest, the fluid being a continuum. Accordingly, we shall consider $\dot{\epsilon}$ rather than ϵ as responsible for the friction. Hence

$$p_{ik} = -2\mu\dot{\epsilon}_{ik}. \tag{9-88}$$

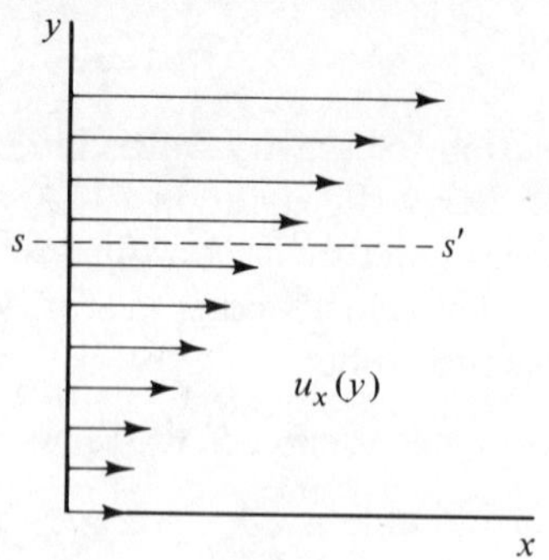

Fig. 9-5

Applying the definition of the ϵ_{ik} [Eq. (7-11)] and differentiating with respect to time for the velocity field given in Eq. (9-86), we find

$$
\dot{\epsilon} = \begin{pmatrix} 0 & \dfrac{1}{2}\dfrac{\partial u_x}{\partial y} & 0 \\[2ex] \dfrac{1}{2}\dfrac{\partial u_x}{\partial y} & 0 & 0 \\[2ex] 0 & 0 & 0 \end{pmatrix}
\tag{9-89}
$$

in support of Newton's assumption.

We were able to show in the theory of elastic solids that σ is a symmetric tensor. The corresponding argument is questionable in the case of the viscous pressure tensor p, since the necessary transition to a sufficiently small volume element is no longer legitimate in the present case. Suppose that we admit the possibility that

$$
p_{ik} \neq p_{ki}.
\tag{9-90}
$$

Then p could always be resolved into a symmetric part $\bar{\mathrm{p}}$ and an anti-symmetric part π such that

$$
\bar{p}_{ik} = \bar{p}_{ki} \quad \text{and} \quad \pi_{ik} = -\pi_{ki}.
\tag{9-91}
$$

The tensor π can now be represented by an axial vector [see, for instance, Eq. (7-12)] and can be invariantly connected only with a quantity of the same character. The vortex frequency

$$
\boldsymbol{\omega} = \tfrac{1}{2}\boldsymbol{\nabla} \times \boldsymbol{u}
\tag{9-92}
$$

is just such a quantity and is the only one that might be used. It would then be necessary to postulate a relation of the form

$$
\pi_{ik} = -\kappa\omega_{ik}.
\tag{9-93}
$$

288

So to admit the possibility of the asymmetry of p amounts to accepting the concept of a "vortex friction."

To test this assumption we fill a drum with some viscous liquid and cause the drum to revolve about its axis. As the liquid begins to rotate at the circumference of the drum, the motion is gradually communicated to the bulk of the liquid. In the steady state the angular frequency ω is a constant throughout the liquid as we know from experience. In contrast to this experience, Eq. (9-93) requires the presence of friction in the steady state.

In addition to the frictional pressures, and superimposed on them, there is a uniform normal pressure of the same type as the pressure in the static case. We denote it by p (it is a scalar, hence no subscripts); it is not, however, identical to the hydrostatic pressure. The total pressure tensor is then

$$\mathsf{p} = \begin{pmatrix} p + p_{xx} & p_{xy} & p_{xz} \\ p_{yx} & p + p_{yy} & p_{yz} \\ p_{zx} & p_{zy} & p + p_{zz} \end{pmatrix}. \tag{9-94}$$

We can now state the equilibrium condition for steady viscous flow as

$$\mathbf{\nabla} \cdot \mathsf{p} = f, \tag{9-95}$$

which, in cartesian coordinates, takes the form

$$\frac{\partial p}{\partial x} + \frac{\partial p_{xx}}{\partial x} + \frac{\partial p_{yx}}{\partial y} + \frac{\partial p_{zx}}{\partial z} = f_x,$$

$$\frac{\partial p}{\partial y} + \frac{\partial p_{xy}}{\partial x} + \frac{\partial p_{yy}}{\partial y} + \frac{\partial p_{zy}}{\partial z} = f_y, \tag{9-96}$$

$$\frac{\partial p}{\partial z} + \frac{\partial p_{xz}}{\partial x} + \frac{\partial p_{yz}}{\partial y} + \frac{\partial p_{zz}}{\partial z} = f_z.$$

These equations determine the flow as soon as we express the six unknowns, p_{ik}, in terms of the velocity components u_x, u_y, u_z by means of Eq. (9-88). For the x component, we obtain

$$\frac{\partial p}{\partial x} - 2\mu \frac{\partial^2 u_x}{\partial x^2} - \mu\left(\frac{\partial^2 u_y}{\partial x\,\partial y} + \frac{\partial^2 u_x}{\partial y^2}\right) - \mu\left(\frac{\partial^2 u_z}{\partial x\,\partial z} + \frac{\partial^2 u_x}{\partial z^2}\right) = f_x, \tag{9-97}$$

which reduces through the use of the incompressibility condition

$$\mathbf{\nabla} \cdot u = 0 = \frac{\partial u_x}{\partial x} + \frac{\partial u_y}{\partial y} + \frac{\partial u_z}{\partial z} \tag{9-98}$$

to

$$\frac{\partial p}{\partial x} - \mu\,\mathbf{\nabla}^2 u_x = f_x. \tag{9-99}$$

The corresponding equations for the y and z directions are

$$\frac{\partial p}{\partial y} - \mu \, \nabla^2 u_y = f_y, \tag{9-100}$$

and

$$\frac{\partial p}{\partial z} - \mu \, \nabla^2 u_z = f_z. \tag{9-101}$$

Equations (9-98) to (9-101) form a four-equation set for the unknowns, u_x, u_y, u_z, and p.

In the inviscid case, only the normal component of the velocity must vanish at the boundary

$$u_n = 0. \tag{9-102}$$

For a viscous fluid, we must also require that the tangential component of the velocity approach zero *continuously* in the neighborhood of the boundary, since a jump in u_t implies an infinitely large gradient in this component and hence infinitely large viscous pressure. Consequently,

$$u_t = 0. \tag{9-103}$$

With this background discussion of viscous flow behind us, we turn to the simplest example, the classical *Poiseuille flow* in a capillary tube. We consider a horizontal tube of circular cross section and radius a, sufficiently small such that the flow proceeds in straight stream lines parallel to the axis of the tube (*laminar* flow). If the axis of the tube coincides with the x axis and the distance from the x axis is $r = \sqrt{y^2 + z^2}$, then the velocity is given by

$$\boldsymbol{u} = [u_x(r), 0, 0]. \tag{9-104}$$

We can neglect gravity, since we have oriented the tube horizontally, and so $\boldsymbol{f} = 0$. From Eqs. (9-100) and (9-101) we see that

$$\frac{\partial p}{\partial y} = \frac{\partial p}{\partial z} = 0; \tag{9-105}$$

hence $p = p(x)$. Since the first term in Eq. (9-99), $\partial p/\partial x$, is a function only of x and the second term, $-\mu \, \nabla^2 u_x$, depends only on r, their difference being zero, either term must be a constant, say $-A$. So

$$\frac{dp}{dx} = -A, \qquad \nabla^2 u_x = -\frac{A}{\mu}, \tag{9-106}$$

where A is the pressure gradient along the tube. Expressing the Laplacian in cylindrical coordinates, the equation for u_x is

$$\frac{1}{r}\frac{d}{dr}\, r\, \frac{du_x}{dr} = -\frac{A}{\mu}. \tag{9-107}$$

Integrating once, we find

$$r \frac{du_x}{dr} = -\frac{A}{\mu}\frac{r^2}{2} + C_1. \tag{9-108}$$

Here the integration constant C_1 must vanish, for otherwise $du_x/dr \to \infty$ as $r \to 0$. Integrating again, we obtain

$$u_x = -\frac{Ar^2}{4\mu} + C_2. \tag{9-109}$$

The integration constant is now determined by the boundary condition

$$u_x(a) = 0, \tag{9-110}$$

and we have

$$u_x = \frac{A}{4\mu}(a^2 - r^2). \tag{9-111}$$

Hence the velocity profile is parabolic. The fluid particles that are on a plane cross section at $t = 0$ lie at $t > 0$ on a paraboloid of revolution, the vertex of which travels in the direction of the flow while the particles in contact with the wall remain fixed.

The volume discharge Q per unit time is

$$Q = \int_0^a 2\pi r u_x \, dr = \frac{\pi A a^4}{8\mu}. \tag{9-112}$$

This formula has been the basis for determining the viscosity coefficient for various liquids since the mid-nineteenth century. The velocity of the flow averaged over the cross section is

$$u_m = \frac{Q}{\pi a^2} = \frac{Aa^2}{8\mu} \tag{9-113}$$

and is just half the maximum velocity. The total pressure drop over a length l of the tube is

$$\Pi = Al = 8\mu l \frac{u_m}{a^2}. \tag{9-114}$$

Note that the pressure drop for laminar flow is found to be proportional to u_m/a^2. In contrast, the pressure drop for *turbulent* flow is approximately proportional to u_m^2/a.

Next, let us examine the magnetohydrodynamic (MHD) equivalent of Poiseuille flow, which is known as *Hartmann flow*. In this case, we shall consider the conductive liquid to flow horizontally between two infinite, horizontal plane walls. We choose our cartesian coordinates so that the boundary walls have the equations $z = \pm L$ and assume laminar flow

$$\mathbf{u} = [u(z), 0, 0], \tag{9-115}$$

and a uniform imposed magnetic field $\mathbf{B}_0$ given by

$$\mathbf{B}_0 = (0, 0, B_0). \tag{9-116}$$

The pressure in the liquid is given by

$$p = p_0(x) + p_1(z) + \rho g z \tag{9-117}$$

so that, apart from the hydrostatic component $p_1(z) = \rho g z$ required to balance the magnetic pressure and the force of gravity, the liquid is subjected to a pressure gradient $-\partial p_0/\partial x$ along the flow direction. We further assume that no quantity varies with y, $\partial/\partial y \equiv 0$, and that, except for p_0, all variables are independent of x. The total magnetic field $\mathbf{B} = \mathbf{B}_0 + \mathbf{b}$, where $\mathbf{b} = [b(z), 0, 0]$, is the field arising from induced currents, is then given by taking the curl of the Ohm's law

$$\mathbf{J} = \frac{1}{\mu_0} \nabla \times \mathbf{b} = \sigma(\mathbf{E} + \mathbf{u} \times \mathbf{B}_0) \tag{9-118}$$

and using the Faraday induction law and Eq. (1-86) to find

$$\frac{1}{\sigma\mu_0} \nabla^2 \mathbf{b} + \nabla \times (\mathbf{u} \times \mathbf{B}_0) = 0. \tag{9-119}$$

The velocity is specified by the stationary ($\partial/\partial t \equiv 0$) form of the equation of motion, including the viscosity term

$$\rho(\mathbf{u} \cdot \nabla)\mathbf{u} = -\nabla(p_0 + p_1) + \frac{1}{\mu_0} (\nabla \times \mathbf{b}) \times \mathbf{B} + \rho\nu \nabla^2 \mathbf{u}, \tag{9-120}$$

where $\rho\nu = \mu$ is the viscosity and ν is called the *kinematic viscosity*.

For the geometry considered here, Eqs. (9-119) and (9-120) reduce to

$$\frac{\partial^2 b}{\partial z^2} + \sigma\mu_0 B_0 \frac{\partial u}{\partial z} = 0, \tag{9-121}$$

$$-\frac{\partial p_0}{\partial x} + \frac{1}{\mu_0} B_0 \frac{\partial b}{\partial z} + \rho\nu \frac{\partial^2 u}{\partial z^2} = 0, \tag{9-122}$$

$$\frac{\partial p_1}{\partial z} + \frac{1}{\mu_0} b \frac{\partial b}{\partial z} = 0. \tag{9-123}$$

Equation (9-122) implies that $-\partial p_0/\partial x$ is a constant, P say, and so for steady laminar flow the pressure gradient in the flow direction is uniform throughout the liquid. Equation (9-123) can be integrated to give

$$p_1 = \text{const} - \frac{b^2}{2\mu_0}, \tag{9-124}$$

and the incompressibility and magnetic field condition equations $\nabla \cdot \boldsymbol{u} = \nabla \cdot \boldsymbol{B} = 0$ are satisfied identically. Equation (9-121) can also be integrated to yield

$$\frac{\partial b}{\partial z} + \sigma\mu_0 u B_0 = A, \tag{9-125}$$

where A is a constant. The components of the current density can then be found, using Eq. (9-118), as

$$J_x = \sigma E_x = 0, \tag{9-126}$$

$$J_y = \frac{1}{\mu_0}\frac{\partial b}{\partial z} = \sigma(E_y - u B_0), \tag{9-127}$$

$$J_z = \sigma E_z = 0, \tag{9-128}$$

implying that $\boldsymbol{E} = (0, E_0, 0)$, where E_0 is constant, and also that the integration constant in Eq. (9-125) is given by $A = \sigma E_0$. Since the electric field is uniform and its component normal to the boundaries vanishes, there can be no volume or surface electric charges and the liquid is not polarized.

Equation (9-122) now assumes the form

$$\rho\nu\frac{\partial^2 u}{\partial z^2} - \sigma u B_0^2 = -(P + \sigma B_0 E_0). \tag{9-129}$$

As in the case of Poiseuille flow, $u_t = 0$ at the walls—that is,

$$u(\pm L) = 0. \tag{9-130}$$

The electromagnetic boundary conditions, assuming the walls to be insulators, must be

$$J_z(\pm L) = 0. \tag{9-131}$$

This condition is satisfied identically by virtue of Eq. (9-128). The solution of Eq. (9-129) that satisfies the boundary condition is easily found to be

$$u = \frac{(P + \sigma B_0 E_0)[\cosh M - \cosh (Mz/L)]}{\sigma B_0^2 \sinh M}, \tag{9-132}$$

where M is the *Hartmann number*, defined by

$$M = B_0 L\sqrt{\frac{\sigma}{\rho\nu}}. \tag{9-133}$$

If no impressed current flows in the liquid, then

$$\int_L^L J_y\, dz = 0, \tag{9-134}$$

which determines the electric field as

$$E_0 = \frac{P(M \coth M - 1)}{\sigma B_0}.$$

(9-135)

Hence

$$u = \frac{PM[\cosh M - \cosh (Mz/L)]}{\sigma B_0^2 \sinh M},$$

(9-136)

or in terms of the mean velocity defined by

$$u_m = \frac{1}{2L} \int_{-L}^{L} u \, dz = \frac{P(M \cosh M - \sinh M)}{\sigma B_0^2 \sinh M},$$

(9-137)

$$u = \frac{u_m M[\cosh M - \cosh (Mz/L)]}{M \cosh M - \sinh M}.$$

(9-138)

We can now integrate Eq. (9-125) to find for the induced magnetic field b

$$b = \frac{u_m PL[\sinh (Mz/L) - (z/L) \sinh M]}{B_0 \sinh M}.$$

(9-139)

Since B is continuous at the plane boundaries, we must have $b = 0$ there and have accordingly chosen the integration constant so that $b(\pm L) = 0$.

The form of the Hartmann velocity profile as given by Eq. (9-138) is shown in Fig. 9-6. The chief effect of the transverse magnetic field is to

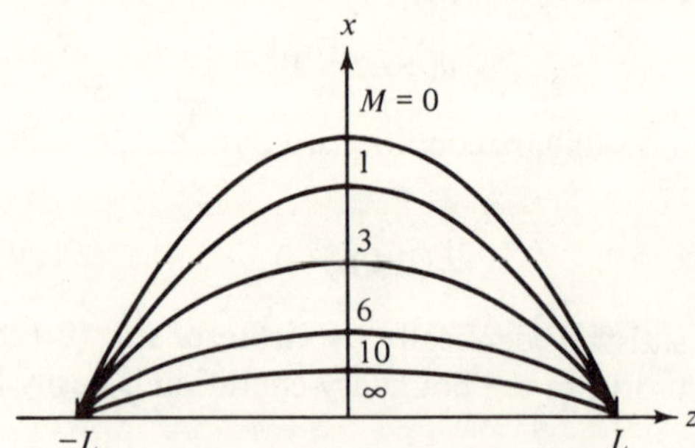

Fig. 9-6 The Hartmann velocity profile.

generate electric currents that retard the fluid flow in the central regions and accelerate the fluid near the boundaries, thereby flattening the Poiseuille parabolic profile realized when $B_0 = 0$. When $M \ll 1$—that is, when the viscous forces are large compared with the electromagnetic forces or the channel is very narrow—Eq. (9-138) tends to the limit

$$u = \frac{3u_m}{2L^2} (L^2 - z^2).$$

(9-140)

When $M \gg 1$—that is, for flows in wide channels or in the presence of strong magnetic fields—Eq. (9-138) tends to the limit

$$u = u_m \left\{ 1 - \exp \frac{[-M(L - |z|)]}{L} \right\}. \tag{9-141}$$

Thus, for this case, the velocity is approximately constant except in two boundary layers whose thickness is of order L/M. This result indicates that the velocity gradient near the walls is of order M/L, and so it might be expected that a large magnetic field could give rise to an instability of the laminar motion near the walls.[1] Experimentally, such is not found to be the case, however; in fact, the presence of a sufficiently strong magnetic field appears to stabilize the motion and inhibit the onset of turbulence.

EXERCISES

9-3.1. Show that any tensor T can be written as the sum of a symmetric tensor $\bar{\mathsf{T}}$ ($\bar{T}_{ij} = \bar{T}_{ji}$) and an antisymmetric tensor τ ($\tau_{ij} = -\tau_{ji}$).

9-3.2. A viscous fluid flows between two infinite parallel plates separated by a distance $2L$. Determine the velocity profile and show that your answer agrees with Eq. (9-140). Determine also the drop Π over a distance l and calculate the force required to keep the plates at rest against the reactive forces of the flow.

9-3.3. Consider a river in laminar flow as shown in Fig. 9-7. The bottom of the stream has a slope angle α with respect to the horizontal and the stream is considered to be infinitely wide. Show that the velocity profile is identical to half the velocity profile obtained in the preceding problem.

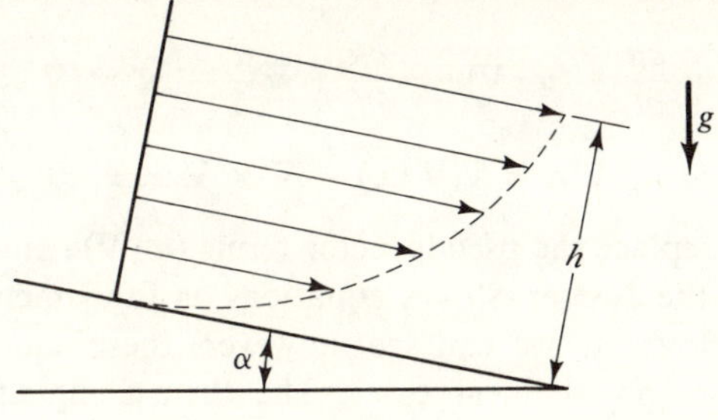

Fig. 9-7

---

[1] Large velocity gradients are commonly unstable, and eddies are usually formed in such cases, as we shall see when we examine the Kármán vortex street.

9-3.4. Verify the solution of Eq. (9-129) given in Eq. (9-132), subject to the boundary conditions $u_t = u_n = 0$ at $z = \pm L$.

9-4 Scaling Relations

The general equation of motion for viscous fluids, known as the *Navier–Stokes equation*, can be written in pseudovector form applicable to cartesian coordinates as

$$\rho \frac{d\mathbf{u}}{dt} + \boldsymbol{\nabla} p - \rho \nu\, \boldsymbol{\nabla}^2 \mathbf{u} = \mathbf{f}, \tag{9-142}$$

This equation must be supplemented in the incompressible case by

$$\boldsymbol{\nabla} \cdot \mathbf{u} = 0. \tag{9-143}$$

In the compressible case, a term, $-(\mu + \lambda)\,\boldsymbol{\nabla}\dot{\Theta} = -(\mu + \lambda)\,\boldsymbol{\nabla}(\boldsymbol{\nabla} \cdot \mathbf{u})$, must be added to the left-hand side of Eq. (9-142) and $\mathbf{f}$ replaced by $\rho\mathscr{F}$. The additional term arises because of the need to replace Eq. (9-88) by

$$p_{ik} = -2\mu\dot{\epsilon}_{ik} - \lambda\,\delta_{ik}\dot{\Theta}, \tag{9-144}$$

as comparison with the stress-strain relation (7-55) suggests. Thus we introduce a second coefficient of viscosity λ associated with the rate of dilation $\dot{\Theta}$. In the literature, one usually finds $\lambda = 2\mu/3$ on the basis of a gas-kinetic argument valid for monatomic gases. For compressible fluids, Eq. (9-143) is replaced by

$$\frac{\partial \rho}{\partial t} + \boldsymbol{\nabla} \cdot (\rho \mathbf{u}) = 0. \tag{9-145}$$

In order to write the Navier–Stokes equation in coordinate-invariant form, the relations

$$\frac{d\mathbf{u}}{dt} = \frac{\partial \mathbf{u}}{\partial t} + (\mathbf{u} \cdot \boldsymbol{\nabla})\mathbf{u} = \frac{\partial \mathbf{u}}{\partial t} + \boldsymbol{\nabla}\frac{u^2}{2} - \mathbf{u} \times (\boldsymbol{\nabla} \times \mathbf{u}) \tag{9-146}$$

and

$$\boldsymbol{\nabla}^2 \mathbf{u} = \boldsymbol{\nabla}(\boldsymbol{\nabla} \cdot \mathbf{u}) - \boldsymbol{\nabla} \times \boldsymbol{\nabla} \times \mathbf{u} \tag{9-147}$$

must be used to replace the pseudovector terms $(\mathbf{u} \cdot \boldsymbol{\nabla})\mathbf{u}$ and $\boldsymbol{\nabla}^2\mathbf{u}$.

We consider the Navier–Stokes equations as fundamental to the entire theory of fluid flow. As we can see, however, these equations cannot be solved except for special linear cases. The British engineer and physicist Osborne Reynolds, near the end of the nineteenth century, carried out a closely knit experimental and theoretical investigation of the scaling properties of these equations and derived a *law of similitude* for similar experiments.

Consider two experimental setups differing only in scale, say bell-mouth fluid intakes as shown in Fig. 9-8. We shall also permit the densities and

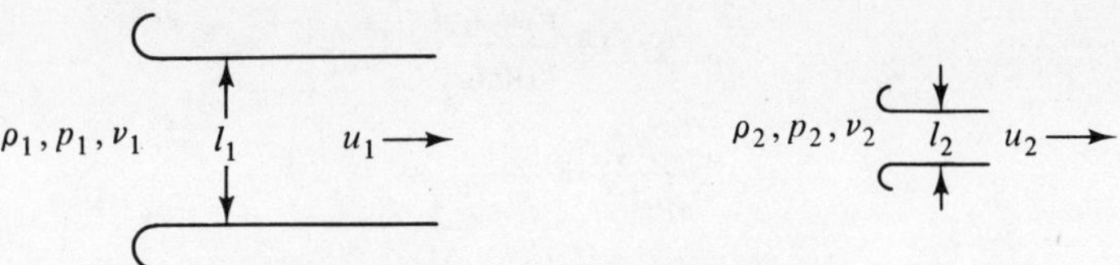

Fig. 9-8 Mechanically similar experiments.

viscosities of the two fluids to differ. We shall assign scaling factors relating the two experiments as follows:

$$l_2 = \alpha l_1, \qquad u_2 = \beta u_1, \qquad p_2 = \gamma p_1. \tag{9-148}$$

Since $u_2 = x_2/t_2 = \beta x_1/t_1$, then $x_2/x_1 = \beta t_2/t_1$ and the characteristic times scale as

$$t_2 = \frac{\alpha}{\beta} t_1. \tag{9-149}$$

The kinematic viscosity and pressure scalings are

$$\nu_2 = \delta \nu_1, \qquad p_2 = \epsilon p_1. \tag{9-150}$$

The Navier–Stokes equation in the form of Eq. (9-142), as applied to experiment 1, is

$$\rho_1 \frac{du_1}{dt_1} + \nabla_1 p_1 - \rho_1 \nu_1 \, \nabla_1^2 u_1 = f. \tag{9-151}$$

Assuming the force per unit volume to be the same in both cases, experiment 2 must be described by

$$\rho_2 \frac{du_2}{dt_2} + \nabla_2 p_2 - \rho_2 \nu_2 \, \nabla_2^2 u_2 = f \tag{9-152}$$

or using the scale factors defined in Eqs. (9-148) to (9-150)

$$\frac{\gamma \beta^2}{\alpha} \rho_1 \frac{du_1}{dt_1} + \frac{\epsilon}{\alpha} \nabla_1 p_1 - \frac{\gamma \, \delta \beta}{\alpha^2} \rho_1 \nu_1 \, \nabla_1^2 u_1 = f. \tag{9-153}$$

Comparing Eqs. (9-151) and (9-153), we see that

$$\frac{\gamma \beta^2}{\alpha} : \frac{\epsilon}{\alpha} : \frac{\gamma \, \delta \beta}{\alpha^2} = 1{:}1{:}1. \tag{9-154}$$

This amounts to the following two independent relations:

$$\frac{\epsilon}{\gamma \beta^2} = 1 \rightarrow \frac{p_2 \rho_1 u_1^2}{p_1 \rho_2 u_2^2} = 1 \tag{9-155}$$

and
$$\frac{\delta}{\alpha\beta} = 1 \rightarrow \frac{\nu_2 u_1 l_1}{\nu_1 u_2 l_2} = 1 \tag{9-156}$$

or
$$\frac{p_2}{\rho_2 u_2^2} = \frac{p_1}{\rho_1 u_1^2} \tag{9-157}$$

and
$$\frac{l_2 u_2}{\nu_2} = \frac{l_1 u_1}{\nu_1}. \tag{9-158}$$

Equations (9-157) and (9-158) are the results of Reynolds' theory of similitude. If these two criteria are satisfied, then flows 1 and 2 are either both laminar or both turbulent, since experiment 2 is simply a *mechanical image* of experiment 1 on a different scale.

The dimensionless number defined by Eq. (9-158) and previously encountered in Section 2-3 is known as the *Reynolds number* and is denoted by R:

$$R = \frac{lu}{\nu}, \tag{9-159}$$

where the length l may designate any linear dimension characteristic of the experiment. The number defined by Eq. (9-157) is denoted by S such that

$$S = \frac{p}{\rho u^2}. \tag{9-160}$$

The transition from laminar to turbulent flow, in particular, is a mechanically *similar* event for the inlet tubes 1 and 2 shown in Fig. 9-6. These transition points are characterized by the same values of R. The critical value of R is quite sensitive to the form of the entrance section. When a bell-shaped entrance is used as in the figure, laminar flow is established and the flow stays laminar up to a value $R \approx 20{,}000$ or so. If the fluid enters the tube via a straight, sharp-edged entrance, then the initial flow is disturbed by lateral components that are not damped out by friction at once. The transition to full turbulence then occurs at a value of R as low as 1200.

We know that viscosity tends to smooth out lateral components of the flow; thus it favors the laminar flow pattern. The inertia of the fluid tends to conserve the lateral components once they arise; thus it acts in favor of turbulence. Note the conflict between viscosity and inertia that is apparent in the definition of the kinematic viscosity $\nu = \mu/\rho$. An increase in μ permits us to increase lu without leaving the laminar range; an increase in ρ diminishes ν and favors turbulence.

The transition to turbulent flow manifests itself not only in the change in the flow pattern but also in a change in the pressure law. Here the number S comes into play. We replace p in Eq. (9-160) by the pressure drop Π between the entrance and exit of the tube. In so doing, the length of the tube L must also be considered. Although L cannot be related to the dimensionless number S defined in Eq. (9-160), we can use the ratio L/l to redefine S as

$$\Pi = \rho u^2 \frac{L}{l} S, \tag{9-161}$$

which is valid for similar tubes 1 and 2 at the same S value but for different values of Π, ρ, u, L, and l. Comparing Eq. (9-161) with Eq. (9-114) for Π found for Poiseuille flow, we have

$$\frac{8\mu Lu}{l^2} = \rho u^2 \frac{L}{l} S \tag{9-162}$$

or
$$S = \frac{8\mu}{\rho u l} = \frac{8}{R}. \tag{9-163}$$

In hydraulic engineering the formula used for the pressure head in a pipe of circular cross section is

$$h = \frac{\lambda u^2 L}{gl}, \tag{9-164}$$

where λ is a pure number. Since $\Pi = \rho g h$, we see that Eqs. (9-164) and (9-161) are consistent if we identify λ with S. Thus

$$\Pi_{\text{turb}} = \rho L S \frac{u^2}{l}. \tag{9-165}$$

If we identify S as $8/R$ as applies to Poiseuille flow, then we find

$$\Pi_{\text{lam}} = 8\mu L \frac{u}{l^2}. \tag{9-166}$$

A more accurate representation of the pressure drop observed in turbulent flow through a smooth pipe is possible within the framework of Eq. (9-161) if a weak dependence of S on R is assumed; that is, S is supposed neither independent of R as in Eq. (9-164) nor proportional to R^{-1} as in Eq. (9-163). The following assumptions have been tested for larger R values:

$$S = \lambda R^{-\kappa} \tag{9-167}$$

and
$$S = \lambda_0 + \lambda_1 R^{-\kappa} \tag{9-168}$$

with $0.2 < \kappa < 0.25$ in the upper formula and somewhat larger values of κ in the lower one.

We found that Poiseuille flow is characterized by a cylindrical parabolic distribution of velocity. If we consider Poiseuille flow between infinite flat plates, the distribution is still parabolic, as we showed for Hartmann flow in the $M = 0$ limit. If we cut along the symmetry plane and consider only the lower half of the flow pattern as shown in Fig. 9-7, we then have the velocity distribution of an infinitely wide river in laminar flow. Hopf found that the critical Reynolds number for such a stream is

$$R_{\text{crit}} = \frac{uh}{\nu} \approx 330. \qquad (9\text{-}169)$$

An even simpler case of laminar flow is the flow generated between two plates, one moving at constant speed U and the other at rest. This is called *Couette flow*, and its velocity profile is shown in Fig. 9-9. An elegant realization of this flow can be made by placing the fluid between two coaxial cylinders of radius $r_2 \gtrsim r_1$. The inner cylinder is kept at rest, and the outer one is rotated at a circumferential speed U. Sir G. I. Taylor has found, for this case, a critical Reynolds number

$$R_{\text{crit}} = \frac{Uh}{\nu} = 1900. \qquad (9\text{-}170)$$

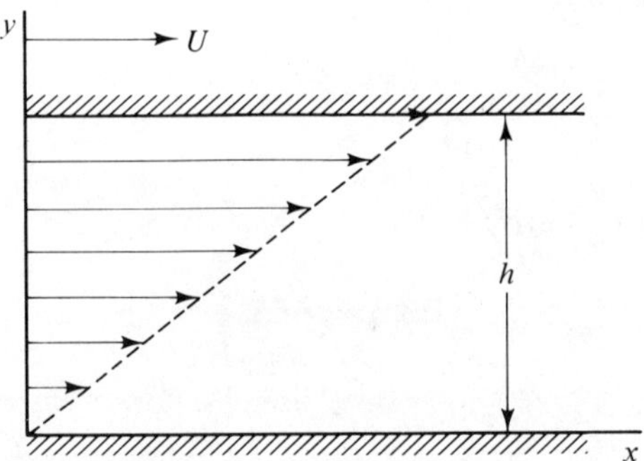

Fig. 9-9 Couette velocity profile.

Next, we wish to derive some dimensionless numbers for scaling magneto-hydrodynamic phenomena. The equation of motion for magnetofluids is the same as that for ordinary fluids except that the force term is always

$$\boldsymbol{J} \times \boldsymbol{B} = \frac{1}{\mu_0} (\boldsymbol{\nabla} \times \boldsymbol{B}) \times \boldsymbol{B}.$$

If we postulate the additional scale factor

$$B_2 = \eta B_1 \qquad (9\text{-}171)$$

then the comparison of the two MHD experiments leads to the relation

$$\frac{\gamma \beta^2}{\alpha} : \frac{\eta^2}{\alpha} = 1{:}1 \qquad (9\text{-}172)$$

or
$$\frac{B^2}{\mu_0 \rho u^2} = S_m, \qquad (9\text{-}173)$$

where the factor μ_0 has been included in order to make the number S_m non-dimensional. Since $B^2/2\mu_0$ plays the role of a pressure in MHD, we see that this number is an analog of the number S previously derived.

In order to complete our specification of the MHD problem, it is convenient to write a somewhat more sophisticated Ohm's law than we have used heretofore:

$$J = \sigma\left[E + u \times B + \frac{\mu}{\sigma}(\nabla p_e - J \times B)\right], \tag{9-174}$$

where

$$\mu = \frac{\sigma}{ne} \tag{9-175}$$

is called the *mobility* and the parenthetical term describes the electric field in a plasma arising from the *Hall effect*. This effect, which does not occur in a totally "liquidized" magnetofluid, occurs because the electrons are much lighter than the ions. When the plasma is subjected to a body force, the electrons tend to respond more quickly than the ions and a charge separation occurs, resulting in the Hall electric field. Taking the curl of Eq. (9-174) and using Eq. (1-86) and the Faraday induction law, we obtain

$$\frac{\partial B}{\partial t} = \nabla \times (u \times B) + \frac{1}{\sigma\mu_0}\nabla^2 B - \frac{\mu}{\sigma\mu_0}\nabla \times [(\nabla \times B) \times B]. \tag{9-176}$$

Defining the scaling factors

$$\sigma_2 = \zeta\sigma_1, \qquad \mu_2 = \theta\mu_1 \tag{9-177}$$

and using Eq. (9-176) to compare two similar experiments, we generate two new numbers

$$R_m = \sigma\mu_0 ul \tag{9-178}$$

known as the *magnetic Reynolds number* and

$$N = \frac{\mu B}{\sigma\mu_0 ul} = \frac{\mu B}{R_m}. \tag{9-179}$$

These numbers, R, S, R_m, S_m, and N, are all mutually independent. Other scaling numbers are in common use in MHD, such as the Hartmann number

$$M = Bl\sqrt{\frac{\sigma}{\rho\nu}} = \sqrt{S_m R R_m} \tag{9-180}$$

and

$$Q = \frac{\sigma B^2 l}{\rho u} = S_m R_m. \tag{9-181}$$

However, these numbers can always be constructed as some combination of the five basic ones.

Neglecting the Hall effect for the moment, we see that the magnetic Reynolds number is effectively the ratio of the orders of magnitude of the first and second terms on the right-hand side of Eq. (9-176). If $R_m \ll 1$, then the first term is negligible in comparison with the second and the equation reduces to

$$\frac{\partial \boldsymbol{B}}{\partial t} = \frac{1}{\sigma\mu_0} \nabla^2 \boldsymbol{B}. \tag{9-182}$$

This is a diffusion equation describing the decay of the magnetic field in a stationary conductor. Dimensional analysis gives the decay time as of order

$$\tau = \sigma\mu_0 l^2. \tag{9-183}$$

If $R_m \gg 1$, the first term is predominant and the equation reduces to

$$\frac{\partial \boldsymbol{B}}{\partial t} = \nabla \times (\boldsymbol{u} \times \boldsymbol{B}). \tag{9-184}$$

This equation implies that the magnetic field lines are "frozen" into the fluid and are convected along with it.

The number $N = \mu B / R_m$ measures the relative magnitudes of the first and third terms; in order for the Hall effect to be significant, this number must exceed unity.

EXERCISES

9-4.1. Referring to problem 9-3.3 and the criterion for the onset of turbulent flow in a wide river given in Eq. (9-169), calculate the stream bed inclination angle α corresponding to the critical Reynolds number.

9-4.2. For pipe flow, let the relation between the average pressure gradient Π/L and the average velocity u_m be

$$\frac{\Pi}{L} = Cu_m^n,$$

where the quantity C is a function of ρ, ν, and l. For laminar flow, $n = 1$; for the turbulent state, $n = 1.72$ to 1.75.

(a) Assume C proportional to a product of powers of ρ, ν, and l and determine the exponents by dimensional analysis.

(b) It should also be possible to represent the preceding relation by the dimensionless quantities R and S and the ratio L/l in the form

$$S = f\left(\frac{L}{l}, R\right).$$

If f is taken as a product of powers and the exponent of L/l is one, show that the result is

$$S = \lambda \frac{L}{l} R^\delta,$$

where λ is a dimensionless constant and δ is related directly to n.

(c) Show that the relation $S = \lambda R^{-1/4}$ is consistent with $n = 1.75$.

9-4.3. Couette flow in straight stream lines as shown in Fig. 9-9 is characterized by

$$\mathbf{u} = \left(\frac{yU}{h}, 0, 0 \right).$$

Show that the equilibrium conditions (9-98) to (9-101) are fulfilled if $p = $ const. Thus no driving pressure is required as for Poiseuille flow in a horizontal tube; however, the upper plate is kept in motion by a force $\mu U/h$.

In the case of cylindrical Couette flow between coaxial cylinders of radius b and a $(b > a)$, moments N_b and N_a are required to maintain the motion of the outer cylinder and keep the inner one at rest. Write the equations for the cylindrical flow. Integrate and compare the result with the linear Couette flow profile in the limiting case $a \to \infty$. Calculate the moments N_b and N_a.

9-4.4. Analyze the Couette flow shown in Fig. 9-9 for the case of an electrically conducting fluid in the presence of an imposed magnetic field $B_0 = (0, B_0, 0)$ in the y direction. Show that the velocity distribution is

$$u_x(y) = \frac{U}{\sinh M} \sinh \left(\frac{Mx}{h} \right),$$

where M is the Hartmann number. Show that this expression reduces to Ux/h in the limit $M \to 0$. Also show that the induced magnetic field is

$$b_z = \frac{\mu_0 \sqrt{\sigma \mu}\, U}{\sinh M} \left[1 - \cosh \left(\frac{Mx}{h} \right) \right].$$

9-5 Hydrodynamic and Hydromagnetic Waves

In this section we renew our acquaintance with waves from the point of view of the powerful theory of characteristics first introduced in Section 3-8. The reader would do well at this point to reread this introduction by way of review.

Let us first consider the one-dimensional time-independent flow of an ordinary gas. The appropriate equations are the equation of continuity, the

Euler equation of motion, and the polytropic relation; they can be written in one dimension as

$$\frac{\partial \rho}{\partial t} + \rho \frac{\partial u}{\partial x} + u \frac{\partial \rho}{\partial x} = 0,$$

$$\rho \left(\frac{\partial u}{\partial t} + u \frac{\partial u}{\partial x} \right) + a^2 \frac{\partial \rho}{\partial x} = 0,$$

$$p = A\rho^n, \qquad A = \text{const.}$$

(9-185)

These equations can easily be solved by linearizing (see problem 9-5.1); we shall, however, treat the nonlinear problem. The quantity a^2 appearing in the last term of the equation of motion is given as

$$\nabla p = \frac{\partial p}{\partial x} = \frac{dp}{d\rho} \frac{\partial \rho}{\partial x} + a^2 \frac{\partial \rho}{\partial x},$$

(9-186)

where a has the units of a velocity.

In order to facilitate the analysis of Eqs. (9-185), we must change from (x, t) as independent variables to φ, t', where the curve $\varphi = 0$ coincides with the wavefront (see Fig. 3-15). The result of this choice is that if we allow time to vary, holding φ constant, then a point can only move along the curve $\varphi = \text{const.}$ So the wavefront $\varphi = 0$ can never be *crossed* by varying time alone, and discontinuities in $\partial u/\partial t$, $\partial \rho/\partial t$ cannot occur as time varies. Thus we can set $t = t'$. The transformation of differential operators is, therefore,

$$\frac{\partial}{\partial t} = \frac{\partial \varphi}{\partial t} \frac{\partial}{\partial \varphi} + \frac{\partial t'}{\partial t} \frac{\partial}{\partial t'} = \frac{\partial \varphi}{\partial t} \frac{\partial}{\partial \varphi} + \frac{\partial}{\partial t},$$

$$\frac{\partial}{\partial x} = \frac{\partial \varphi}{\partial x} \frac{\partial}{\partial \varphi} + \frac{\partial t'}{\partial x} \frac{\partial}{\partial t'} = \frac{\partial \varphi}{\partial x} \frac{\partial}{\partial \varphi}.$$

(9-187)

In terms of this transformation of variables, the continuity and dynamic equations become

$$\frac{\partial \rho}{\partial t} + \frac{\partial \varphi}{\partial t} \frac{\partial \rho}{\partial \varphi} + \frac{\partial \varphi}{\partial x} \left(\rho \frac{\partial u}{\partial \varphi} + u \frac{\partial \rho}{\partial \varphi} \right) = 0,$$

$$\frac{\partial u}{\partial t} + \frac{\partial \varphi}{\partial t} \frac{\partial u}{\partial \varphi} + \frac{\partial \varphi}{\partial x} \left(u \frac{\partial u}{\partial \varphi} + \frac{a^2}{\rho} \frac{\partial \rho}{\partial \varphi} \right) = 0.$$

(9-188)

Along any of the curves $\varphi(x, t) = \text{const}$, the total derivative of φ is zero; therefore

$$d\varphi = \frac{\partial \varphi}{\partial x} dx + \frac{\partial \varphi}{\partial t} dt = 0$$

(9-189)

and
$$\lambda \equiv \frac{dx}{dt} = -\frac{\partial \varphi/\partial t}{\partial \varphi/\partial x}.$$

(9-190)

The quantity λ is the gradient of the $\varphi = $ const curves in the x, t plane. So, on $\varphi = 0$, λ corresponds to the propagation speed of the wavefront. At points behind the wavefront λ denotes the local propagation speeds of the continuous wave.

We shall denote the *jump* in a quantity X by $[X]$. If we let P_I and P_II be points arbitrarily close to $\varphi = 0$ on opposite sides of the wavefront, then

$$[X] = X(P_\mathrm{II}) - X(P_\mathrm{I}). \tag{9-191}$$

Since, by our choice of coordinates, no discontinuities occur in time derivatives, $[\partial \rho/\partial t] = [\partial u/\partial t] = 0$.

Next, let us apply the jump operator to Eqs. (9-188), bearing in mind the *weak wave* assumption that u and ρ must be continuous across the wavefront:

$$\frac{\partial \varphi}{\partial t}\left[\frac{\partial \rho}{\partial \varphi}\right] + \frac{\partial \varphi}{\partial x}\left(\rho\left[\frac{\partial u}{\partial \varphi}\right] + u\left[\frac{\partial \rho}{\partial \varphi}\right]\right) = 0,$$

$$\frac{\partial \varphi}{\partial t}\left[\frac{\partial u}{\partial \varphi}\right] + \frac{\partial \varphi}{\partial x}\left(u\left[\frac{\partial u}{\partial \varphi}\right] + \frac{a^2}{\rho}\left[\frac{\partial \rho}{\partial \varphi}\right]\right) = 0. \tag{9-192}$$

Dividing by $\partial \varphi/\partial x$ and using Eq. (9-190), we obtain

$$(u - \lambda)\left[\frac{\partial \rho}{\partial \varphi}\right] + \rho\left[\frac{\partial u}{\partial \varphi}\right] = 0,$$

$$\frac{a^2}{\rho}\left[\frac{\partial \rho}{\partial \varphi}\right] + (u - \lambda)\left[\frac{\partial u}{\partial \varphi}\right] = 0. \tag{9-193}$$

These equations can only have a nontrivial solution (that is, one in which the jumps are nonzero) when the determinant of their coefficients vanishes:

$$\begin{vmatrix} u - \lambda & \rho \\ \dfrac{a^2}{\rho} & u - \lambda \end{vmatrix} = 0. \tag{9-194}$$

This is called the *characteristic determinant*. Its vanishing implies

$$(u - \lambda)^2 - a^2 = 0, \tag{9-195}$$

which has as its solutions

$$\lambda = u \pm a. \tag{9-196}$$

Hence the wavefront and the continuous disturbance behind it propagate at fluid speed $\pm$ local sound speed.[2]

---

[2] The quantity $\sqrt{dp/d\rho} = \sqrt{np/\rho} \approx 331$ m/sec in air under standard conditions at sea level.

Since the $\varphi = 0$ curve advances into a constant state and since u and a are continuous across the wavefront, $\lambda = $ const and the wavefront trace, $\varphi = 0$, must be a *straight* line with slope equal to its propagation speed. For waves propagating into a nonconstant state, the slope of the wavefront trace will be equal to the local value of λ and the trace will not, in general, be straight.

The curves $\varphi = $ const are determined by combining Eq. (9-190) and (9-196) to yield the *characteristic equations*

$$\frac{dx}{dt} = u + a, \qquad \frac{dx}{dt} = u - a. \tag{9-197}$$

Before proceeding from the simple case of *acoustic waves* to the more complex waves of magnetohydrodynamics, let us note that either of the difference Eqs. (9-193) can be used to determine the relationship that must exist between $[\partial\rho/\partial\varphi]$ and $[\partial u/\partial\varphi]$ in one-dimensional flow when use is made of the permissible values of λ given in Eq. (9-196). We find

$$-a\left[\frac{\partial\rho}{\partial\varphi}\right] + \rho\left[\frac{\partial u}{\partial\varphi}\right] = 0,$$

$$a\left[\frac{\partial\rho}{\partial\varphi}\right] + \rho\left[\frac{\partial u}{\partial\varphi}\right] = 0, \tag{9-198}$$

corresponding to the two characteristics of the acoustic wave.

If the wave is propagating into a constant state, these equations have a convenient interpretation in terms of the infinitesimal increments experienced by ρ and u in the neighborhood of the wavefront $\varphi = 0$. In this case, all derivatives with respect to φ ahead of the wavefront vanish, and so the jump experienced by a quantity $(\partial/\partial\varphi)X(\varphi, t)$ when crossing $\varphi = 0$ reduces to $\partial X/\partial\varphi$ evaluated immediately behind the wavefront. If we approximate $\partial X/\partial\varphi$ behind the wavefront $\varphi = 0$ at time t_0 by the difference equation

$$\frac{\partial X}{\partial\varphi} = \frac{X(0, t_0) - X(\delta\varphi_0, t_0)}{\delta\varphi_0}, \tag{9-199}$$

where $\varphi = \delta\varphi_0$ is a characteristic curve close behind the wavefront, then we write, approximately,

$$\frac{\partial X}{\partial\varphi} = \frac{\delta X}{\delta\varphi_0}, \tag{9-200}$$

where $$\delta X = X(0, t_0) - X(\delta\varphi_0, t_0). \tag{9-201}$$

This result can be applied to Eqs. (9-198), which then assume the alternate form

$$-a\,\delta\rho + \rho\,\delta u = 0,$$

$$a\,\delta\rho + \rho\,\delta u = 0. \tag{9-202}$$

This notation has the advantage that when the wave propagates into a constant state, the differences $\delta\rho$ and δu in Eqs. (9-202) can be replaced by the differentials $d\rho$ and du, which are, of course, infinitesimals. For wave propagation into a nonconstant state, the δ notation must be regarded as equivalent to the earlier notation for the jump in the normal derivative of the variable in question.

Let us turn next to the equations of magnetohydrodynamics and examine them for wave solutions

$$\frac{\partial\rho}{\partial t} + \nabla\cdot(\rho u) = 0,$$

$$\frac{\partial u}{\partial t} + (u\cdot\nabla)u - \frac{1}{\rho\mu_0}(\nabla\times B)\times B + \frac{a^2}{\rho}\nabla\rho = 0, \tag{9-203}$$

$$\frac{\partial B}{\partial t} - \nabla\times(u\times B) = 0,$$

$$\nabla\cdot B = 0,$$

where use has been made of Eq. (9-186). We can avoid considerable algebraic complication if we choose the x axis normal to the wavefront at some given point P. In the neighborhood of this point, the problem is locally one-dimensional and Eqs. (9-203) can be simplified by writing them in (x, t) space. We find

$$\frac{\partial\rho}{\partial t} + \frac{\partial}{\partial x}(\rho u_x) = 0,$$

$$\frac{\partial u}{\partial t} + u_x\frac{\partial u}{\partial x} - \frac{1}{\rho\mu_0}\left(\hat{n}\times\frac{\partial B}{\partial x}\right)\times B + \frac{a^2}{\rho}\frac{\partial\rho}{\partial x}\hat{n} = 0,$$

$$\frac{\partial B}{\partial t} - \hat{n}\times\frac{\partial}{\partial x}(u\times B) = 0, \tag{9-204}$$

$$\frac{\partial B_x}{\partial x} = 0,$$

where $\hat{n}$ is the unit normal to the wavefront directed along the positive x axis. In component form, we have

$$\frac{\partial\rho}{\partial t} + u_x\frac{\partial\rho}{\partial x} + \rho\frac{\partial u_x}{\partial x} = 0, \tag{9-205}$$

$$\frac{\partial u_x}{\partial t} + u_x\frac{\partial u_x}{\partial x} + \frac{1}{\rho\mu_0}B_y\frac{\partial B_y}{\partial x} + \frac{1}{\rho\mu_0}B_z\frac{\partial B_z}{\partial x} + \frac{a^2}{\rho}\frac{\partial\rho}{\partial x} = 0, \tag{9-206}$$

$$\frac{\partial u_y}{\partial t} + u_x\frac{\partial u_y}{\partial x} - \frac{1}{\rho\mu_0}B_x\frac{\partial B_y}{\partial x} = 0, \tag{9-207}$$

$$\frac{\partial u_z}{\partial t} + u_x \frac{\partial u_z}{\partial x} - \frac{1}{\rho\mu_0} B_x \frac{\partial B_z}{\partial x} = 0, \tag{9-208}$$

$$\frac{\partial B_x}{\partial t} = 0, \tag{9-209}$$

$$\frac{\partial B_y}{\partial t} + \frac{\partial}{\partial x}(B_y u_x - B_x u_y) = 0, \tag{9-210}$$

$$\frac{\partial B_z}{\partial t} + \frac{\partial}{\partial x}(B_z u_x - B_x u_z) = 0, \tag{9-211}$$

$$\frac{\partial B_x}{\partial x} = 0. \tag{9-212}$$

We see from Eqs. (9-209) and (9-212) that, in the neighborhood of the point P of the wavefront, $B_x = B_{0x} = \text{const}$. Since we shall look only for weak wave solutions, ρ, $\boldsymbol{u}$, and $\boldsymbol{B}$ are constant across the entire wavefront. This result must also be true for $B_x = B_{0x}$ in the case of weak waves. So, using the delta notation, we can write

$$\delta B_x = 0. \tag{9-213}$$

The remaining six equations can be further simplified by choosing the z axis in such a way that B_z behind the wavefront vanishes. When B_z is set equal to zero in the equations, the independent variables are changed from (x, t) to (φ, t), and the resulting equations are differenced across the wavefront, the following set of MHD characteristic equations is obtained:

$$\mp c_n\, \delta\rho + \rho\, \delta u_x = 0, \tag{9-214}$$

$$\mp c_n\rho\, \delta u_x + a^2\, \delta\rho + \frac{1}{\mu_0} B_y\, \delta B_y = 0, \tag{9-215}$$

$$\mp c_n\rho\, \delta u_y - \frac{1}{\mu_0} B_x\, \delta B_y = 0, \tag{9-216}$$

$$\mp c_n\rho\, \delta u_z - \frac{1}{\mu_0} B_x\, \delta B_z = 0, \tag{9-217}$$

$$\mp c_n\, \delta B_y + B_y\, \delta u_x - B_x\, \delta u_y = 0, \tag{9-218}$$

$$\mp c_n\, \delta B_z - B_x\, \delta u_z = 0, \tag{9-219}$$

where, if Eq. (9-190) is used as the definition of λ,

$$c_n = |\lambda - u_x|, \tag{9-220}$$

defines the speed of the wavefront relative to the fluid speed. The minus and plus signs associated with c_n correspond, respectively, to the negative and positive values of $u_x - \lambda$. If the terms are arranged in the order $\delta\rho$, δu_x, δu_y, δu_z, δB_y, δB_z, then the characteristic determinant of their coefficients becomes

$$\begin{vmatrix} \mp c_n & \rho & 0 & 0 & 0 & 0 \\ \dfrac{a^2}{\rho} & \mp c_n & 0 & 0 & \dfrac{B_y}{\rho\mu_0} & 0 \\ 0 & 0 & \mp c_n & 0 & -\dfrac{B_x}{\rho\mu_0} & 0 \\ 0 & 0 & 0 & \mp c_n & 0 & -\dfrac{B_x}{\rho\mu_0} \\ 0 & B_y & -B_x & 0 & \mp c_n & 0 \\ 0 & 0 & 0 & -B_x & 0 & \mp c_n \end{vmatrix} = 0. \tag{9-221}$$

It is convenient here to regard c_n rather than λ as the fundamental parameter. If we define the *Alfvén speed b* as

$$b = \sqrt{\dfrac{B^2}{\rho\mu_0}}, \tag{9-222}$$

and the Alfvén speed in the x direction b_x as

$$b_x = \sqrt{\dfrac{B_{0x}^2}{\rho\mu_0}}, \tag{9-223}$$

then the expansion of the characteristic determinant becomes expressible as

$$(c_n^2 - b_x^2)[(c_n^2 - a^2)(c_n^2 - b_x^2) - c_n^2(b^2 - b_x^2)] = 0. \tag{9-224}$$

From this important equation we shall obtain a great deal of information on waves occurring in an MHD fluid.

The first root of Eq. (9-224) results from setting the expression in parentheses (as opposed to the square bracket term) equal to zero:

$$c_n = b_x. \tag{9-225}$$

Expanding the quantity in square brackets

$$c_n^4 - c_n^2 b_x^2 - c_n^2 a^2 + a^2 b_x^2 - c_n^2 b^2 + c_n^2 b_x^2 = 0$$

or

$$c_n^4 - c_n^2(a^2 + b^2) + a^2 b_x^2 = 0,$$

we find two roots

$$c_n = c_f = (\tfrac{1}{2}\{(a^2 + b^2) + [(a^2 + b^2)^2 - 4a^2 b_x^2]^{1/2}\})^{1/2} \tag{9-226}$$

and

$$c_n = c_s = (\tfrac{1}{2}\{(a^2 + b^2) - [(a^2 + b^2)^2 - 4a^2 b_x^2]^{1/2}\})^{1/2}. \tag{9-227}$$

Using the obvious inequality

$$(a^2 + b^2)^2 - 4a^2 b_x^2 > (a^2 - b^2)^2 > 0, \tag{9-228}$$

we see that the roots c_f and c_s of Eq. (9-224) are always real, as are the λ given by

$$\lambda = u_x \pm c_f, \tag{9-229}$$

$$\lambda = u_x \pm c_s, \tag{9-230}$$

$$\lambda = u_x \pm b_x. \tag{9-231}$$

The $+$ and $-$ signs in Eqs. (9-229) to (9-230) correspond, respectively, to the $-$ and $+$ signs of c_n in the characteristic equations. It is obvious that c_f as defined in Eq. (9-226) is greater than c_s. For this reason, the wave corresponding to Eq. (9-229) is called the *fast wave*, and the wave described by Eq. (9-230) is called the *slow wave*. For reasons to be given below, Eq. (9-231) is said to describe the *transverse wave*.

We shall first consider the fast and slow waves, known collectively as *magnetoacoustic* waves. We assume that $c_n \neq 0$ and $c_n \neq b_x$; by Eq. (9-224), we are then left with

$$(c_n^2 - a^2)(c_n^2 - b_x^2) = c_n^2(b^2 - b_x^2). \tag{9-232}$$

Here the right-hand side is positive definite, and so we must have either

$$c_n^2 \geq a^2 \quad \text{and} \quad c_n^2 \geq b_x^2 \tag{9-233}$$

or
$$c_n^2 \leq a^2 \quad \text{and} \quad c_n^2 \leq b_x^2. \tag{9-234}$$

Since $c_f > c_s$, conditions (9-233) are equivalent to

$$c_f \geq a \quad \text{and} \quad c_f \geq b_x, \tag{9-235}$$

whereas conditions (9-234) are equivalent to

$$c_s \leq a \quad \text{and} \quad c_s \leq b_x. \tag{9-236}$$

If $b = b_x$, for which the transverse magnetic field vanishes, then $c_s \to a$, the local speed of sound, and $c_f \to b_x$, the Alfvén speed.

Using Eqs. (9-217) and (9-219), we eliminate δu_z between them and obtain

$$\mp c_n \rho \left(\mp c_n \frac{\delta B_z}{B_x} \right) - \frac{1}{\mu_0} B_x \, \delta B_z = 0,$$

or
$$(c_n^2 - b_x^2) \, \delta B_z = 0. \tag{9-237}$$

Since we have assumed, for magnetoacoustic waves, that $(c_n^2 - \cdots - b_x^2) \neq 0$, it follows that $\delta B_z = 0$ and hence, from Eqs. (9-217) and (9-219), that $\delta u_z = 0$. The value of the other jumps, $\delta \rho$, δu_x, δu_y, and δB_y, can be expressed in terms of a dimensionless parameter characterizing the jump of one of these quantities. If we express $\delta \rho$ as

$$\delta \rho = \epsilon \rho, \tag{9-238}$$

where ϵ is a dimensionless parameter that must be assigned different values for fast and slow waves, then other jumps are

$$\delta u_x = \pm \epsilon c_n, \tag{9-239}$$

$$\delta u_y = \frac{\mp \epsilon c_n b_x b_y}{(c_n^2 - b_x^2)} \, \text{sgn} \, (B_x B_y), \tag{9-240}$$

$$\delta B_y = \frac{\epsilon c_n^2 B_y}{(c_n^2 - b_x^2)}, \tag{9-241}$$

where $b_y = \sqrt{B_y^2/\rho\mu_0}$ and $\text{sgn} \, (B_x B_y) \equiv B_x B_y/|B_x B_y|$ is a unit multiplier bearing the sign of its argument. In Eqs. (9-239) to (9-241) c_n can be either c_s or c_f.

Let us look into the jump properties of the transverse wave. As we have seen, these waves are characterized by $c_n = b_x$. We shall assume that c_f and c_s are both distinct from c_n. Setting $c_n = b_x$ in the characteristic equations leads easily to the results

$$\delta u_x = \delta u_y = \delta B_y = \delta\rho = 0, \tag{9-242}$$

and

$$\delta u_z = \mp \frac{\text{sgn} \, (B_x)}{\sqrt{\rho\mu_0}} \, \delta B_z. \tag{9-243}$$

Since the vector $\delta \boldsymbol{B}$ has δB_z as its only nonzero component, it is directed transversely with respect to the wavefront normal $\hat{\boldsymbol{n}}$. So it can be expressed in terms of a dimensionless parameter ϵ as

$$\delta \boldsymbol{B} = \epsilon \hat{\boldsymbol{n}} \times \boldsymbol{B}. \tag{9-244}$$

The quantity $\delta \boldsymbol{u}$, which has δu_z as its only nonzero component, can be written

$$\delta \boldsymbol{u} = \mp \epsilon \, \frac{\text{sgn} \, (B_x)}{\sqrt{\rho\mu_0}} \, \hat{\boldsymbol{n}} \times \boldsymbol{B}. \tag{9-245}$$

Thus changes across the wavefront occur only in the transverse components of $\delta \boldsymbol{B}$ and $\delta \boldsymbol{u}$; this fact justifies our use of the term *transverse wave*.

Since $\delta p = a^2 \, \delta\rho$ and $\delta\rho = 0$ for transverse waves, we see that these waves are noncompressive. Additional information is revealed by considering the jump in B^2:

$$\delta(B^2) = 2\boldsymbol{B} \cdot \delta \boldsymbol{B}. \tag{9-246}$$

Since $\delta \boldsymbol{B} = \epsilon \hat{\boldsymbol{n}} \times \boldsymbol{B}$ and $2\epsilon \boldsymbol{B} \cdot \boldsymbol{n} \times \boldsymbol{B} = 0$, then $\delta(B^2) = 0$ and the total pressure $p + B^2/2\mu_0$ must also have zero jump value.

For a more physical look at the hydromagnetic waves, we shall consider a linearized solution of the MHD equations. Imagine an unbounded, incompressible conductive fluid in which is embedded a homogeneous magnetic field $\boldsymbol{B}_0$. We perturb the static equilibrium by causing a column of fluid to move in the y direction with velocity $\boldsymbol{u}$. This perturbation, as shown in Fig. 9-10, gives rise to an induced electric field, $\boldsymbol{u} \times \boldsymbol{B}_0$. This electric field drives eddy currents that interact with the magnetic field to impede the motion of the column and accelerate adjacent fluid. In this way, motion is transferred and a wave arises. This type of wave is called an *Alfvén wave* after H. Alfvén, who postulated their existence in 1942.

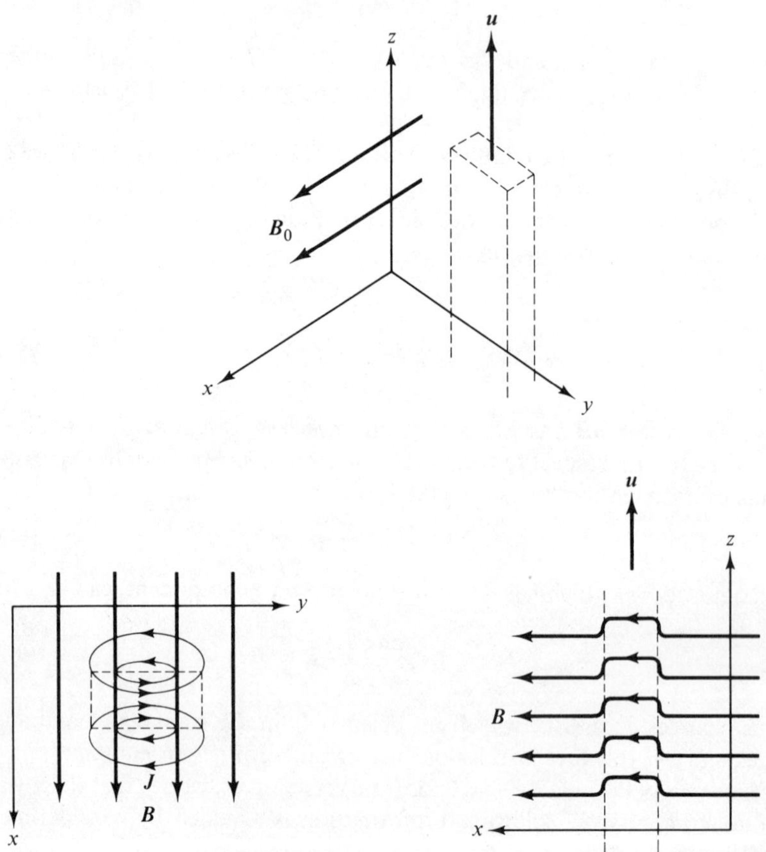

Fig. 9-10 Excitation of Alfvén waves.

We must first linearize the equations by assuming that the wave constitutes only a *small* perturbation to the static equilibrium. The net magnetic field is written as the sum of the imposed homogeneous field and a small induced field

$$\boldsymbol{B} = \boldsymbol{B}_0\boldsymbol{i} + \boldsymbol{b}, \tag{9-247}$$

where $|b| \ll |B|$. The velocity u is also considered to be a small quantity. The linearized equations can be written

$$\frac{\partial b}{\partial t} = \nabla \times (u \times B_0), \tag{9-248}$$

$$\frac{\partial u}{\partial t} = -\nabla\left(\frac{p}{\rho} + \frac{B^2}{2\rho\mu_0}\right) + \frac{1}{\rho\mu_0}(B \cdot \nabla)b, \tag{9-249}$$

$$\nabla \cdot u = 0, \qquad \nabla \cdot B = 0. \tag{9-250}$$

Using identity (1-95), Eqs. (9-250), and

$$\frac{\partial}{\partial y} = \frac{\partial}{\partial z} = 0, \qquad B_0 = (B_0, 0, 0), \tag{9-251}$$

Eq. (9-248) reduces to

$$\frac{\partial b}{\partial t} = B_0 \frac{\partial u}{\partial x}, \tag{9-252}$$

and Eqs. (9-249) and (9-250) become

$$\frac{\partial u}{\partial t} = \frac{B_0}{\rho\mu_0} \frac{\partial b}{\partial x} - \frac{\partial}{\partial x}\left(\frac{p}{\rho} + \frac{B^2}{2\rho\mu_0}\right)i \tag{9-253}$$

and

$$\frac{\partial u_x}{\partial x} = \frac{\partial b_x}{\partial x} = 0. \tag{9-254}$$

Thus u_x and b_x are both constant in space and we can avoid trivial (nonwave) solutions by considering both constants to be zero. Since the first two terms in Eq. (9-253) have no x components,

$$\frac{\partial}{\partial x}\left(\frac{p}{\rho} + \frac{B^2}{2\rho\mu_0}\right) = 0, \tag{9-255}$$

and Eq. (9-253) reduces to

$$\frac{\partial u}{\partial t} = \frac{B_0}{\rho\mu_0} \frac{\partial b}{\partial x}. \tag{9-256}$$

Equations (9-252) and (9-256) can be combined to produce wave equations in b and u:

$$\frac{\partial^2 b}{\partial t^2} = \frac{B_0^2}{\rho\mu_0} \frac{\partial^2 b}{\partial x^2}, \tag{9-257}$$

$$\frac{\partial^2 u}{\partial t^2} = \frac{B_0^2}{\rho\mu_0} \frac{\partial^2 u}{\partial x^2}. \tag{9-258}$$

These equations represent plane waves traveling in the x direction with phase velocity

$$c_n = \pm \frac{B_0}{\sqrt{\rho\mu_0}}. \qquad (9\text{-}259)$$

They are transverse waves, since b and u are free of x components. These Alfvén waves are a restricted form of the general transverse wave that we derived with the use of the theory of characteristics. We note that the propagation velocity in the linear limit is the Alfvén speed, just as we found by using the original nonlinear equations.

The most physical way of looking at Alfvén waves is to make an analogy with wave motion along a massive string under tension. The phase velocity of such waves was found to be (see Section 3-7)

$$c_n = \pm \sqrt{\frac{\tau}{\sigma}}, \qquad (9\text{-}260)$$

where τ is the tension in the string and σ is the mass per unit length. The magnetic force

$$f = J \times B = \frac{1}{\mu_0} (\nabla \times B) \times B = \frac{1}{\mu_0} (B \cdot \nabla)B - \nabla \frac{B^2}{2\mu_0} \qquad (9\text{-}261)$$

can also be expressed as the divergence of a magnetic stress tensor T such that

$$f_i = -\frac{\partial T_{ij}}{\partial x_j} = \frac{\partial}{\partial x_j} \left(\frac{B_i B_j}{\mu_0} - \frac{B^2}{2\mu_0}\, \delta_{ij} \right). \qquad (9\text{-}262)$$

This amounts to a scalar pressure $B^2/2\mu_0$ transverse to the magnetic field lines and a tension B/μ_0 per field line. The fluid mass per unit length of a field line is ρ/B. So if we view a magnetic field line as an elastic string with tension $\tau = B/\mu_0$ and mass per unit length $\sigma = \rho/B$, then, by Eq. (9-260), the phase velocity is

$$c_n = \pm \sqrt{\frac{B}{\mu_0} \cdot \frac{B}{\rho}} = \frac{B}{\sqrt{\rho\mu_0}}, \qquad (9\text{-}263)$$

the Alfvén speed.

In order to include compressibility in our linearized wave analysis, we replace the relations $\nabla \cdot u = 0$ and $\rho = \text{const}$ by the linearized continuity and polytropic relations

$$\frac{\partial \rho}{\partial t} + \rho_0\, \nabla \cdot u = 0 \qquad (9\text{-}264)$$

and

$$\frac{\partial \rho}{\partial t} = \frac{\rho_0}{p_0 n} \frac{\partial p}{\partial t}, \qquad (9\text{-}265)$$

where $\rho - \rho_0 \ll \rho$ and n is the polytropic index. Combining these equations, we obtain

$$\nabla \cdot \boldsymbol{u} + \frac{1}{\rho_0 n}\frac{\partial p}{\partial t} = 0. \tag{9-266}$$

Equations (9-248), (9-249), (9-266), and $\nabla \cdot \boldsymbol{B} = 0$ constitute the linearized equations for a compressible fluid. If we again look for plane waves traveling along $\boldsymbol{B}_0$, we find the transverse Alfvén waves as before, as well as longitudinal waves unaffected by the magnetic field. These longitudinal waves are simply ordinary sound waves. We also find waves propagating normal to the field. Such waves are longitudinal in $\boldsymbol{u}$ and transverse in $\boldsymbol{b}$; they are the magneto-acoustic waves revealed earlier by the theory of characteristics. In this linearized analysis, however, there are no fast and slow waves as described by Eqs. (9-226) and (9-227); only the phase velocity

$$c_n = \left(\frac{np_0 + B^2/\mu_0}{\rho}\right)^{1/2} \tag{9-267}$$

is found.

The inclusion of finite but large electrical conductivity leads to damped waves. As smaller conductivities are contemplated, the solutions become aperiodic and the fluid mechanics finally decouples altogether from the electromagnetics as $\sigma \to 0$.

For large perturbation energies, the wave profiles of weak waves can steepen until an infinite gradient develops. The processes involved are revealed by taking into account heat conduction and viscosity effects that become more and more pronounced as the wavefront steepens. The equilibrium state involves a balance between the effects of nonlinearity and dissipation. In waves of this type, the physical quantities, as well as their derivatives, are discontinuous across the wavefront. Such waves are called *shock waves*. The analysis of shock phenomena involves recasting the fluid or hydromagnetic equations into the form of conservation relations of the form

$$\frac{\partial U}{\partial t} + \nabla \cdot \boldsymbol{F} = G, \tag{9-268}$$

from which a set of equations known as the *Rankine–Hugoniot* relations can be formulated in terms of the jumps. Space limitations preclude our dealing further with shocks. The interested reader is referred to other textbooks.[3]

---

[3] See, for instance, L. D. Landau and E. M. Lifshitz, *Fluid Mechanics*, Reading, Mass.: Pergamon Press, 1959; and Alan Jeffrey, *Magnetohydrodynamics*, London: Oliver & Boyd, 1966.

EXERCISES

9-5.1. Linearize the first two of Eqs. (9-185) by assuming that deviations of the density from the equilibrium value ρ_0 are small. Thus $\partial(\rho u)/\partial x \approx \rho_0 \, \partial u/\partial x$ and $\rho \, \partial u/\partial t \approx \rho_0 \, \partial u/\partial t$. Neglecting the nonlinear term in u, show that the resulting equations can be combined to yield a wave equation.

9-5.2. Show that in a polytropic gas

$$\frac{dp}{d\rho} = \frac{np}{\rho} = a^2.$$

For experimental values of the sound speed, density, and pressure of

$$a = 331 \text{ m/sec}$$

$$\rho = 1.29 \text{ kg/m}^3$$

$$p = 1.01 \cdot 10^5 \text{ nt/m}^2$$

for air under standard conditions, calculate the value of the polytropic index, n. Show that your answer is compatible with the formula

$$n = \frac{f + 2}{f},$$

where f is the number of degrees of freedom of the molecule.

9-5.3. Verify the expansion of Eq. (9-221) as given by Eq. (9-224).

9-5.4. Show that the pressure changes across the wavefront of a magneto-acoustic wave is given by

$$\delta p = \epsilon a^2 \rho.$$

9-5.5. Show that Eqs. (9-225) to (9-227) can be rewritten in the form

$$\frac{b_x}{b} = \cos \theta, \tag{9-269}$$

$$\frac{c_f}{b} = (\tfrac{1}{2}\{(1 + s) + [(1 + s)^2 - 4s \cos^2 \theta]^{1/2}\})^{1/2}, \tag{9-270}$$

$$\frac{c_s}{b} = (\tfrac{1}{2}\{(1 + s) - [(1 + s)^2 - 4s \cos^2 \theta]^{1/2}\})^{1/2}, \tag{9-271}$$

where $s = a^2/b^2$ and θ is the angle between the wavefront normal and the magnetic field vector $\mathbf{B}$. By regarding b_x/b, c_f/b, and c_s/b as radial coordinates and θ as the azimuthal angle with respect to $\mathbf{B} = (B, 0, 0)$, construct a polar plot of the normal velocity of the transverse, fast, and slow waves. How does the value of s affect the nature of these curves?

9-5.6. Consider the Maxwell equations

$$\nabla \times B = \frac{1}{c^2}\frac{\partial E}{\partial t}$$

and
$$\nabla \times E = -\frac{\partial B}{\partial t},$$

appropriate for a vacuum. Change from the coordinates x, y, z, t to ϕ, y, z, t, where $\phi = 0$ is the wavefront trace. Show that the characteristic equations can be written

$$c^2 \hat{n} \times \delta B + \lambda\, \delta E = 0,$$

$$\hat{n} \times \delta E - \lambda\, \delta B = 0,$$

where $\hat{n} = \nabla\phi/|\nabla\phi|$ is the unit normal to $\phi = 0$ and $\lambda = -(\partial\phi/\partial t)/|\nabla\phi|$ is the normal velocity of the wavefront. Show that δE, δB, and $\hat{n}$ are mutually orthogonal, and so electromagnetic waves are transverse waves. Show that the characteristic roots are

$$\lambda = \pm c.$$

9-5.7. Write the linearized equations for a hydromagnetic plane wave traveling along B_0 in a polytropic fluid. Show that the component of these equations parallel to B_0 yields an acoustical wave and that the component perpendicular to B_0 yields a transverse (Alfvén) wave.

9-5.8. Write and solve the linearized equations for a magnetoacoustic wave propagating at right angles to B_0. Find the polarization of the vectors u and b and show that the phase velocity of the wave is $c_n = [(np_0 + B^2/\mu_0)/\rho_0]^{1/2}$.

9-5.9. Reconsider the equations for a plane Alfvén wave by including finite conductivity. Show that solutions of the form $\exp(i\omega t + \alpha x)$ satisfy the linearized equations. Determine the damping constant α.

9-6 Vorticity

Vorticity in real fluids is a common phenomenon. Isolated vortices are formed when a basin of water is drained or a cup of tea is stirred; however, vortices are formed in nature more often by an *instability* that occurs in flows with large transverse velocity gradient (shear). Perhaps the most common occurrence of vortices is in turbulent flow wherein an entire spectrum of vortices takes place. Some of the problems involved in describing turbulence are examined in Section 9-7.

The first comprehensive mathematical description of vorticity was given by Helmholtz in 1858. The most important result of Helmholtz's paper can be expressed as follows.

> *Vortices can neither be created nor destroyed in an inviscid, incompressible fluid, provided that the external forces acting on the fluid are conservative.*

In order to establish this theorem, we write Euler's equation in a coordinate invariant form

$$\frac{\partial u}{\partial t} - u \times (\nabla \times u) = -\nabla\left(\frac{u^2}{2} + \frac{p + \phi}{\rho}\right), \qquad (9\text{-}272)$$

where ϕ is the scalar potential corresponding to the conservative external force density f. We now take the curl of this equation, writing ω for $\frac{1}{2}\nabla \times u$ and recalling that the curl of any gradient vanishes:

$$\frac{\partial \omega}{\partial t} - \nabla \times (u \times \omega) = 0. \qquad (9\text{-}273)$$

The second term of this equation can be recast by using identity (1-95):

$$\nabla \times (u \times \omega) = u(\nabla \cdot \omega) - \omega(\nabla \cdot u) + (\omega \cdot \nabla)u - (u \cdot \nabla)\omega. \quad (9\text{-}274)$$

In the present case, this relation is simplified because of our assumption of incompressibility ($\nabla \cdot u = 0$) and the fact that $\nabla \cdot \omega = 0$ by virtue of identity (1-99). Thus Eq. (9-273) becomes

$$\frac{\partial \omega}{\partial t} + (u \cdot \nabla)\omega = \frac{d\omega}{dt} = (\omega \cdot \nabla)u. \qquad (9\text{-}275)$$

From it we see that the rate of change of ω is zero if ω itself is zero. In other words, if at any time a fluid is vortex-free, it will remain so forever. The motion of an inviscid fluid cannot generate vorticity.

In order to make a rigorous statement, we must develop an important vector theorem. Let A be some vector associated with the moving fluid (a "particle" function as opposed to a "point" function) and let $d\sigma$ be a surface element that moves with the fluid, varying in size, shape, and position in the course of time. We wish to consider the flux of A through $d\sigma$ at successive times t_1 and t_2.

The development of $d\sigma$ is shown in Fig. 9-11. The line elements ds_1 and ds_2 form the boundaries of $d\sigma_1$ and $d\sigma_2$ and are correlated by the displacement $\delta = u\, dt$. The volume swept out by the motion of $d\sigma$ in time dt is $d\tau = \delta \cdot d\sigma$. The definition of the divergence operator or Gauss's theorem enables us to write

$$\oint A \cdot d\sigma = \int \nabla \cdot A \, d\tau \approx \nabla \cdot A \, d\tau = (\nabla \cdot A)\, \delta \cdot d\sigma. \qquad (9\text{-}276)$$

The integral on the left refers to the *closed* surface bounding the volume swept out by $d\sigma$ in time dt and so must include flux through the "side wall" as well

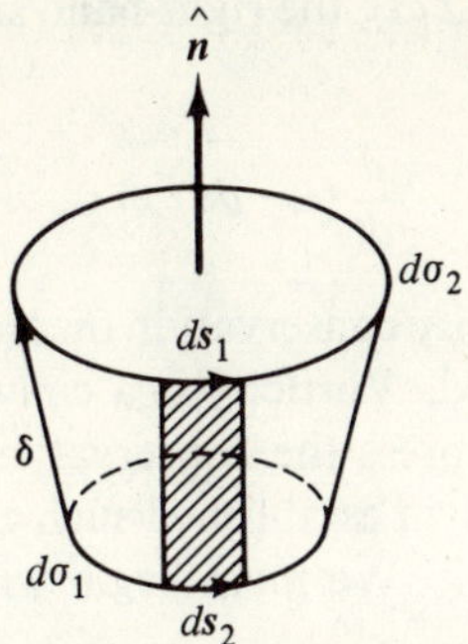

Fig. 9-11 Development of a surface element $d\sigma$ that moves with the fluid.

as the ends. The change in vector A over time dt must also be considered; it can, however, be dealt with separately. For the moment we shall consider A as "frozen" at the instant t_1.

The crosshatched portion of the side wall of $d\tau$ has an area $ds \times \delta$; hence its contribution to $A \cdot d\sigma$ is

$$A \cdot (ds \times \delta) = (\delta \times A) \cdot ds. \tag{9-277}$$

The integral over the side wall can therefore be written

$$\oint (\delta \times A) \cdot ds = \int \nabla \times (\delta \times A) \cdot d\sigma, \tag{9-278}$$

where Stokes' theorem has been invoked.

The contributions to the flux of A through the bottom and top surface are simply $(A \cdot d\sigma)_2 - (A \cdot d\sigma)_1$. Using Eqs. (9-276) and (9-278), the net change in the flux of A due to motion and distortion of $d\sigma$ can be written

$$(A \cdot d\sigma)_2 - (A \cdot d\sigma)_1 = [\delta(\nabla \cdot A) - \nabla \times (\delta \times A)] \cdot d\sigma. \tag{9-279}$$

Finally, we must consider the contribution caused by changes in A itself. The change in the flux due to time changes of A is merely $(\partial A/\partial t) \cdot d\sigma \, dt$. Therefore we can write the total change in the flux as

$$\frac{d}{dt}(A \cdot d\sigma) = \left[\frac{\partial A}{\partial t} + u(\nabla \cdot A) - \nabla \times (u \times A)\right] \cdot d\sigma. \tag{9-280}$$

This result was found by Helmholtz but in connection with electrodynamics rather than hydrodynamics.

If we now identify the vector A as the vortex frequency ω, then we find

$$\frac{d}{dt}(\omega \cdot d\sigma) = \left[\frac{\partial \omega}{\partial t} - \nabla \times (u \times \omega)\right] \cdot d\sigma, \tag{9-281}$$

319

since $\nabla \cdot \boldsymbol{\omega} = 0$. By Eq. (9-273), the right-hand side of this equation vanishes and we have

$$\frac{d}{dt}(\boldsymbol{\omega} \cdot d\boldsymbol{\sigma}) = 0, \tag{9-282}$$

and so the Helmholtz vorticity conservation theorem as stated at the beginning of this section is established. Vorticity is a convective quantity of the flow.

It is also possible to express the conservation of vorticity as an integral theorem. In 1869 Lord Kelvin established such a theorem for the purpose of clarifying Helmholtz's work. We must begin with the definition of *circulation* Γ:

$$\Gamma \equiv \oint \boldsymbol{u} \cdot d\boldsymbol{s}. \tag{9-283}$$

We let $\mathscr{C}$ be the integration circuit at time t and $\mathscr{C}'$ the integration circuit at time $t + \Delta t$ as shown in Fig. 9-12. Let $d\boldsymbol{s}$ be an arc element of $\mathscr{C}$ and $d\boldsymbol{s}'$ an

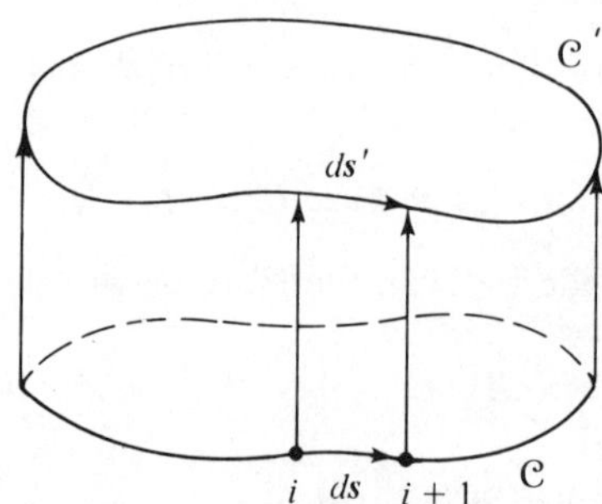

Fig. 9-12 Development of a convected circulation path.

arc element of $\mathscr{C}'$. The vertical arrows represent the paths of neighboring particles i and $i + 1$ during the displacement. These arrows simply correspond to the vector $\boldsymbol{\delta}$ in the similar previous figure—that is,

$$\boldsymbol{\delta}_i = \boldsymbol{u}_i \Delta t, \qquad \boldsymbol{\delta}_{i+1} = \boldsymbol{u}_{i+1} \Delta t. \tag{9-284}$$

We see from Fig. 9-12 that

$$\boldsymbol{u}_i \Delta t + d\boldsymbol{s}' = \boldsymbol{u}_{i+1} \Delta t + d\boldsymbol{s} \tag{9-285}$$

or

$$\frac{d\boldsymbol{s}' - d\boldsymbol{s}}{\Delta t} = \boldsymbol{u}_{i+1} - \boldsymbol{u}_i. \tag{9-286}$$

We now wish to calculate

$$\frac{d\Gamma}{dt} = \lim_{\Delta t \to 0} \frac{\Gamma' - \Gamma}{\Delta t} = \frac{d}{dt}\oint \boldsymbol{u} \cdot d\boldsymbol{s} = \oint \frac{d\boldsymbol{u}}{dt} \cdot d\boldsymbol{s} + \int \boldsymbol{u} \cdot \frac{d}{dt}d\boldsymbol{s}, \tag{9-287}$$

where Γ' and Γ are the circulations about $\mathscr{C}'$ and $\mathscr{C}$, respectively. The quantity

$$\frac{d}{dt}\,ds = \lim_{\Delta t \to 0} \frac{ds' - ds}{\Delta t} = d\boldsymbol{u};\tag{9-288}$$

so the last integral on the right-hand side of Eq. (9-287) becomes

$$\oint_{\mathscr{C}} \boldsymbol{u} \cdot d\boldsymbol{u} = \frac{u^2}{2}\bigg|_{A=B}^{B} = 0.\tag{9-289}$$

Let us examine the first integral

$$\oint \frac{d\boldsymbol{u}}{dt} \cdot d\boldsymbol{s} = \oint - \nabla\left(\frac{p + \phi}{\rho}\right) \cdot d\boldsymbol{s},$$

which also vanishes. Thus we find

$$\frac{d\Gamma}{dt} = 0 \to \Gamma = \text{const},\tag{9-290}$$

and

the circulation evaluated about an arbitrary circuit, every element of which is convected with the flow, is constant.

The connection between the circulation Γ and the vortex frequency $\boldsymbol{\omega}$ is easily established by Stokes' theorem:

$$\Gamma = \oint \boldsymbol{u} \cdot d\boldsymbol{s} = \int \nabla \times \boldsymbol{u} \cdot d\boldsymbol{\sigma} + 2\int \boldsymbol{\omega} \cdot d\boldsymbol{\sigma}.\tag{9-291}$$

We see by comparing Eqs. (9-282), (9-290), and (9-291) that the contents of Kelvin's circulation theorem and Helmholtz's vorticity conservation theorem are identical.

By virtue of the relation $\nabla \cdot \boldsymbol{\omega} = 0$, we also see that the vortex frequency is a solenoidal vector. Thus a vortex tube or filament must either be closed in a ring or end only at the boundaries of the fluid.

The dynamics of vortices is very strange indeed and differs markedly from the dynamics of mass points. An isolated vortex is not subject to forces but remains at rest with respect to the fluid. A uniform rectilinear motion can only be acquired by interaction with a second vortex of equal strength but opposite sense of rotation or under the action of a wall at rest. For this reason, the relativity principle of classical mechanics, according to which the state of rest and the uniform motion are equivalent, no longer applies; the fluid to which the vortex belongs plays the role of a preferred reference frame.

The behavior of vortices in regard to the second of Newton's laws is even stranger. The action of one vortex on another does not determine the *acceleration* but the *velocity*.

A remarkable analogy exists between the equations of electrodynamics and those of vortical fluid dynamics. If we define the *magnetic field intensity* H by the equation

$$B = \mu H, \tag{9-292}$$

where μ is called the *permeability* of the material, then, for the relation between the magnetic field and its source, we can write the current density

$$\nabla \times H = J. \tag{9-293}$$

The field H is a solenoidal one, as is J:

$$\nabla \cdot H = \nabla \cdot J = 0, \tag{9-294}$$

and the electromagnetic Lorentz force density can be written

$$J \times H = (\nabla \times H) \times H = \frac{f}{\mu}. \tag{9-295}$$

In analogy with Eqs. (9-293) to (9-295), we have for fluid dynamics

$$\nabla \times u = \zeta, \tag{9-296}$$

where $\zeta = 2\omega$ is known as the *vorticity* and serves as the source vector for the vortex velocity field just as J does for H. In an incompressible fluid

$$\nabla \cdot u = \nabla \cdot \zeta = 0. \tag{9-297}$$

The superposition of a vortex on a velocity field results in a force known as the *Magnus force*

$$\zeta \times u = (\nabla \times u) \times u = -\frac{f}{\rho}. \tag{9-298}$$

The most common manifestation of the Magnus force is in aerodynamics. The unequal flow over the top and bottom of an aeroplane wing is equivalent to a vortex bound in the wing. The interaction of this vortex with the velocity field induced by the forward motion of the aircraft provides the lifting force to balance gravity. Note that the one-to-one analogy between electrodynamics and incompressible vortex flow is violated by the fact that the Lorentz force and the Magnus force have *opposite* signs.

Closed vortices or vortex rings are analogous to the magnetic field produced by a circular current loop. They can be generated by pulsing a small volume of air (or smoke) through a sharp-edged circular aperture. The smaller the aperture, the faster the propagation speed of the ring through the background air. If a square aperture is substituted for the circular one, then the emergent vortex is square. However, regular pulsations develop that cause the vortex to evolve rapidly toward a circular shape.

Two coaxial vortex rings moving in the same direction interact in a most interesting fashion, as shown by the sequence depicted in Fig. 9-13. The rings are attractive, just as current loops having the same sense are. The trailing ring will accelerate and the leading ring will decelerate. The increase in the propagation speed of the trailing ring occurs at the expense of its poloidal motion. The vortex filament therefore shrinks and the trailing ring becomes smaller. Meanwhile, the decelerating leading ring increases its diameter. The smaller ring catches up with the larger one and passes through it. The roles are then reversed; the trailing ring becomes the leading one and vice versa

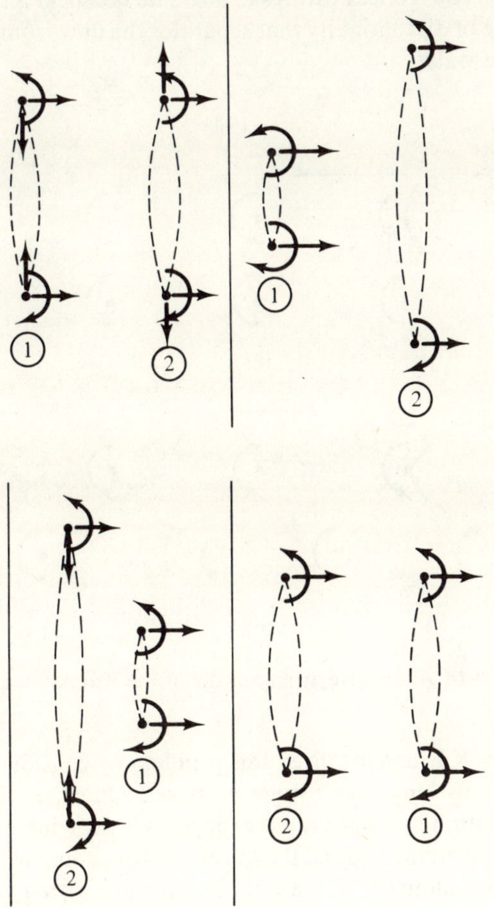

Fig. 9-13 Alternate threading of two coaxial vortex rings.

and the process is repeated. This phenomenon is most easily demonstrated by dropping colored drops from a pipet on the surface of water. When a drop hits the surface, it creates a ring that penetrates the water, expanding at the same time. The next drop creates a vortex ring that is initially smaller than the first one and that will overtake it. Several mutual passthroughs can be observed with this arrangement.

In Fig. 9-14 we show two versions of the flow about a plate situated obliquely in a uniform velocity field. The situation on the left—smooth potential flow—is not realized in practice. The usual situation is for a flow separation to occur at the edges of the plate and a turbulent wake to develop behind the plate. We can expect vortices to be formed by instability as a result of the high transverse velocity gradient occurring at the edge of the plate. These vortices serve as roller bearings to alleviate the shear and are distributed along the surface of discontinuity that separates the flow from the more or less dead fluid in the wake.

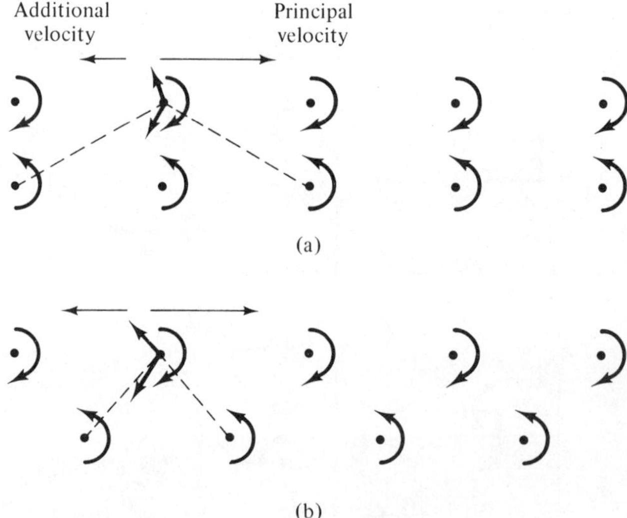

Fig. 9-14 Alternative arrangements of the Kármán vortex street.

Theodore von Kármán idealized this problem by considering the vortices to form an infinitely long, straight *vortex street*. On either side of the street the sense of rotation is opposite; the upper vortices are clockwise if the principal velocity is to the left and the lower ones are counterclockwise. If the additional velocity induced by the mutual influence of the vortices on each other is parallel to the street, then the vortices must either be opposite one another as shown in Fig. 9-14(a) or staggered as shown in Fig. 9-14(b).

Von Kármán was able to show, by analyzing the stability of both arrangements, that the vortices do indeed alternate. It is this alternate formation of vortices that causes a plate placed obliquely in a flow to *flutter* as vortices are formed alternately at opposite edges.

Turning next to magnetohydrodynamics, let us examine the possibilities for vortices in the presence of a uniform magnetic field. In Fig. 9-15 we show

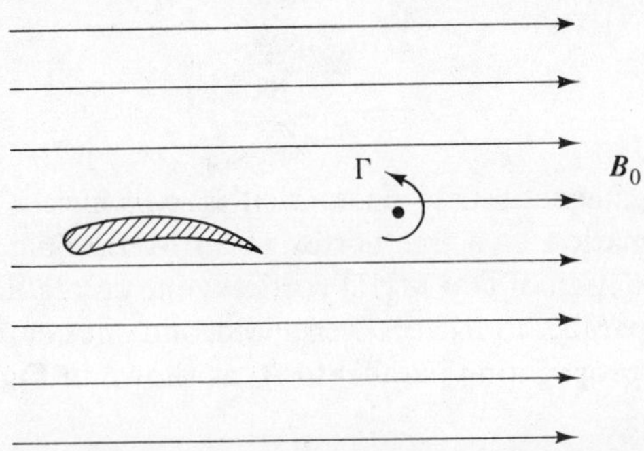

Fig. 9-15 Generation of a vortex in a magneto-fluid.

an aerofoil that has just been accelerated to the left. The creation of a clockwise vortex bound in the wing causes a counterclockwise vortex to be ejected from the trailing edge. We shall assume that the fluid conductivity is infinite and hence that the magnetic field must be "wound up" by the vortex. No matter how weak the magnetic field, the field cannot be wound up forever. The final configuration must feature magnetic field lines coinciding with the vortex velocity field vector. The condition for dynamic (stationary) equilibrium in an incompressible MHD fluid

$$\nabla\left(\frac{\rho u^2}{2} + p\right) = \boldsymbol{J} \times \boldsymbol{B} - \rho(\boldsymbol{\zeta} \times \boldsymbol{u}), \tag{9-299}$$

becomes with the assumption

$$\boldsymbol{u} = \alpha\boldsymbol{B}, \qquad \boldsymbol{\zeta} = \alpha\mu_0\boldsymbol{J}, \tag{9-300}$$

$$\nabla\left(\frac{\rho u^2}{2} + p\right) = (\boldsymbol{J} \times \boldsymbol{B})(1 - \alpha^2\rho\mu_0),$$

or

$$[\nabla \times (\boldsymbol{J} \times \boldsymbol{B})](1 - \alpha^2\rho\mu_0) = 0. \tag{9-301}$$

In order to interpret Eq. (9-301), we should recall that Helmholtz's Eq. (9-273) implies that the vorticity be frozen into the fluid. This condition serves the requirement that the vortex filament be a *force-free line singularity*. We see that such an equilibrium is possible for the vortex depicted in Fig. 9-15, not only for the cases where $J \times B$ is conservative and its potential adds to the pressure or where the magnetic field is force free but also for the case

$$\alpha^2 = \frac{1}{\rho\mu_0}. \tag{9-302}$$

The first two cases for which $\nabla \times (J \times B) = 0$ correspond to a vortex filament frozen into the flow. The third possibility represents vortices that move at a speed

$$u = \pm \frac{B^2}{\sqrt{\rho\mu_0}} \tag{9-303}$$

up or down the magnetic field lines—that is, a *nonlinear Alfvén wave*. In practice, the formation of a free vortex in an MHD fluid such that $\zeta \perp B$ results in the formation of two MHD vortices: one corotating with u parallel to B moving antiparallel to B_0 at Alfvén speed, and one contrarotating with u antiparallel to B propagating parallel to B_0 as shown in Fig. 9-16.

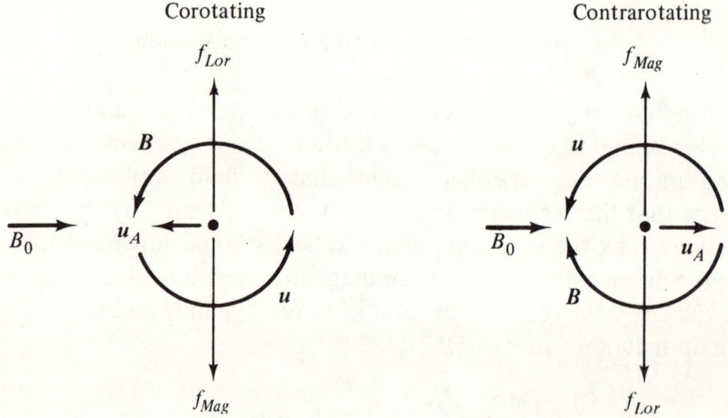

Fig. 9-16 Force-free line singularities in MHD.

In the case of the flow of an MHD fluid past a blunt body, the vortex street depends on the speed of the flow. For $u < u_A = B_0/\sqrt{\rho\mu_0}$, a front street, as well as a rear one, is formed. For $u > u_A$, the front street disappears and the rear street is double, consisting of both co- and contrarotating vortices. (See Fig. 9-17.)

Other types of MHD vortices are possible in correspondence with Helmholtz's condition. In this case, $\zeta \times B = 0$, and the magnetic field is either force-free and the vortex resembles an ordinary hydrodynamic vortex or $J \times B$ is conservative, in which case a column of rotating fluid is confined by an axial magnetic field. This latter configuration, known as a Θ pinch, has

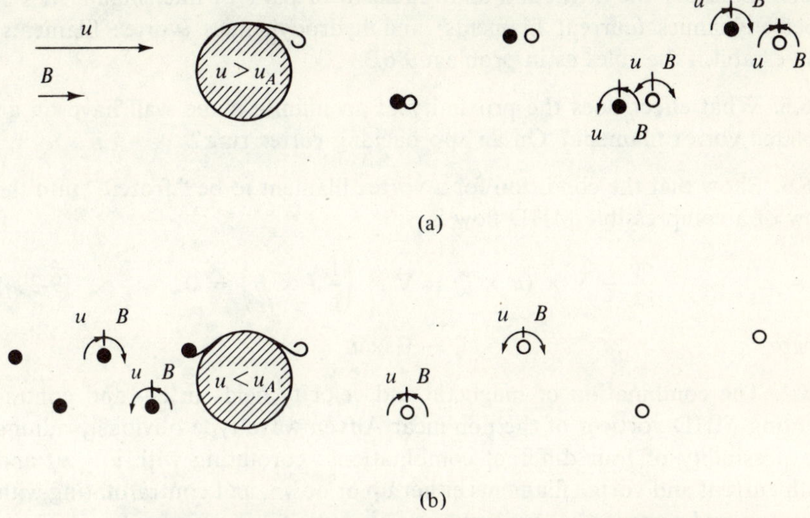

(a)

(b)

Fig. 9-17 MHD Kármán vortex street, (a) super-Alfvénic,
(b) sub-Alfvénic.

received considerable experimental attention as a possible environment for controlled thermonuclear fusion reactions.

EXERCISES

9-6.1. A vortex ideally consists of a *vortex filament* (line singularity) surrounded by an irrotational circular flow. If the flow surrounding the filament is written in the form $u_\theta = cr_n$, show that all flows for which $n \neq -1$ are rotational.

9-6.2. Show that two vortex filaments of equal strength

(a) and opposite sense move at equilibrium with a common velocity u in a straight line.

(b) and like sense move at equilibrium on a circle at opposite ends of a diameter.

9-6.3. In electrodynamics it is well known that opposite point singularities (plus and minus charges and north and south magnetic poles) attract. In hydrodynamics the point singularities are *sources* and *sinks*, in which velocity stream lines begin and end, respectively. Show by means of Bernoulli's equation that a pair of point singularities in hydrodynamics repels if dissimilar and attracts if similar.

9-6.4. Consider the attraction and repulsion of pairs of line singularities in electrodynamics (current filaments) and hydrodynamics (vortex filaments) and establish the rules as in problem 9-6.3.

9-6.5. What effect does the proximity of an infinite plane wall have on an isolated vortex filament? On an approaching vortex ring?

9-6.6. Show that the condition for a vortex filament to be "frozen" into the flow of a compressible MHD flow is

$$\frac{\partial \zeta}{\partial t} - \nabla \times (u \times \zeta) = \nabla \times \left(\frac{1}{\rho} J \times B\right) = 0, \qquad (9\text{-}304)$$

where
$$\zeta = \nabla \times u.$$

9-6.7. The combination of magnetic and velocity fields in co- and contra-rotating MHD vortices of the nonlinear Alfvén wave type obviously admits the possibility of four different combinations: corotating with $u = u_A$ and both current and vortex filaments either up or down, and contrarotating with $u = -u_A$ and current up, vorticity down or vorticity up, current down. Show by considering all possibilities that corotating vortices ignore the presence of other corotating vortices of either orientation and that contrarotating vortices similarly ignore other contrarotating vortices. Show also that a co interacts with a contra. Find the sense of the interaction (attraction or repulsion) for all possibilities.

9-7 Turbulence

We have seen that the flow of a viscous fluid through a small tube or between narrowly spaced plates is laminar for small u. As the speed increases, the flow becomes unstable, transverse components of velocity arise, and vortical fluid motions are generated. In steady laminar flow the motion is governed by

$$\nabla p = \rho\nu\, \nabla^2 u = \rho\nu[\nabla(\nabla \cdot u) - \nabla \times \zeta], \qquad (9\text{-}305)$$

in which the inertial terms $\rho(u \cdot \nabla)u = \tfrac{1}{2}\rho\, \nabla u^2 - \rho u \times \zeta$ are ignored. When these inertial terms become comparable with the right-hand side of Eq. (9-305), the flow pattern begins to change. If l is a length representative of the dimension of the flow region, then the ratio of the inertial terms to the viscous terms is ul/ν, which is the Reynolds number. For R greater than some number that varies with the details of the experiment, a state of fully established turbulent flow is reached. This flow is characterized by a spectrum of *turbulent eddies*, the largest of which are of the same order-of-magnitude size as the flow region. Moreover, there is a continual exchange of energy between eddies of different sizes. The energy input is largely through the larger eddies. These eddies, being characterized by large Reynolds numbers, experience very little viscous dissipation and transfer energy in some manner to smaller eddies and they, in turn, to smaller ones yet, until the energy is finally dissipated as heat by viscosity in the smallest eddies. The main problem in this most difficult area of fluid mechanics is to determine the distribution of energy according to the size of the eddies and to describe the mechanism by which energy transfer between the eddies takes place.

In a turbulent flow the fluid velocity u experiences fluctuations in position and time. With Fourier analysis in mind, we can write $u(r, t)$ as

$$u(r, t) = \sum_k u(k, t) \exp(ik \cdot r), \tag{9-306}$$

where the summation is taken over wave number k. The equation of motion

$$\frac{\partial u}{\partial t} + (u \cdot \nabla)u + \nabla \mathscr{P} + \nu \nabla^2 u = 0, \tag{9-307}$$

and for $u \ll a$, the sound speed

$$\nabla \cdot u = 0, \tag{9-308}$$

we find using Eq. (9-306) and

$$\mathscr{P} = \sum_k \mathscr{P}(k, t) \exp(ik \cdot r), \tag{9-309}$$

$$\frac{\partial u(k)}{\partial t} = -i \sum_{k'} [u(k') \cdot k]u(k - k') - i\mathscr{P}(k)k - \nu|k|^2 u(k), \tag{9-310}$$

and
$$k \cdot u(k) = 0. \tag{9-311}$$

Multiplying Eq. (9-310) by $u^*(k)$, the complex conjugate velocity, and making use of the fact that

$$k \cdot u^*(k) = 0, \tag{9-312}$$

we obtain

$$\frac{1}{2}\frac{\partial}{\partial t}|u(k)|^2 = -i \sum_{k'} [u(k') \cdot k][u(k - k') \cdot u(-k)] - \nu|k|^2. \tag{9-313}$$

We can draw several important conclusions from Eqs. (9-310) and (9-313). First, we see that viscosity plays no role in the energy transfer from large eddies to small eddies, since k is not coupled to k' in the last term of (9-313). Viscosity only serves to reduce the amplitude of each wave-number component independently. Next, we note that there is no pressure term in Eq. (9-313). This fact implies that pressure effects similarly are not responsible for energy transfer from large to small eddies. Pressure is operative, however, in exchanging energy in different directions and therefore acts to promote isotropy. We see that, as expected, the coupling occurs in the inertial terms.

As the critical Reynolds number is reached and the onset of turbulence commences, energy is initially fed into the largest eddies. Because of the transfer of energy to smaller and smaller eddies, the turbulence loses its anisotropic character at higher Reynolds numbers and in the fully developed state is both isotropic and homogeneous. This tendency to isotropy is independent of the dimensions of the flow region and of the manner in which the energy is supplied. We can therefore expect the fully developed turbulent structure to depend solely on ϵ and ν, the rate of energy dissipation per unit mass and the kinematic viscosity, respectively. We should also expect that in any proper mathematical formulation the limit $\nu \rightarrow 0$ would exist and that in this limit the turbulence would depend only on ϵ. This assumption is reasonable because, as $\nu \rightarrow 0$, the size of an energy-dissipating eddy also approaches zero. This principle, first suggested by Kolmogoroff, enables us to determine the *spectrum* of the turbulence $F(k)$ as defined by

$$\langle |\boldsymbol{u}|^2 \rangle = \int_0^\infty F(k)\,dk. \tag{9-314}$$

It follows from this definition that $F(k)$ has the dimensions of $(\text{length})^3/(\text{time})^2$. The only combination of ν and ϵ having the units of length is $(\nu^3/\epsilon)^{1/4}$, and the only combination having units of time is $(\nu/\epsilon)^{1/2}$. Thus

$$F(k) = (\nu^5 \epsilon)^{1/4} \Phi(k\nu^{3/4} \epsilon^{-1/4}), \tag{9-315}$$

where Φ is a general function of its dimensionless argument. Since $F(k)$ must be independent of ν as $\nu \rightarrow 0$, we find

$$F(k) = \text{const} \times \epsilon^{2/3} k^{-5/3}, \tag{9-316}$$

which is the Kolmogoroff spectrum. Unfortunately, the integral in Eq. (9-314) is divergent for this choice, and so it cannot be valid in the limit as $k \rightarrow 0$.

Since diffusive action is an element of turbulent motion, weak magnetic fields in highly conducting turbulent fluids will tend to be "stretched" by the diffusion. The intensity of the magnetic field is a measure of the number of field lines passing through a given area. Consequently, the action of turbulent diffusion tends to increase the intensity of a magnetic field initially embedded in it.

As we have seen, the magnetic field in a conducting fluid satisfies

$$\frac{\partial \boldsymbol{B}}{\partial t} - \nabla \times (\boldsymbol{u} \times \boldsymbol{B}) = \frac{1}{\sigma \mu_0} \nabla^2 \boldsymbol{B} \tag{9-317}$$

and

$$\nabla \cdot \boldsymbol{B} = 0, \tag{9-318}$$

which we can combine by using identity (1-95) and the assumption of incompressibility to find

$$\frac{\partial \boldsymbol{B}}{\partial t} + (\boldsymbol{u} \cdot \nabla)\boldsymbol{B} - (\boldsymbol{B} \cdot \nabla)\boldsymbol{u} = \frac{1}{\sigma \mu_0} \nabla^2 \boldsymbol{B}. \tag{9-319}$$

Multiplying this equation scalarly by B and averaging, we find for the average rate of increase of magnetic energy per unit volume

$$\frac{1}{2}\frac{\partial}{\partial t}\langle B^2\rangle = \langle B \cdot [(B \cdot \nabla)u]\rangle - \frac{1}{2}\langle (u \cdot \nabla)B^2\rangle$$

$$+ \frac{1}{\sigma\mu_0}\langle B \cdot \nabla^2 B\rangle. \tag{9-320}$$

By Fourier analyzing the magnetic field into a superposition of plane waves, the last term in Eq. (9-320) can be put in the form $-\langle |\nabla B|^2\rangle/\sigma\mu_0$. Similarly, the first two terms on the right-hand side of Eq. (9-320) can be recast as $\langle B^2\, \partial u_\parallel/\partial s\rangle$, where $u_\parallel$ is the component of u along a field line and s is the arc length along the same line. So Eq. (9-320) is simplified to

$$\frac{1}{2}\frac{\partial}{\partial t}\langle B^2\rangle = \left\langle B^2\,\frac{\partial u_\parallel}{\partial s}\right\rangle - \frac{1}{\sigma\mu_0}\langle |\nabla B|^2\rangle. \tag{9-321}$$

Thus B^2 will increase or decrease, depending on the relative magnitudes of the two terms on the right-hand side. The second term is always negative and describes the average rate of conversion of magnetic energy to joule heat. The first term represents the flow of energy to and from turbulent motion, depending on whether the field lines are extended or contracted by the turbulence.

Using the Navier–Stokes equation and identity (1-99), we see that the vorticity $\zeta = \nabla \times u$ satisfies equations

$$\frac{\partial \zeta}{\partial t} - \nabla \times (u \times \zeta) = \nu\,\nabla^2\zeta, \tag{9-322}$$

$$\nabla \cdot \zeta = 0, \tag{9-323}$$

entirely similar to Eqs. (9-317) to (9-318) for the magnetic field. So in the absence of a magnetic field we must find, in analogy to Eq. (9-321),

$$\frac{1}{2}\frac{\partial}{\partial t}\langle \zeta^2\rangle = \left\langle \zeta^2\,\frac{\partial u_\parallel}{\partial s}\right\rangle - \nu\langle |\nabla \zeta|^2\rangle. \tag{9-324}$$

For large Reynolds numbers, it is found experimentally that the two terms on the right-hand side of Eq. (9-324) are comparable, as might be expected for a state of fully developed turbulence. Since turbulence tends to diffuse apart any two fluid particles that are initially close together, the field lines are stretched and the magnetic field energy tends to be amplified. This process is quite similar to the one in which vortex filaments become elongated. Since small-scale turbulence (high k) is more effective at magnetic field amplification than the large eddies, besides contributing most of the vorticity, it seems plausible that there is an equipartition of energy between magnetic field and motion, on the high-k end of the spectrum at least. This factor implies the existence of a steady state, at which point the tension in the field lines balances the tendency to turbulent diffusion. Kolmogoroff's principle implies that $\langle \zeta^2 \rangle$ for large wave numbers must go as $(\epsilon \nu)^{1/2}$, which is also the order of magnitude of the kinetic energy per unit mass. Since we assume that in the steady state

$$\left\langle \frac{B^2}{2\mu_0} \right\rangle \approx \left\langle \frac{\rho u^2}{2} \right\rangle, \tag{9.325}$$

then

$$\frac{\langle B^2 \rangle}{\rho \mu_0} \approx (\epsilon \nu)^{1/2}. \tag{9-326}$$

In the preceding discussion we showed how a large turbulent magnetic field is generated from a weak initial field. If a strong field is present initially, it will tend to inhibit the onset of turbulence if the motion is sub-Alfvénic:

$$\frac{B_0^2}{\rho \mu_0} \gg \langle u^2 \rangle. \tag{9-327}$$

EXERCISE

9-7.1. Show that the mean velocity gradient varies with the cube root of distance for eddies obeying the Kolmogoroff spectrum. This is the *Kolmogoroff–Obukhov law*.

SOME ASPECTS
OF THE
CALCULUS OF VARIATIONS

In the next chapter we shall examine alternative methods of formulating classical mechanics in such a way that problems involving unknown forces of constraint can be solved. These formulations, which also allow fields to be treated, require that we learn the rules of a branch of mathematics known as the *calculus of variations*. The development of the calculus of variations was begun by Newton; however, Johann and Jakob Bernoulli, Euler, Legendre, Lagrange, Hamilton, and Jacobi all made important contributions.

10-1 The Euler Equation

The basic problem of the calculus of variations is to determine the function $y(x)$ such that the integral

$$J = \int_{x_1}^{x_2} f(y, y', x)\, dx \qquad (10\text{-}1)$$

of the functional $f(y, y' = dy/dx, x)$ between the limits x_2 and x_1 is *extremal*—that is, J is a maximum, minimum, or saddlepoint. The determination of $y(x)$ is equivalent to establishing the *path* of integration that causes J to be extremal with respect to any neighboring path. In many cases of physical interest, the extremum will be a minimum.

This problem is rather more involved than the corresponding problem in differential calculus. In differential calculus the extremal value of some function $y(x)$ is determined by comparing values of y for a range of values of x. Here we shall compare values of J for various neighboring integration paths

$y(x)$ connecting the fixed endpoints.[1] In Fig. 10-1 we show two possible paths out of an infinite number of possibilities. The difference between these two paths for a given x is δy, which we shall call the variation of y. This difference will be described by introducing a new function $\eta(x)$ and a scaling factor α. The function $\eta(x)$ is arbitrary except that

$$\eta(x_1) = \eta(x_2) = 0, \tag{10-2}$$

and $\eta(x)$ must be differentiable. The path of integration can therefore be written

$$y(x, \alpha) = y(x, 0) + \alpha\eta(x), \tag{10-3}$$

and the difference between neighboring paths is

$$\delta y = y(x, \alpha) - y(x, 0) = \alpha\eta(x). \tag{10-4}$$

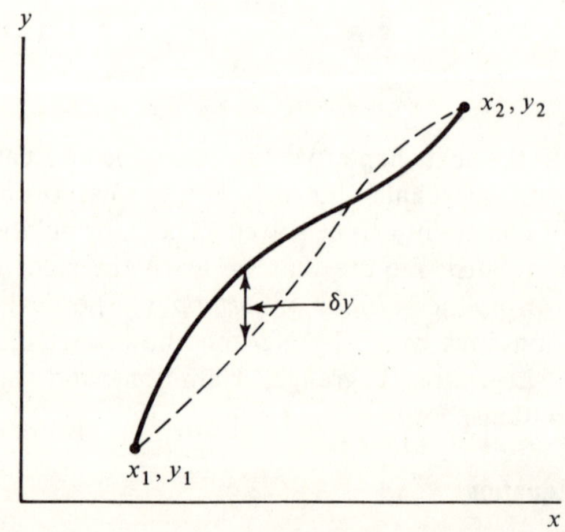

Fig. 10-1

We shall select $y(x, \alpha = 0)$ as the path that will extremize J; $y(x, \alpha)$ then describes some neighboring path. Thus J is a function of α and we rewrite Eq. (10-1) as

$$J(\alpha) = \int_{x_1}^{x_2} f[y(x, \alpha), y'(x, \alpha), x]\, dx, \tag{10-5}$$

and our condition for an extreme value must be

$$\left.\frac{\partial J}{\partial \alpha}\right|_{\alpha=0} = 0 \tag{10-6}$$

---

[1] The endpoints x_2, x_1 need not be considered fixed; however, for all problems of present interest the assumption of fixed endpoints is convenient.

in analogy to the similar condition $dy/dx = 0$ necessary to extremize y in differential calculus. Regarding $y(x, \alpha)$ and $y'(x, \alpha) = \partial y(x, \alpha)/\partial x$ as independent variables, we find

$$\frac{\partial J(\alpha)}{\partial \alpha} = \int_{x_1}^{x_2} \left[\frac{\partial f}{\partial y} \frac{\partial y}{\partial \alpha} + \frac{\partial f}{\partial y'} \frac{\partial y'}{\partial \alpha} \right] dx. \tag{10-7}$$

From Eq. (10-3) we see that

$$\frac{\partial y(x, \alpha)}{\partial \alpha} = \eta(x), \qquad \frac{\partial y'(x, \alpha)}{\partial \alpha} = \frac{d\eta(x)}{dx}, \tag{10-8}$$

and so

$$\frac{\partial J(\alpha)}{\partial \alpha} = \int_{x_1}^{x_2} \left(\frac{\partial f}{\partial y} \eta(x) + \frac{\partial f}{\partial y'} \frac{d\eta(x)}{dx} \right) dx. \tag{10-9}$$

Integrating the second term in Eq. (10-9) by parts yields

$$\int_{x_1}^{x_2} \frac{d\eta(x)}{dx} \frac{\partial f}{\partial y'} \, dx = \eta(x) \frac{\partial f}{\partial y'} \bigg|_{x_1}^{x_2} - \int_{x_1}^{x_2} \eta(x) \frac{d}{dx} \frac{\partial f}{\partial y'} \, dx. \tag{10-10}$$

The first term on the right-hand side of Eq. (10-10) vanishes as a result of assumption (10-2), and Eq. (10-9) becomes

$$\frac{\partial J(\alpha)}{\partial \alpha} = \int_{x_1}^{x_2} \left(\frac{\partial f}{\partial y} - \frac{d}{dx} \frac{\partial f}{\partial y'} \right) \eta(x) \, dx = 0, \tag{10-11}$$

which must vanish as the condition for extrema to exist. Since $\eta(x)$ is arbitrary except at the endpoints where it vanishes, we can specify that it have the same sign as the parenthetical part of the integrand whenever the latter is nonzero. Thus the integrand is positive definite, and Eq. (10-11) can only be satisfied if

$$\frac{\partial f}{\partial y} - \frac{d}{dx} \frac{\partial f}{\partial y'} = 0. \tag{10-12}$$

This equation, known as the *Euler equation*, is the condition for J to have an extreme value.

An alternative form of Euler's equation is quite useful in cases where f does not depend explicitly on x. In order to derive this equation, we note that

$$\frac{df}{dx} = \frac{d}{dx} f(y, y', x) = \frac{\partial f}{\partial x} + \frac{\partial f}{\partial y} \frac{dy}{dx} + \frac{\partial f}{\partial y'} \frac{dy'}{dx}$$

$$= \frac{\partial f}{\partial x} + y' \frac{\partial f}{\partial y} + y'' \frac{\partial f}{\partial y'}, \tag{10-13}$$

and $\quad\quad\quad\quad \dfrac{d}{dx}\left(y' \dfrac{\partial f}{\partial y'} \right) = y'' \dfrac{\partial f}{\partial y'} + y' \dfrac{d}{dx} \dfrac{\partial f}{\partial y} \tag{10-14}$

or eliminating $y''(\partial f/\partial y')$ with the use of Eq. (10-13), we find

$$\frac{d}{dx}\left(y'\frac{\partial f}{\partial y'}\right) = \frac{df}{dx} - \frac{\partial f}{\partial x} - y'\frac{\partial f}{\partial y} + y'\frac{d}{dx}\frac{\partial f}{\partial y'}$$

$$= \frac{df}{dx} - \frac{\partial f}{\partial x} - y'\left(\frac{\partial f}{\partial y} - \frac{d}{dx}\frac{\partial f}{\partial y'}\right)$$

$$= \frac{df}{dx} - \frac{\partial f}{\partial x}, \tag{10-15}$$

where the Euler equation (10-12) was used to eliminate the parenthetical term. Therefore

$$\frac{\partial f}{\partial x} - \frac{d}{dx}\left(f - y'\frac{\partial f}{\partial y'}\right) = 0 \tag{10-16}$$

is equivalent to the Euler equation (10-12) and is especially useful when $\partial f/\partial x = 0$. Then

$$f - y'\frac{\partial f}{\partial y'} = \text{const.} \tag{10-17}$$

The original problem that engaged the interest of the Bernoullis and led to the development of the calculus of variations in its present form was the *brachistochrone* (or shortest time) problem. A particle slides freely along a curve from (x_1, y_1) to the origin under the action of a uniform gravitational field. We wish to find the curve for which the time of descent is a minimum. The descent time τ is given by

$$\tau = \int_0^{x_1,y_1}\frac{ds}{u}. \tag{10-18}$$

Since

$$\tfrac{1}{2}mu^2 = mg(y_1 - y)$$

or

$$u = \sqrt{2g(y_1 - y)}, \tag{10-19}$$

and

$$ds = \sqrt{1 + y'^2}\,dx, \tag{10-20}$$

then

$$\tau = \frac{1}{\sqrt{2g}}\int_0^{x_1}\left(\frac{1 + y'^2}{y_1 - y}\right)^{1/2}dx. \tag{10-21}$$

We see that the integrand function f does not depend explicitly on x. So using the alternate form of the Euler equation, Eq. (10-17), we find with a bit of algebra

$$y' = \frac{dy}{dx} = \left[\frac{a - (y_1 - y)}{y_1 - y}\right]^{1/2}, \tag{10-22}$$

where a^{-2} is the constant associated with Eq. (10-17). This differential equation is most easily solved by introducing the parameter θ as

$$y = y_1 - \frac{a}{2}(1 + \cos \theta), \tag{10-23}$$

with the use of which we obtain, with Eq. (10-22),

$$dx = \frac{a}{2}(1 + \cos \theta)\, d\theta. \tag{10-24}$$

Integrating Eq. (10-24), we find for x

$$x + c = \frac{a}{2}(\theta + \sin \theta), \tag{10-25}$$

where c is the constant of integration. Equation (10-23) for y can be rewritten

$$y + a - y_1 = \frac{a}{2}(1 - \cos \theta). \tag{10-26}$$

Equations (10-25) and (10-26) are the equations for a *cycloid* having its vertex at the origin. The integration constants a and c are determined by the boundary condition at the origin

$$x(0) = y(0) = 0 \tag{10-27}$$

to be

$$c = 0, \qquad a = y_1.$$

Thus

$$x = \frac{y_1}{2}(\theta + \sin \theta),$$

$$\tag{10-28}$$

$$y = \frac{y_1}{2}(1 - \cos \theta).$$

This cycloid is shown in Fig. 10-2.

It is a further interesting property of cycloidal trajectories that the time of descent from any point (x_2, y_2) to the origin is invariant.

Another physical problem of interest is the question of *geodesics*, curves representing extremal paths between any two points, the paths being constrained to lie on a given surface. For the case of a sphere of radius R, the element of surface length is

$$ds = R(d\theta^2 + \sin^2 \theta\, d\varphi^2). \tag{10-29}$$

The distance s from point 1 to point 2 is therefore

$$s = R \int_1^2 \left[\left(\frac{d\theta}{d\varphi}\right)^2 + \sin^2 \theta \right]^{1/2} d\varphi. \tag{10-30}$$

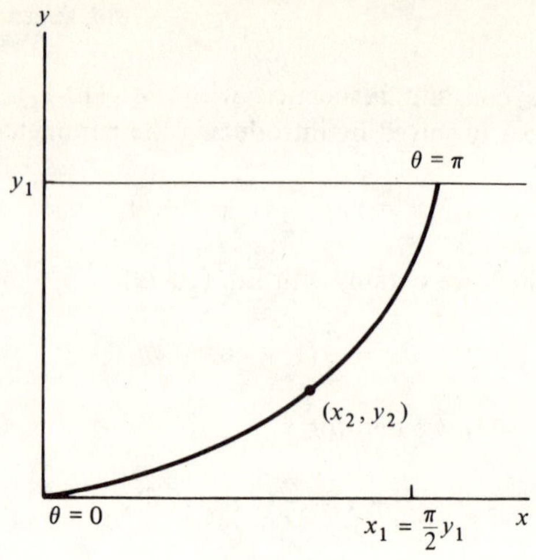

Fig. 10-2

We note that the integrand function f is independent of φ, and so, using Eq. (10-17), we find

$$(\theta'^2 + \sin^2 \theta)^{1/2} - \theta' \frac{\partial}{\partial \theta'} (\theta'^2 + \sin^2 \theta)^{1/2} = a \qquad (10\text{-}31)$$

or carrying out the differentiation and multiplying the equation by $f = (\theta'^2 + \sin^2 \theta)^{1/2}$ yields

$$\sin^2 \theta + a(\theta'^2 + \sin^2 \theta)^{1/2} = 0 \qquad (10\text{-}32)$$

This equation can be solved algebraically for θ'

$$\theta' = \frac{d\theta}{d\varphi} = \frac{\sin^2 \theta}{a} (1 - a^2 \csc^2 \theta)^{1/2} \qquad (10\text{-}33)$$

and integrated to yield

$$\varphi = \arcsin \left[\frac{a \cot \theta}{(1 - a^2)^{1/2}} \right] + \psi$$

or

$$\cot \theta = \frac{\sqrt{1 - a^2}}{a} \sin (\varphi - \psi), \qquad (10\text{-}34)$$

where ψ is the constant of integration. It is left as an exercise to show that in cartesian coordinates Eq. (10-34) becomes

$$Ay - Bx = z, \qquad (10\text{-}35)$$

where

$$A = \frac{\sqrt{1 - a^2}}{a} \cos \psi,$$

$$\qquad (10\text{-}36)$$

$$B = \frac{\sqrt{1 - a^2}}{a} \sin \psi.$$

This is the equation of a plane passing through the center of the sphere. Hence the geodesic defined by the intersection of the surface of the sphere with the plane described by Eq. (10-35) is a *great circle*. Note that the great circle route between two points on a sphere is the maximum as well as the minimum path, depending on which way you go.

EXERCISES

10-1.1. Derive Eq. (10-22), using Eq. (10-17).

10-1.2. Show that the time required for a particle to slide without friction along a cycloid from any initial point (x_2, y_2) to the vertex is

$$\tau = \pi\sqrt{\frac{y_1}{2g}}.$$

10-1.3. Determine the curve passing through the fixed points (x_1, y_1) and (x_2, y_2) such that the surface generated by rotating the curve about the x axis has minimum area.

10-1.4. Show that Eq. (10-34) describes a plane containing the center of the sphere as given by Eqs. (10-35) and (10-36).

10-1.5. Find the geodesic on a circular cylinder.

10-1.6. Fermat's principle states that a beam of light propagating in a medium having a variable index of refraction will follow a path that corresponds to an extremal time of propagation between two given points. Show that this statement corresponds to extremizing the integral

$$\int_{x_1}^{x_2} \frac{(1 + y'^2)^{1/2}}{n(x, y)} \, dx,$$

where $n(x, y)$ is the index of refraction. So for a constant index of refraction, light propagates along geodesics.

10-2 Generalization for Several Variables

It is possible to generalize the integral J to be extremized in several ways. We shall first suppose that more than one dependent variable is involved in the functional f. We shall label these dependent variables y_i, where i runs from 1 to n, and assume that they are all functions of the same independent variable x.

As before, we shall determine the extreme value of J by comparing neighboring integration paths. We let

$$y_i(x, \alpha) = y_i(x, 0) + \alpha\eta_i(x), \qquad i = 1, \ldots, n, \qquad (10\text{-}37)$$

where the η_i are independent of one another except for the fact that they all vanish at the endpoints. Since Eq. (10-6) still applies as the condition for an extreme value to exist, we find upon differentiating

$$J = \int_{x_1}^{x_2} f[y_1(x), y_2(x), \ldots, y_n(x), y_1'(x), y_2'(x), \ldots, y_n'(x), x]\, dx \quad (10\text{-}38)$$

with respect to α and evaluating for $\alpha = 0$,

$$\int_{x_1}^{x_2} \sum_i \left(\frac{\partial f}{\partial y_i} \eta_i + \frac{\partial f}{\partial y_i'} \frac{d\eta_i}{dx} \right) dx = 0. \quad (10\text{-}39)$$

Again we integrate the second term by parts and obtain

$$\int_{x_1}^{x_2} \sum_i \left(\frac{\partial f}{\partial y_i} - \frac{d}{dx} \frac{\partial f}{\partial y_i'} \right) \eta_i\, dx = 0. \quad (10\text{-}40)$$

Since the η_i are arbitrary and mutually independent, each term in the sum must vanish independently and we obtain an entire set of Euler equations

$$\frac{\partial f}{\partial y_i} - \frac{d}{dx} \frac{\partial f}{\partial y_i'} = 0, \qquad i = 1, 2, \ldots, n, \quad (10\text{-}41)$$

each of which must be satisfied in order that J have an extreme value.

Consider, next, the case of an integral J having a single dependent function of several independent variables $y = y(x_1, x_2, \ldots, x_n)$

$$J = \int\!\!\int \cdots \int f[y, y_{x_1}, y_{x_2}, \ldots, y_{x_n}, x_1, x_2, \ldots, x_n]\, dx_1\, dx_2 \ldots dx_n, \quad (10\text{-}42)$$

where $y_{x_i} \equiv \partial y / \partial x_i$. The variational problem in this case is to find the function $y(x_1, x_2, \ldots, x_n)$ for which J is extremal. In analogy with Eq. (10-3), we shall set

$$y(x_1, x_2, \ldots, x_n, \alpha) = y(x_1, x_2, \ldots, x_n, \alpha = 0) + \alpha \eta(x_1, x_2, \ldots, x_n), \quad (10\text{-}43)$$

where $y(x_1, x_2, \ldots, x_n, \alpha = 0)$ is the function for which

$$\left. \frac{\partial J}{\partial \alpha} \right|_{\alpha = 0} = 0. \quad (10\text{-}44)$$

The η function of the independent variables must be differentiable and must vanish at the endpoints just as in the previously considered cases. Carrying out the operation indicated in Eq. (10-44) upon Eq. (10-42), we find

$$\left. \frac{\partial J}{\partial \alpha} \right|_{\alpha = 0} = \int\!\!\int \cdots \int \left[\frac{\partial f}{\partial y} \eta + \sum_{i=1}^{n} \left(\frac{\partial f}{\partial y_{x_i}} \eta_{x_i} \right) \right] dx_1\, dx_2 \ldots dx_n = 0. \quad (10\text{-}45)$$

Integrating each of the terms $(\partial f/\partial y_{x_i})\eta_{x_i}$ by parts gives

$$\int\int\cdots\int\left[\frac{\partial f}{\partial y}-\sum_i\left(\frac{\partial}{\partial x_i}\frac{\partial f}{\partial y_{x_i}}\right)\right]\eta(x_1,x_2,\ldots,x_n)\,dx_1\,dx_2\ldots dx_n=0.\quad(10\text{-}46)$$

Because of the arbitrariness of η, the term in square brackets must vanish, and we find for the Euler equation appropriate to n independent variables

$$\frac{\partial f}{\partial y}-\sum_{i=1}^{n}\left(\frac{\partial}{\partial x_i}\frac{\partial f}{\partial y_{x_i}}\right)=0.\quad(10\text{-}47)$$

Finally, if the integrand f contains more than one dependent variable (say N) and more than one independent variable (say n), then the condition for an extremum to exist will be the satisfaction of a set of N Euler equations:

$$\frac{\partial f}{\partial y_i}-\sum_{j=1}^{n}\left\{\frac{\partial}{\partial x_j}\left[\frac{\partial f}{\partial(\partial y_i/\partial x_j)}\right]\right\}=0,\qquad i=1,2,\ldots,N.\quad(10\text{-}48)$$

EXERCISE

10-2.1. Show that a function $\phi(x,y,z)$ having a minimum average value of the square of its gradient over some region of space obeys Laplace's equation.

10-3 The Inclusion of Auxiliary Conditions

Let us turn now to the problem of extremizing a function f or an integral J, subject to certain constraints. Considering the differential calculus problem first, we note that the condition for a function $f(x,y,z)$ to be extreme is

$$df=\frac{\partial f}{\partial x}\,dx+\frac{\partial f}{\partial y}\,dy+\frac{\partial f}{\partial z}\,dz=0.\quad(10\text{-}49)$$

The necessary and sufficient condition for the satisfaction of Eq. (10-49) is

$$\frac{\partial f}{\partial x}=\frac{\partial f}{\partial y}=\frac{\partial f}{\partial z}=0.\quad(10\text{-}50)$$

Quite often the independent variables x, y, z are subject to constraint so that they are no longer independent and Eq. (10-50) does not obtain. In principle, it is possible to use each constraint equation to eliminate a variable and thereby reduce the equation set. In many cases, doing so is undesirable or inconvenient, and so a technique known as *Lagrange multipliers* is used.

Let us take as our equation of constraint

$$\varphi(x,y,z)=\text{const}\quad(10\text{-}51)$$

from which we can write

$$d\varphi = \frac{\partial\varphi}{\partial x}\,dx + \frac{\partial\varphi}{\partial y}\,dy + \frac{\partial\varphi}{\partial z}\,dz = 0. \tag{10-52}$$

In this case, there are only two independent variables. If we take x and y to be independent, then dz is no longer arbitrary. We can, however, write, using Eqs. (10-49) and (10-52),

$$df + \lambda\,d\varphi = \left(\frac{\partial f}{\partial x} + \lambda\frac{\partial\varphi}{\partial x}\right) dx + \left(\frac{\partial f}{\partial y} + \lambda\frac{\partial\varphi}{\partial y}\right) dy + \left(\frac{\partial f}{\partial z} + \lambda\frac{\partial\varphi}{\partial z}\right) dz = 0, \tag{10-53}$$

where the Lagrange multiplier λ is chosen so that

$$\frac{\partial f}{\partial z} + \lambda\frac{\partial\varphi}{\partial z} = 0, \qquad \frac{\partial\varphi}{\partial z} \neq 0. \tag{10-54}$$

With this condition, Eq. (10-53) becomes

$$\left(\frac{\partial f}{\partial x} + \lambda\frac{\partial\varphi}{\partial x}\right) dx + \left(\frac{\partial f}{\partial y} + \delta\frac{\partial\varphi}{\partial y}\right) dy = 0. \tag{10-55}$$

Since dx and dy are arbitrary and independent,

$$\frac{\partial f}{\partial x} + \lambda\frac{\partial\varphi}{\partial x} = \frac{\partial f}{\partial y} + \lambda\frac{\partial\varphi}{\partial y} = 0. \tag{10-56}$$

When Eqs. (10-54) and (10-56) are satisfied, $f(x, y, z)$ is an extremum. Note that there are four unknowns in the problem (x, y, z, λ) and four equations [(10-54), (10-56), and (10-51)] for their solution. We need only find x, y, and z in order to solve the physical problem, and λ is usually not determined. For this reason, λ is sometimes called Lagrange's *undetermined multiplier*. If $\partial\varphi/\partial x = \partial\varphi/\partial y = \partial\varphi/\partial z = 0$ at the extremum, the Lagrange multiplier technique fails.

Turning to the calculus of variations problem, we seek, as before, the path of integration that will make the integral

$$J = \int f\left(y_i, \frac{\partial y_i}{\partial x_j}, x_j\right) dx_j \tag{10-57}$$

an extremum. Here we consider x_j to represent a set of independent variables and y_i a set of dependent variables. Again, the condition for J to be an extremum is

$$\delta J = \frac{\partial J}{\partial \alpha}\bigg|_{\alpha=0} = 0, \tag{10-58}$$

where we have reintroduced the δ operator as a notational convenience. Suppose that one or more constraints are imposed, having the form

$$\varphi_k(y_i, x_j) = 0. \tag{10-59}$$

Clearly, we can multiply Eq. (10-59) by a function of x_j, say $\lambda_k(x_j)$, and integrate over the same range as in J to find

$$\Phi = \int \lambda_k(x_j)\varphi_k(y_i, x_j)\, dx_j = 0 \tag{10-60}$$

from which, obviously,

$$\delta\Phi = 0. \tag{10-61}$$

Alternatively, the constraint may appear in the form of an integral

$$\int \varphi_k(y_i, x_j)\, dx_j = \text{const.} \tag{10-62}$$

We can introduce a *constant* Lagrange multiplier in this case and again obtain Eq. (10-61).

Either for *point* or for *integral* constraints, we can add Eqs. (10-58) and (10-61) to obtain

$$\delta \int \left[f\left(y_i, \frac{\partial y_i}{\partial x_j}, x_j\right) + \sum_k \lambda_k \varphi_k(y_i, x_j) \right] dx_j = 0, \tag{10-62}$$

where λ_k may be a function of x_j for a point constraint. If the entire integrand of Eq. (10-62) is treated as a new functional

$$g\left(y_i, \frac{\partial y_i}{\partial x_j}, x_j\right) = f + \sum_k \lambda_k \varphi_k, \tag{10-63}$$

then g must satisfy the Euler equation

$$\frac{\partial g}{\partial y_i} - \sum_j \frac{\partial}{\partial x_j} \frac{\partial g}{\partial(\partial y_i/\partial x_j)} = 0. \tag{10-64}$$

EXERCISES

10-3.1. Find the ratio of radius r to height h of a circular cylinder such that the surface area is minimized subject to the constraint of constant volume.

10-3.2. Find the closed plane curve of fixed perimeter having maximum area.

10-3.3. Find the curve of fixed length l drawn between the origin and the point $(x, y) = (L, 0)$ such that the area between the curve and the x axis is maximal.

ALTERNATIVE FORMULATIONS
OF CLASSICAL MECHANICS

We have seen that Newton's law of motion

$$F = \frac{dp}{dt} \tag{11-1}$$

describes the dynamics of a particle in an inertial reference frame. If cartesian coordinates are employed and the particle is not subject to external constraints, the equations of motion are usually straightforward. If either or both conditions are not met, however, the equations may be difficult to formulate or solve. For instance, suppose that a particle is constrained to move along a certain path or in a given surface. The equations of motion must then take into account the *forces of constraint* that maintain the particle in the given path or surface. If the surface is a smooth horizontal plane, then the constraint force is the gravitational force and no complications arise. On the other hand, if the particle is a bead constrained to slide along a wire bent into some complicated shape, then the force of constraint may be difficult or impossible to express. Since the force entering into Eq. (11-1) is the vector sum of *all* the forces acting on the particle, we may well find ourselves in the position of being unable to write Newton's second law.

Since we know that Newton's theory is correct (in the nonrelativistic limit), the problem is not one of devising a *new* theory of mechanics but rather of reformulating the Newtonian theory in such a way that the complications mentioned can be circumvented. Such a reformulation is contained in *Hamilton's principle*, and the equations of motion that result from its application are called *Lagrange's equations*. These equations are entirely equivalent to Newton's equations, as we shall show. Moreover, Hamilton's principle can be applied successfully to a broader range of physical phenomena involving *fields* that are inaccessible to analysis by Newton's equations.

11-1 Hamilton's Principle

The intuitive notion that nature always acts in such a way as to extremize certain physical quantities goes back a long way. The first principle of this type was stated by Hero in the second century B.C. Hero found that the law of reflection of light (equal angles of incidence and reflection) could be obtained by postulating that a light ray, in traveling from one point to another by reflection from a plane mirror, always takes the shortest path. In 1657 Pierre de Fermat reformulated Hero's principle by postulating that a light ray always travels the path requiring an extremal *time* and in this way was also able to derive Snell's law of refraction.

Following the development of the calculus of variations in the latter years of the seventeenth century, a general variational principle of dynamics was formulated by Maupertuis. He postulated that mechanical motion occurs in such a way as to minimize action. This principle, known as the *principle of least action*, was justified only on theological grounds and was later given a firm mathematical foundation by Lagrange. This principle and others that were later proposed are less general than Hamilton's principle, and we shall forego a detailed discussion of them.

In 1834 Hamilton published a dynamical principle on which all of mechanics and indeed most of classical physics can be based. Hamilton's principle, in terms of the calculus of variations, can be stated

$$\delta \int_{t_1}^{t_2} (T - V)\, dt = 0; \tag{11-2}$$

that is, the motion of a dynamical system between two points in a specific time interval is such as to extremize the time integral of the difference between the kinetic and potential energies. In most applications of importance, the extremum turns out to be a minimum.

Since the kinetic energy of a particle expressed in fixed cartesian coordinates is a function only of the velocities $\dot{x}_i$ and, for conservative forces, the potential energy is a function of position x_i, we can define

$$L(\dot{x}_i, x_i) \equiv T(\dot{x}_i) - V(x_i), \tag{11-3}$$

in terms of which Hamilton's principle becomes

$$\delta \int_{t_1}^{t_2} L\, dt = 0. \tag{11-4}$$

The functional $L(\dot{x}_i, x_i)$ can be identified with the functional $f[y_i(x), y_i'(x), x]$ used in Chapter 10 by means of the substitutions

$$x \to t$$

$$y(x) \to x(t)$$

$$y'(x) \to \dot{x}(t)$$

and the Euler equations (10-41) then go over into the *Euler–Lagrange equations*

$$\frac{\partial L}{\partial x_i} - \frac{d}{dt}\frac{\partial L}{\partial \dot{x}_i} = 0. \tag{11-5}$$

These equations are the dynamical equations of Lagrange, and L is known as the *Lagrangian* for the problem.

As a simple example, consider the one-dimensional harmonic oscillator for which

$$L = \tfrac{1}{2}m\dot{x}^2 - \tfrac{1}{2}kx^2. \tag{11-6}$$

Then

$$\frac{\partial L}{\partial x} = kx, \qquad \frac{\partial L}{\partial \dot{x}} = m\dot{x}, \tag{11-7}$$

and Eq. (11-5) yields

$$m\ddot{x} + kx = 0, \tag{11-8}$$

which is identical with Newton's second law. Note, however, that in deriving Eq. (11-8) by the application of Hamilton's principle, the concept of *force* does not enter. This factor suggests that as long as energy can be defined without recourse to Newtonian concepts, Hamilton's principle allows the analysis of the dynamics.

EXERCISES

11-1.1. In Fig. 11-1 we show a light ray passing from point 1 to point 2 via a reflection. Use Fermat's principle to show that the angle of incidence θ equals the angle of reflection φ. What is the nature of the extremum?

11-1.2. Use Fig. 11-2, showing a light ray being refracted in passing from point 1 to point 2, and Fermat's principle to establish Snell's law of refraction. What is the nature of the extremum?

11-1.3. A simple pendulum with a mass m and string length l oscillates about its equilibrium position. If θ is the angular deflection of the pendulum from

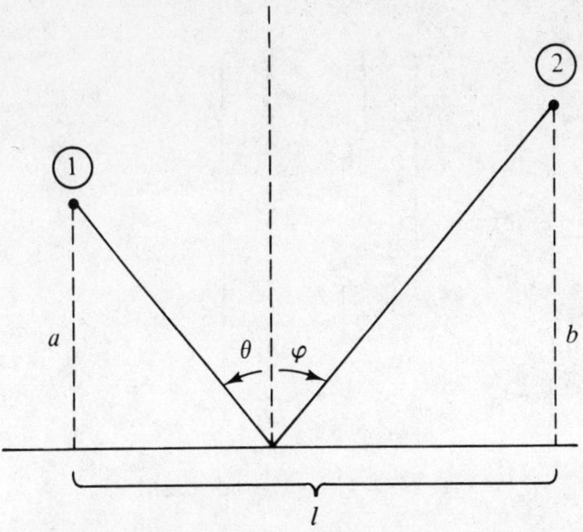

Fig. 11-1 Reflection from a plane mirror.

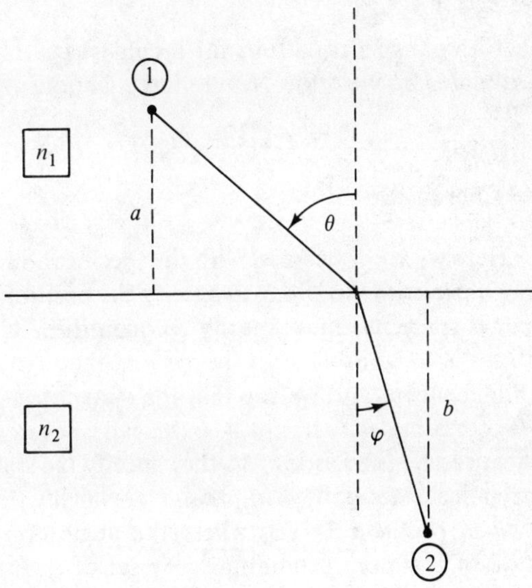

Fig. 11-2 Refraction at the interface between media having different indices of refraction.

equilibrium ($\theta = 0$), write the Lagrangian of the system in terms of θ and use the Euler–Lagrange equations to show that

$$\ddot{\theta} + \frac{g}{l}\sin\theta = 0.$$

11-1.4. Set up the Lagrangian for the Atwood machine shown in Fig. 11-3.

347

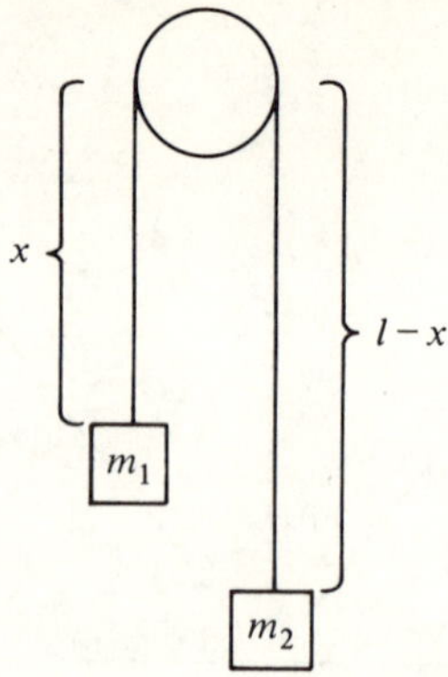

Fig. 11-3 The Atwood machine.

Assume the pulley to be massless, frictionless, and of negligible size. Use the Euler–Lagrange equation to derive the equation of motion for the system. (Suggestion: Let $V = 0$ for $x = 0$.)

11-1.5. Show that Lagrange's equations for a conservative system, in terms of cartesian coordinates x_i, yield the Newtonian equations of motion

$$F_i = -\nabla_i V = \dot{p}_i.$$

11-2 Generalized Coordinates

Here we are concerned with the specification of a mechanical system containing n particles. In order to specify the position of the particles in three-dimensional space, we must specify $3n$ quantities. If a number N of constraint equations is to be applied to the system, then only $3n - N$ of the coordinates are independent and we say that the system has $3n - N$ *degrees of freedom*. These coordinates need not be cartesian or even curvilinear; *any* $3n - N$ parameters can be used as long as they specify the state of the system completely. The choice is not limited to parameters having the dimensions of length, as we saw in problem 11-1.3, where the angle $\theta(t)$ was chosen to represent the state of a simple pendulum. Any set of quantities that completely describes the state of a system is called a set of *generalized coordinates*. We shall write these coordinates as q_i, where $i = 1, 2, \ldots, 3n - N$. In certain cases, we can use a set of generalized coordinates whose number exceeds the number of degrees of freedom and take the constraint equations into account with the use of Lagrange multipliers in order to be able to calculate the forces of constraint.

Unfortunately, there are no hard and fast rules governing the selection of generalized coordinates for a particular problem. The criteria for judging the suitability of the choice are the simplicity and manageability of the resulting

348

Lagrange equations. The development of intuition in this matter is largely a function of experience.

In addition to the generalized coordinates q_i, we also define a set of *generalized* velocities $\dot{q}_i$ (which need not have the dimensions of velocity). Allowing for the possibility that the equations connecting the rectangular coordinates of the particles x_j ($j = 1, 2, \ldots, 3n$) and the q_i may contain the time t explicitly, the transformation equations will have the form

$$x_j = x_j(q_1, q_2, \ldots, q_M, t), \tag{11-9}$$

where M may be equal to or greater than $3n - N$, depending on whether the Lagrange multiplier technique is to be employed. In general, the velocities of the particles will depend on the generalized coordinates, generalized velocities, and time, and so

$$\dot{x}_j = \dot{x}_j(q_1, q_2, \ldots, q_M, \dot{q}_1, \dot{q}_2, \ldots, \dot{q}_M, t). \tag{11-10}$$

The N constraint equations are written

$$\varphi_k = \varphi_k(x_1, x_2, \ldots, x_{3n}, t). \tag{11-11}$$

The state of an n-particle system subject to N constraints can be completely specified by a point in a $(3n - N)$-dimensional space known as *configuration space.* Each dimension of this space corresponds to one of the q_i. The evolution of the system in time is therefore given by a curve in configuration space, each point of which specifies the state (or configuration) of the system at a particular instant. The path of the system through configuration space automatically satisfies the constraints, since the coordinates are chosen to correspond only to allowable motions of the system.

In terms of the foregoing, Hamilton's principle can be restated as follows.

> *The path followed by a dynamical system moving through configuration space within a specified time interval is that path that extremizes the time integral of the appropriate Lagrangian.*

Since energy is a scalar and scalars, together with other tensors, are invariant to coordinate transformation, the *value* of the Lagrangian will be unique for a given state of the system. We recognize, of course, that the potential energy V is undetermined to within an additive constant; hence L is similarly indefinite. The Lagrangian only enters the problem in terms of its derivative, however, and in this sense it is immaterial whether we express L in terms of $x_j, \dot{x}_j$ or $q_i, \dot{q}_i$. Thus

$$\begin{aligned} L &= T(\dot{x}_j) - V(x_j) \\ &= T(q_i, \dot{q}_i, t) - V(q_i, t) \\ &= L(q_i, \dot{q}_i, t), \end{aligned} \tag{11-12}$$

and the Lagrange equations, in terms of generalized coordinates, become

$$\frac{\partial L}{\partial q_i} - \frac{d}{dt}\frac{\partial L}{\partial \dot{q}_i} = 0. \tag{11-13}$$

There are $3n - N$ of these equations. Together with $2(3n - N)$ initial conditions and the N equations of constraint, they provide a complete description of the dynamical evolution of the system subject to the conditions (1) that the forces acting on the system (except the constraint forces) must be conservative and (2) that the constraint equations must be functions only of the coordinates of the system and, in general, time. Constraints that satisfy this condition are termed *holonomic*.

To illustrate, consider the motion of a particle of mass m that is constrained to move on the surface of a hemisphere of radius R centered at the origin and subject to a uniform gravitational field in the minus z direction as shown in Fig. 11-4. The particle has two degrees of freedom, and a logical choice of generalized coordinates is the angles λ and θ corresponding to latitude and longitude. In terms of these coordinates, the square of the particle velocity is

$$u^2 = R^2(\dot{\theta}^2 \cos^2 \lambda + \dot{\lambda}^2). \tag{11-14}$$

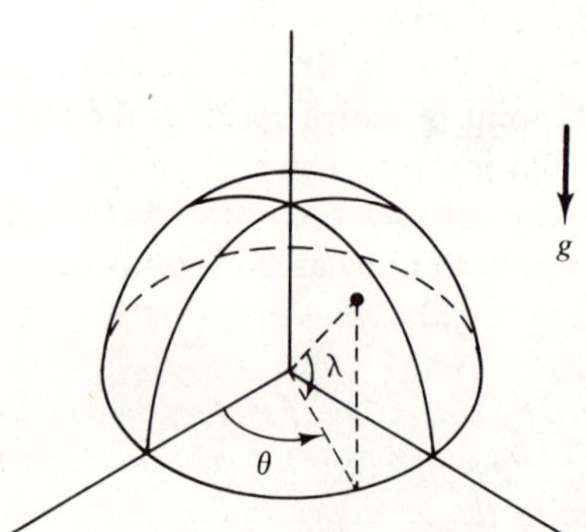

Fig. 11-4

The potential energy is (choosing $V = 0$ at $z = 0$)

$$V = mgR \sin \lambda, \tag{11-15}$$

and so the Lagrangian can be written

$$L = \tfrac{1}{2}mR^2(\dot{\theta}^2 \cos^2 \lambda + \dot{\lambda}^2) - mgR \sin \lambda. \tag{11-16}$$

We note right away that L does not explicitly contain θ. Therefore $\partial L/\partial \theta = 0$, and the Lagrange equation for the longitude coordinate is

$$\frac{d}{dt}\frac{\partial L}{\partial \dot{\theta}} = 0. \tag{11-17}$$

Hence

$$\frac{\partial L}{\partial \dot\theta} = mR^2 \cos^2 \lambda \dot\theta = J = \text{const.} \tag{11-18}$$

This equation expresses the conservation of the angular momentum of the particle about the z axis.

The Lagrange equation for the latitude coordinate

$$\frac{\partial L}{\partial \lambda} - \frac{d}{dt}\frac{\partial L}{\partial \dot\lambda} = 0, \tag{11-19}$$

yields, after calculating the derivatives,

$$\ddot\lambda + \frac{J^2 \sin \lambda}{m^2 R^4 \cos^3 \lambda} + \frac{g}{R} \cos \lambda = 0. \tag{11-20}$$

EXERCISES

11-2.1. A bead moves on a smooth wire bent into the form of a vertical circle of radius a. The wire loop rotates about its vertical diameter with a uniform angular frequency ω. The angle θ measures the angular distance of the bead from the lowest point on the loop—that is, the lower end of the rotation axis. Assuming a uniform gravitational acceleration g, find the Lagrangian for the system and find the equation of motion.

11-2.2. A disk of mass m and radius a rolls without slipping down a plane of mass M and inclination angle ϕ. The inclined plane is free to slide without friction along a horizontal surface. Describe the motion by the Lagrangian technique.

11-2.3. A bead of mass m is constrained to move along a smooth wire bent into a parabolic shape described by $y = bx^2$; $z = \text{const.}$ Find the Lagrangian for the system and determine the equation of motion under the action of a conservative force.

11-2.4. A particle of mass m moves on a smooth horizontal table under the action of an attached string that passes through a hole and is attached at its other end to a spring of constant k as shown in Fig. 11-5. Analyze the motion of the particle.

11-2.5. The spring in the problem depicted in Fig. 11-5 is replaced by a mass M acted on by gravity. How does the particle on the table move now?

11-2.6. A double pendulum consists of two equal-length, equal-mass pendula, one suspended from the bob of the other. The system is constrained to move in a plane. Find the equations of motion of the system.

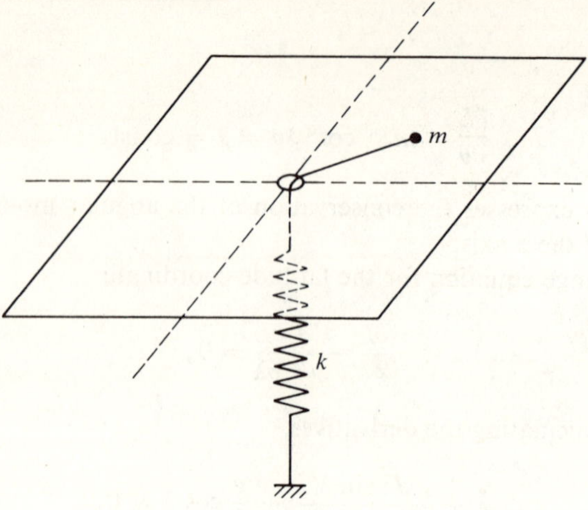

Fig. 11-5

11-2.7. Find the Lagrangian for a simple pendulum of mass m_2 and length l supported from a bead of mass m_1 that is free to slide in a horizontal line within the plane in which m_2 moves. What is the most convenient set of generalized coordinates for this problem?

11-2.8. A simple pendulum of mass m and length l has a support point that moves uniformly with constant frequency ω on a vertical circle of radius a. Determine the motion of the pendulum bob.

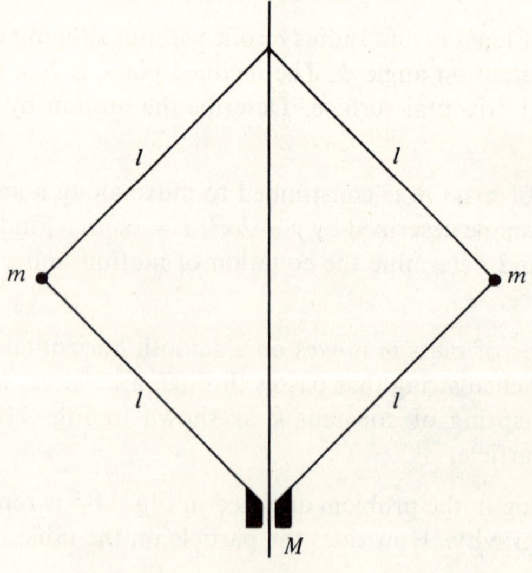

Fig. 11-6

11-2.9. A flyball governor for a steam engine is shown schematically in Fig. 11-6. Two particles, each of mass m, are attached by four arms of equal length l hinged at either end. The mass M is tubular and slides without friction on the vertical rod. The entire assembly rotates with constant angular frequency ω. Set up the Lagrangian and derive the equation of motion.

11-3 Lagrange Multipliers and the Forces of Constraint

As noted, constraints that can be expressed as coordinate relations are holonomic. If a system is subject only to such constraints, then a set of generalized coordinates can always be found, in terms of which the problem can be formulated without explicit reference to the constraints. This situation is not true of nonholonomic constraints that also involve the particle velocities, such as

$$\varphi_k(x_j, \dot{x}_j, t) = 0, \tag{11-21}$$

unless the constraint equations can be integrated to reduce them to holonomic form.

For example, suppose that we have a constraint relation of the form

$$a\dot{x} + b\dot{y} + c\dot{z} + e = 0, \tag{11-22}$$

where a, b, c, e are functions of x, y, z, t. In general, this equation is not integrable and the constraint is nonholonomic. If, however,

$$a = \frac{\partial \varphi}{\partial x}, \qquad b = \frac{\partial \varphi}{\partial y}, \qquad c = \frac{\partial \varphi}{\partial z}, \qquad e = \frac{\partial \varphi}{\partial t}, \tag{11-23}$$

where $\varphi = \varphi(x, y, z, t)$, then Eq. (11-22) becomes

$$\frac{\partial \varphi}{\partial x}\frac{dx}{dt} + \frac{\partial \varphi}{\partial y}\frac{dy}{dt} + \frac{\partial \varphi}{\partial z}\frac{dz}{dt} + \frac{\partial \varphi}{\partial t} = \frac{d\varphi}{dt} = 0, \tag{11-24}$$

which can be integrated to yield

$$\varphi(x, y, z, t) - \text{const} = 0, \tag{11-25}$$

and the constraint is revealed to be holonomic after all. From this example we conclude that constraints expressible in the form

$$\sum_i \frac{\partial \varphi_k}{\partial q_i}\, dq_i + \frac{\partial \varphi_k}{\partial t}\, dt = 0 \tag{11-26}$$

are reducible to holonomic form.

If the constraints for a problem are expressed in differential rather than algebraic form, then, as we saw in Section 10-3, they can be incorporated directly into Lagrange's equations by means of Lagrange multipliers. It

follows directly from Eqs. (10-63) and (10-64) that if the constraints are of
the form

$$\sum_i \frac{\partial \varphi_k}{\partial q_i}\, dq_i = 0,\tag{11-27}$$

then Lagrange's equations assume the form

$$\frac{\partial L}{\partial q_i} - \frac{d}{dt}\frac{\partial L}{\partial \dot{q}_i} + \sum_k \lambda_k(t)\,\frac{\partial \varphi_k}{\partial q_i} = 0.\tag{11-28}$$

It is left for the reader to show that, in fact, constraints like Eq. (11-26), where
a term $(\partial \varphi_k/\partial t)\, dt$ is included, also lead to Lagrange's equations expressed
in Eq. (11-28).

The main advantage of the Lagrangian formulation is, of course, that
explicit knowledge of the constraint forces is unnecessary. In certain cases,
however, the calculation of these forces may be required. By writing
Lagrange's equations in the form of Eq. (11-28), these forces are obtained.
The kth Lagrange multiplier is simply equal to the force exerted by the kth
constraint.

To illustrate, let us consider a disk of mass m rolling without slipping
down an inclined plane of inclination angle ϕ as depicted in Fig. 11-7. The
kinetic energy of the disk consists of translational and rotational energy:

$$T = \tfrac{1}{2}m\dot{x}^2 + \tfrac{1}{2}I\dot{\theta}^2$$

$$= \tfrac{1}{2}m\dot{x}^2 + \tfrac{1}{4}mR^2\dot{\theta}^2,\tag{11-29}$$

where $I = \tfrac{1}{2}m\dot{\theta}^2$ is the moment of inertia of the disk. The potential energy is

$$V = mg(l - x)\sin\phi,\tag{11-30}$$

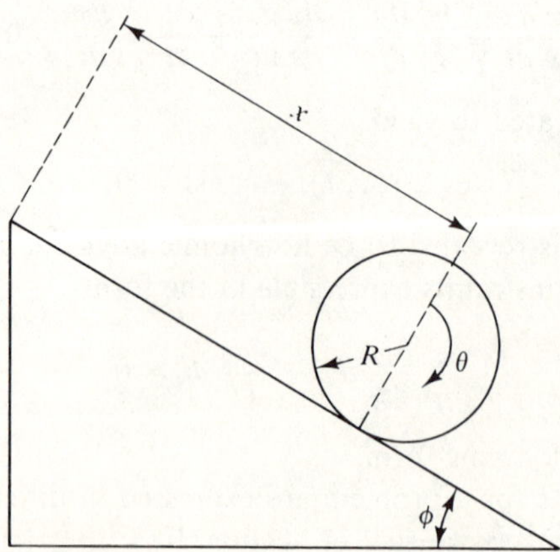

Fig. 11-7

where l is the length of the inclined plane and V is taken to be zero at $x = l$. The Lagrangian for the problem is therefore

$$L = \tfrac{1}{2}m\dot{x}^2 + \tfrac{1}{4}mR^2\dot{\theta}^2 - mg(l - x)\sin\phi. \tag{11-31}$$

The constraint on the problem is the requirement that the disk roll without slippage; it is expressed by

$$\varphi(x, \theta) = x - R\theta = 0. \tag{11-32}$$

Because of the constraint, the system is thus seen to have only one degree of freedom. We can choose either x or θ as our generalized coordinate and use Eq. (11-32) to eliminate the other. If, however, we wish to know the constraint force, we can use *both* x and θ as generalized coordinates and incorporate the constraint via the Lagrange multiplier technique. The Lagrange equations in this case are

$$\frac{\partial L}{\partial x} - \frac{d}{dt}\frac{\partial L}{\partial \dot{x}} + \lambda\frac{\partial \varphi}{\partial x} = 0 \tag{11-33}$$

and

$$\frac{\partial L}{\partial \theta} - \frac{d}{dt}\frac{\partial L}{\partial \dot{\theta}} + \lambda\frac{\partial \varphi}{\partial \theta} = 0. \tag{11-34}$$

The calculation of the derivatives is straightforward, and we find for the equations of motion

$$m\ddot{x} - mg\sin\phi - \lambda = 0, \tag{11-35}$$

$$\tfrac{1}{2}mR^2\ddot{\theta} + \lambda R = 0. \tag{11-36}$$

These equations, together with Eq. (11-32), constitute a complete set of equations for the unknowns x, θ, and λ. Differentiating Eq. (11-32) twice

$$\ddot{x} - R\ddot{\theta} = 0 \tag{11-37}$$

and eliminating $\ddot{\theta}$ between Eqs. (11-36) and (11-37), we find

$$\lambda = -\tfrac{1}{2}m\ddot{x}. \tag{11-38}$$

Substituting this result into Eq. (11-35) yields

$$\ddot{x} = \tfrac{2}{3}g\sin\phi. \tag{11-39}$$

Hence

$$\lambda = -\tfrac{1}{3}mg\sin\phi \tag{11-40}$$

and

$$\ddot{\theta} = \frac{2g\sin\phi}{3R}, \tag{11-41}$$

which solves the problem.

We see from Eq. (11-35) that if the plane was frictionless and the disk were to slide without rolling, then $\ddot{x} = g\sin\phi$. Therefore the effect of the rolling constraint is to reduce the linear acceleration by a third. The friction force that causes the disk to roll is merely $(mg/3) \sin \phi$, directed up the plane as given by Eq. (11-40).

If it has been unnecessary to calculate the constraint force, then $\dot{\theta}$ might have been eliminated from the Lagrangian with the use of Eq. (11-32). Then

$$L = \tfrac{3}{4}m\dot{x}^2 - mg(l - x) \sin \phi. \tag{11-42}$$

In this case, there is only one generalized coordinate x and one Lagrange equation:

$$\frac{\partial L}{\partial x} - \frac{d}{dt}\frac{\partial L}{\partial \dot{x}} = 0, \tag{11-43}$$

and the equation of motion obtained is identical with Eq. (11-39). Although a simpler procedure, it does not yield the force of constraint.

EXERCISES

11-3.1. Show that constraints of the form of Eq. (11-26) can be incorporated by the use of Lagrange multipliers to yield Eq. (11-28). Why does the explicit time derivative not appear in this equation?

11-3.2. Reformulate problem 11-1.4 on the Atwood machine, using Lagrange multipliers, and determine the tension in the connecting string.

11-3.3. A particle of mass m is constrained to move in a smooth plane that rotates with constant angular frequency ω about a horizontal axis. A constant gravitational acceleration g acts vertically downward. Solve the problem to find the particle motion in the plane and the reaction of the plane on the particle.

11-3.4. A particle slides down a smooth stationary sphere under the action of gravity, starting from rest very near the top. Find the reaction of the sphere on the particle as a function of λ, the latitude angle. Find the value of λ for which the particle leaves the spherical surface.

11-3.5. A bead of mass m moves under gravity along a smooth wire bent into the form of a vertical circle of radius R. Calculate the reaction force of the wire on the bead as a function of the total energy of the bead and its position on the loop.

11-4 Conservation of Energy

The kinetic energy of a system of n particles written in terms of fixed cartesian coordinates is a homogeneous quadratic function of the velocity components:

$$T = \frac{1}{2}\sum_{i=1}^{n}\sum_{j=1}^{3} m_i \dot{x}_{ij}^2. \tag{11-44}$$

We wish to rewrite T in terms of the generalized coordinates and velocities. Using the transformation equations

$$x_{ij} = x_{ij}(q_k, t), \qquad k = 1, \ldots, 3n - N = f, \tag{11-45}$$

we find for the velocity components

$$\dot{x}_{ij} = \sum_{k=1}^{f} \frac{\partial x_{ij}}{\partial q_k} \dot{q}_k + \frac{\partial x_{ij}}{\partial t}. \tag{11-46}$$

Squaring this equation, we obtain

$$\dot{x}_{ij}^2 = \sum_{k,l} \frac{\partial x_{ij}}{\partial q_k} \frac{\partial x_{ij}}{\partial q_l} \dot{q}_k \dot{q}_l + 2 \sum_{k} \frac{\partial x_{ij}}{\partial q_k} \frac{\partial x_{ij}}{\partial t} \dot{q}_k + \left(\frac{\partial x_{ij}}{\partial t} \right)^2, \tag{11-47}$$

and the most general expression for the kinetic energy in terms of generalized variables becomes

$$T = \frac{1}{2} \sum_{i} \sum_{j,k,l} m_i \frac{\partial x_{ij}}{\partial q_k} \frac{\partial x_{ij}}{\partial q_l} \dot{q}_k \dot{q}_l + \sum_{i} \sum_{j,k} m_i \frac{\partial x_{ij}}{\partial q_k} \frac{\partial x_{ij}}{\partial t} \dot{q}_k$$

$$+ \frac{1}{2} \sum_{i} \sum_{j} m_i \left(\frac{\partial x_{ij}}{\partial t} \right)^2. \tag{11-48}$$

We now wish to specialize this result for the case in which time does not enter explicitly into the transformation, Eq. (11-45). Such systems involving only fixed constraints are said to be *scleronomic*. Here only the first group of summations in Eq. (11-48) survives, and the kinetic energy, in terms of generalized variables, is a homogeneous quadratic function, just as in the case of cartesian variables:

$$T = \sum_{k,l} \alpha_{kl} \dot{q}_k \dot{q}_l, \tag{11-49}$$

where the quantity α_{kl} is given by

$$\alpha_{kl} = \frac{1}{2} \sum_{i}^{n} \sum_{j}^{f} m_i \frac{\partial x_{ij}}{\partial q_k} \frac{\partial x_{ij}}{\partial q_l}. \tag{11-50}$$

Differentiating Eq. (11-49) with respect to q_m, where m is a summation index not to be confused with the mass of the particles, we find

$$\frac{\partial T}{\partial \dot{q}_m} = \sum_{i} \alpha_{ml} \dot{q}_l + \sum_{k} \alpha_{km} \dot{q}_k. \tag{11-51}$$

Multiplying this result by q_m and summing over m, we have

$$\sum_{m} \dot{q}_m \frac{\partial T}{\partial \dot{q}_m} = \sum_{l,m} a_{ml} \dot{q}_l \dot{q}_m + \sum_{k,m} a_{km} \dot{q}_k \dot{q}_m. \tag{11-52}$$

In perusing this equation, we note that, since all the indices are dummies, the two terms on the right-hand side must be identical and we obtain the useful result

$$\sum_i \dot{q}_i \frac{\partial T}{\partial \dot{q}_i} = 2T. \tag{11-53}$$

This relation is a special case of *Euler's theorem*, which states that if $f(x_i)$ is a homogeneous function of the x_i of degree n, then

$$\sum_i x_i \frac{\partial f}{\partial x_i} = nf. \tag{11-54}$$

Suppose that we are considering a closed system—that is, a system that does not interact with anything exterior to itself. In this case, the Lagrangian of the system cannot depend explicitly on time.[1] Thus $L = L(q_i, \dot{q}_i)$ and

$$\frac{dL}{dt} = \sum_i \frac{\partial L}{\partial q_i} \dot{q}_i + \sum_i \frac{\partial L}{\partial \dot{q}_i} \ddot{q}_i. \tag{11-55}$$

Using Lagrange's equations

$$\frac{\partial L}{\partial q_i} = \frac{d}{dt} \frac{\partial L}{\partial \dot{q}_i} \tag{11-56}$$

to eliminate $\partial L / \partial q_i$ in Eq. (11-55), we find

$$\frac{dL}{dt} = \sum_i \dot{q}_i \frac{d}{dt} \frac{\partial L}{\partial \dot{q}_i} + \sum_i \frac{\partial L}{\partial \dot{q}_i} \ddot{q}_i$$

$$= \sum_i \frac{d}{dt} \left(\dot{q} \frac{\partial L}{\partial \dot{q}_i} \right)$$

or

$$\frac{d}{dt} \left(L - \sum_i \dot{q}_i \frac{\partial L}{\partial \dot{q}_i} \right) = 0. \tag{11-57}$$

This is easily integrated to yield

$$L - \sum_i \dot{q}_i \frac{\partial L}{\partial \dot{q}_i} = -H = \text{const.} \tag{11-58}$$

If the potential energy of the system is neither velocity nor time dependent, then $V = V(q_i)$, since we are still considering a scleronomic system for which $x_{ij} = x_{ij}(q_k)$ or $q_k = q_k(x_{ij})$. Thus

$$\frac{\partial L}{\partial \dot{q}_i} = \frac{\partial (T - V)}{\partial \dot{q}_i} = \frac{\partial T}{\partial \dot{q}_i}, \tag{11-59}$$

---

[1] This conclusion also holds for systems embedded in a uniform force field.

which enables us to rewrite Eq. (11-58) as

$$T - V - \sum_i \dot{q}_i \frac{\partial T}{\partial \dot{q}_i} = -H. \qquad (11\text{-}60)$$

Using the Euler theorem result given in Eq. (11-53), we therefore obtain

$$T - V - 2T = -H$$

or $\qquad\qquad T + V = E = H = \text{const}; \qquad\qquad (11\text{-}61)$

that is, the total energy of the system E is a constant of the motion. The function H defined in Eq. (11-58) is called the *Hamiltonian* of the system and is equal to the total energy if the system is scleronomic and the potential energy is velocity independent. A system can be described by generalized coordinates that are in motion with respect to fixed cartesian axes and still be conservative; in this case, however, the Hamiltonian is no longer equal to the total energy.[2]

EXERCISES

11-4.1. Devise a mechanical system for each of the following cases such that a judicious choice of generalized coordinates results in
   (a) $H \neq E \neq \text{const}.$
   (b) $H = E \neq \text{const}.$
   (c) $E \neq H = \text{const}.$
   (d) $E = H = \text{const}.$

11-4.2. A simple pendulum consists of a bob of mass m attached by a string to a fixed point. The length of the string is $l = a - bt$, where a and b are constants. Compare the Hamiltonian function and the total energy of the system and discuss your conclusions.

11-5 Hamilton's Equations

Clearly, the Lagrangian of a system expressed in the fixed cartesian coordinates of an inertial frame is invariant to a translation of the entire system in that space. For simplicity, we shall consider a one-particle system and write the Lagrangian as $L(x_j, \dot{x}_j)$. The change in L due to an infinitesimal displacement of the particle $\delta r = \sum_j \delta x_j \, \hat{e}_j$ is

$$\delta L = \sum_j \frac{\partial L}{\partial x_j} \delta x_j + \sum_j \frac{\partial L}{\partial \dot{x}_j} \delta \dot{x}_j = 0. \qquad (11\text{-}62)$$

---

[2] Systems for which $x_{ij} = x_{ij}(q_k, t)$ are termed *rheonomic*.

For a simple displacement, the δx_j are not explicit or implicit functions of time. Hence

$$\delta \dot{x}_j = \delta \frac{dx_j}{dt} = \frac{d}{dt} \delta x_j = 0 \tag{11-63}$$

and δL reduces to

$$\delta L = \sum_j \frac{\partial L}{\partial x_j} \delta x_j = 0. \tag{11-64}$$

Since each of the δx_j is independent, Eq. (11-64) implies that

$$\frac{\partial L}{\partial x_j} = \frac{d}{dt} \frac{\partial L}{\partial \dot{x}_j} = 0 \tag{11-65}$$

by virtue of Lagrange's equations. So

$$\frac{\partial L}{\partial \dot{x}_j} = \frac{\partial}{\partial \dot{x}_j}(T - V) = \frac{\partial T}{\partial \dot{x}_j} = \frac{\partial}{\partial \dot{x}_j}\left(\tfrac{1}{2}m \sum_i \dot{x}_i^2\right)$$

$$= m\dot{x}_j = p_j = \text{const}, \tag{11-66}$$

and we see that the homogeneity of space in an inertial frame implies that the linear momentum of a closed system is constant in time.

By analogy, we can extrapolate this result for the case in which the Lagrangian is expressed in generalized coordinates and define the generalized momenta as

$$p_i = \frac{\partial L}{\partial \dot{q}_i}. \tag{11-67}$$

Using this relation, the definition of the Hamiltonian function, Eq. (11-58), can be written

$$H = \sum_i p_i \dot{q}_i - L. \tag{11-68}$$

As we have seen, the Lagrangian is taken to be a function of q_i, $\dot{q}_i$, and (possibly) t. Thus a transformation of the form

$$\dot{q}_i = \dot{q}_i(q_k, p_k, t) \tag{11-69}$$

can be derived from Eq. (11-67) and the Hamiltonian is then written

$$H = (q_k, p_k, t) = \sum_i p_i \dot{q}_i - L(q_k, \dot{q}_k, t). \tag{11-70}$$

The Hamiltonian is always considered as a function of q_k, p_k, t, just as the Lagrangian is always written in terms of $q_k, \dot{q}_k, t$.

The equations of motion associated with the Hamiltonian $H(q_k, p_k, t)$ are readily derived by rewriting Hamilton's principle, using Eq. (11-70),

$$\delta \int_{t_1}^{t_2} \left(\sum_i p_i \dot{q}_i - H \right) dt = 0. \tag{11-71}$$

Carrying out the variation, we find

$$\int_{t_1}^{t_2} \sum_i \left(p_i\, \delta \dot{q}_i + \dot{q}_i\, \delta p_i - \frac{\partial H}{\partial q_i}\, \delta q_i - \frac{\partial H}{\partial p_i}\, \delta p_i \right) dt = 0. \tag{11-72}$$

In the Hamiltonian formulation, the q_i and the p_i are considered to be independent. The $\dot{q}_i$ are *not* independent of the q_i. Therefore, in the term $\delta \dot{q}_i$, the variation and time differentiation operations can be commuted, and the first term in Eq. (11-72) becomes

$$\int_{t_1}^{t_2} \sum_i p_i\, \delta \dot{q}_i\, dt = \int_{t_1}^{t_2} \sum_i p_i \frac{d(\delta q_i)}{dt}\, dt. \tag{11-73}$$

Integrating the right-hand side by parts, the integrated term vanishes and we find

$$\int_{t_1}^{t_2} \sum_i p_i\, \delta \dot{q}_i\, dt = -\int_{t_1}^{t_2} \sum_i \dot{p}_i\, \delta q_i\, dt. \tag{11-74}$$

Using this result, Eq. (11-72) becomes

$$\int_{t_1}^{t_2} \sum_i \left[\left(\dot{q}_i - \frac{\partial H}{\partial p_i} \right) \delta p_i - \left(\dot{p}_i + \frac{\partial H}{\partial q_i} \right) \delta q_i \right] dt = 0. \tag{11-75}$$

Since we consider δq_i and δp_i to be independent variations in the Hamiltonian formalism, the expressions in parentheses must both vanish and we have Hamilton's equations of motion

$$\dot{q}_i = \frac{\partial H}{\partial p_i}, \tag{11-76}$$

$$-\dot{p}_i = \frac{\partial H}{\partial q_i}, \tag{11-77}$$

which, because of their symmetry, are also known as the *canonical equations of motion*. It should be explained that the coordinates and momenta are not "independent" in the strictest sense of the word. If the time dependence of the coordinates is known, the problem is, of course, solved. Since q_i and $\dot{q}_i$ are related by a simple time derivative that is independent of the behavior of the system

$$\dot{q}_i = \frac{d}{dt} q_i(t), \tag{11-78}$$

the relation between the q_i and the p_i,

$$p_i = \frac{\partial L}{\partial \dot{q}_i}(q_i, \dot{q}_i, t), \tag{11-79}$$

are the equations of motion. Therefore the elimination of the assumed independence of q_i and p_i only occurs when the problem is solved.

If f is the number of degrees of freedom of the system in question, the canonical equations constitute a set of $2f$ first-order equations that replaces the f second-order Lagrangian set. In order to formulate the canonical equations for a problem, it is necessary to write the appropriate Hamiltonian as a function of the generalized coordinates and momenta. Sometimes, notably when $H = E$, it is possible to do so directly. In more complicated cases, it is necessary first to write the Lagrangian and then to calculate the generalized momenta by using Eq. (11-67).

As an example of the formulation of a problem via the Hamiltonian technique, we shall again consider the disk rolling without slipping down a plane of length l and inclination angle ϕ. The Lagrangian for this case was found to be

$$L = \tfrac{3}{4}m\dot{x}^2 - mg(l - x)\sin\phi \tag{11-80}$$

from which we find for the generalized momentum p_x:

$$p_x = \frac{\partial L}{\partial \dot{x}} = \tfrac{3}{2}m\dot{x}. \tag{11-81}$$

So using Eq. (11-70), we find for the Hamiltonian

$$H = p_x\dot{x} - L$$

$$= p_x\left(\frac{2}{3}\frac{p_x}{m}\right) - \tfrac{3}{4}m\left(\frac{2}{3}\frac{p_x}{m}\right)^2 + mg(l - x)\sin\phi$$

$$= \frac{p_x^2}{3m} + mg(l - x)\sin\phi. \tag{11-82}$$

Hamilton's equations of motion then yield

$$\frac{\partial H}{\partial p_x} = \frac{2}{3}\frac{p_x}{m} = \dot{x} \tag{11-83}$$

and

$$\frac{\partial H}{\partial x} = -mg\sin\phi = -\dot{p}_x. \tag{11-84}$$

Taking the time derivative of Eq. (11-83) and eliminating p_x between the resulting equation and Eq. (11-84), we find

$$\ddot{x} = \tfrac{2}{3}g\sin\phi, \tag{11-85}$$

in agreement with Eq. (11-39).

Here the solution of the problem by the Hamiltonian method proved to be somewhat less straightforward than with the Lagrangian formulation, as is often true. However, because there is greater freedom in choosing the variable in the Hamiltonian formulation (the q_i and p_i are independent, whereas the q_i and the $\dot{q}_i$ are not), an advantage often results from using the Hamiltonian method. For example, in celestial mechanics, particularly when the motion of the body of interest is perturbed by the presence of other bodies, the Hamiltonian approach is superior.

Inspection of Eq. (11-77) shows that if q_i does not appear in the Hamiltonian, then $\dot{p}_i = 0$, and the conjugate momentum p_i is a constant of the motion. Such coordinates as are not contained explicitly in the Hamiltonian are said to be *cyclic*. If a coordinate is cyclic in H, it is also cyclic in L. Even though q_k may not appear in L, the corresponding generalized velocity $\dot{q}_k$ will still, in general, be present. Thus

$$L = L(q_1, q_2, \ldots, q_{k-1}, q_{k+1}, \ldots, q_f, \dot{q}_1, \ldots, \dot{q}_f, t), \qquad (11\text{-}86)$$

and no reduction in the number of degrees of freedom of the system results from the cyclic nature of the coordinate; there are still f second-order equations to be solved. In the Hamiltonian formulation, however, if q_k is cyclic, then $p_k = c_k = $ const and

$$H = H(q_1, \ldots, q_{k-1}, q_{k+1}, \ldots, q_f, p_1, \ldots, p_{k-1}, c_k, p_{k+1}, \ldots, p_f, t). \quad (11\text{-}87)$$

In this case, there are $2f - 2$ first-order equations to be solved, and the number of degrees of freedom of the system has effectively been reduced to $f - 1$. Thus the Hamiltonian formulation is particularly well suited for dealing with problems in which one or more of the coordinates are cyclic. Obviously, the best possible formulation of a problem would be to arrange matters so that *all* the coordinates are cyclic. Then each coordinate could be described trivially as

$$q_k(t) = \int \frac{\partial H}{\partial c_k}\, dt. \qquad (11\text{-}88)$$

There are, in fact, transformations known as *canonical transformations* that do exactly that. The system of dynamics that follows from this approach is known as *Hamilton–Jacobi theory* and is particularly useful in modern field theories. We shall not pursue this line of development here, however. The curious reader is referred to more advanced books.[3]

---

[3] See, for example, Herbert Goldstein, *Classical Mechanics*. Reading, Mass.: Addison-Wesley Publishing Company, 1959, Chapter 9.

EXERCISES

11-5.1. The principle of least action can be stated as

$$\delta \int_{t_1}^{t_2} \sum_i \dot{q}_i p_i \, dt = 0.$$

For what type of mechanical system does this principle hold?

11-5.2. If $\varphi(p_k, q_k)$ and $\psi(p_k, q_k)$ are any two continuous functions of the generalized coordinates and momenta, then the *Poisson brackets* of φ and ψ are defined by

$$[\varphi, \psi] \equiv \sum_k \left(\frac{\partial \varphi}{\partial q_k} \frac{\partial \psi}{\partial p_k} - \frac{\partial \varphi}{\partial p_k} \frac{\partial \psi}{\partial q_k} \right).$$

Based on this definition, show that

$$\frac{d\varphi}{dt} = [\varphi, H] + \frac{\partial \varphi}{\partial t}$$

and hence that

$$\frac{dH}{dt} = \frac{\partial H}{\partial t}.$$

11-5.3. Write the Hamiltonian function for a particle of mass m moving freely in a conservative force field of potential V. Show that if cartesian coordinates are used, the canonical equations reduce to Newton's equations.

11-5.4. A particle of mass m moves under the influence of a central force given by

$$F = -\frac{k}{r^2} e^{-\alpha t} \hat{r},$$

where k and α are positive constants. Does $H = E$? Is H a constant?

11-5.5. Reexamine problem 11-2.5 from the Hamiltonian point of view.

11-5.6. Calculate the Hamiltonian for the double pendulum described in problem 11-2.6. Find the canonical equations of motion for the system.

11-6 Additional Comments on Hamilton's Principle

In this section we contrast the points of view represented by the Lagrangian and Newtonian formulations of classical mechanics. First, and most importantly, we note that the physical content of both theories is the same. The results of a Newtonian or Lagrangian (or Hamiltonian) analysis of a given problem must agree.

The Newtonian approach, as we have seen, emphasizes the effect of an external agency (the *force*) acting on a body. In the formalisms derivable from Hamilton's principle, only quantities associated with the body (the kinetic and potential *energies*) enter into consideration. The concept of force never arises. Since energy is a scalar quantity, the Lagrangian and Hamiltonian functions are invariant to coordinate transformation. The scope of such coordinate transformations is not limited to transformations between orthogonal coordinate systems in ordinary space; they may include transformations between ordinary and generalized coordinates. Thus a problem that may involve very complicated equations of motion in ordinary coordinates or, because of a lack of knowledge of the forces of constraint, a problem for which we are unable to write Newton's second law may be transformed into a configuration space chosen so as to accommodate automatically the constraints and yield the maximum simplification of the problem.

A more subtle difference in the differential Newton's laws approach and the integral Hamilton's principle approach to mechanics is that the former takes a *cause* (force) *and effect* (motion) point of view, whereas, according to Hamilton's principle, the motion of a body results from the tendency of nature to achieve a given *purpose*—namely, the extremization of the time integral of the difference between the kinetic and potential energies.

THE STABILITY
OF EQUILIBRIUM

We live in a world in which all equilibria are subject to fluctuations of one sort or another. We must, therefore, judge the quality of an equilibrium by its response to the perturbations that it experiences. We say that systems for which the perturbation does not grow beyond certain bounds and for which the equilibrium condition is maintained, at least in an average sense, are *stable*. Any other response to a perturbation identifies an *unstable* (exponentially growing perturbation) or *overstable* (oscillation with exponentially growing envelope) equilibrium.

12-1 Stability in a Conservative Force Field

Let us consider an n-dimensional mechanical system in a conservative force field. The requirement that the external force on the system vanish in order that a state of equilibrium result is equivalent to requiring that the potential energy of the system have an extremal value:

$$\frac{\partial V}{\partial x_i} = 0, \qquad i = 1, 2, \ldots, n. \tag{12-1}$$

We shall now demonstrate that if this extremum is a minimum, then the equilibrium is stable.

The system under consideration is scleronomic. Consequently, the total energy is conserved

$$T + V = T_0 + V_0 \tag{12-2}$$

and the kinetic energy of the system is a homogeneous quadratic function and so is positive definite. Equation (12-2) implies that if the change in the potential energy is positive—$\Delta V = V - V_0 > 0$—then the change in the kinetic energy must be equal in magnitude and opposite in sign:

$$\Delta T = -\Delta V. \tag{12-3}$$

Let us consider the system at its equilibrium position q_{0i} to be suddenly endowed with a virtual kinetic energy τ. If the potential energy increases in *all* directions from equilibrium, then a displacement of the system in velocity space must lead to simple oscillation about the equilibrium.

As a simple example, consider a particle in a gravitational field. The particle is constrained to move only on the curve shown in Fig. 12-1. Since the potential is mgz and the coordinates x and z are not independent, owing to the constraint, we can also label the z axis as $V(x)$. It is possible to visualize this situation directly as a particle required to move over rough terrain in one horizontal direction. Obviously, A, B, C, D are all equilibrium points. The equilibria at A and D are unstable, corresponding to local maxima of $V(x)$.

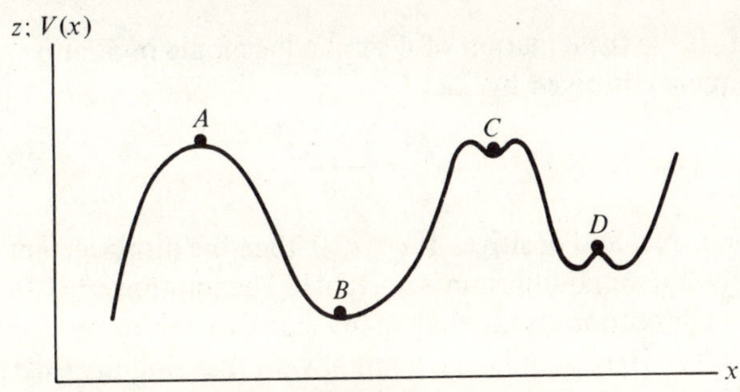

Fig. 12-1

Any displacement in velocity or position will cause the ball to roll down the hill, so to speak. The equilibria at B and C are stable, corresponding as they do to local potential minima. Note that the equilibrium at C is only stable for small perturbation energies; for perturbation energies exceeding the depth of the local potential well, the system is unstable. We shall limit ourselves to small perturbations and so will regard equilibria B and C as equivalent.

If, in addition to forces derivable from the potential V, the system is also subject to forces at right angles to the velocities, such as the Coriolis or magnetic forces, the conclusion is the same. The presence of friction forces acting in a direction opposed to the velocity similarly does not alter the conclusion (due to Dirichlet) that potential minima correspond to stable equilibria.

Next, let us adopt a dynamic rather than an energetic point of view and discuss a generalization due to Liapunov. For simplicity, we consider a one-dimensional spatial displacement of a single particle. The equation of motion of the particle is

$$m \frac{d^2x}{dt^2} = -V'(x).$$ (12-4)

If we take the displacement $\xi = x - x_0$ to be small and limit the equation of motion to apply only in the neighborhood of the equilibrium, we can replace the right-hand side of Eq. (12-4) by the first nonzero term of the Taylor expansion of $-V'(x)$ about the equilibrium point:

$$-V'(x) \approx -V'(x_0) - V''(x_0) \cdot (x - x_0) - \cdots$$

$$= -V''(x_0) \cdot \xi,$$ (12-5)

since $-V'(x_0) = F(x_0)$, which vanishes by definition. The equation of motion in the neighborhood of equilibrium then becomes

$$m \frac{d^2\xi}{dt^2} + V''(x_0) \cdot \xi = 0.$$ (12-6)

We see that this is the equation of a simple harmonic oscillator in which the angular frequency is given by

$$\omega^2 = \frac{V''(x_0)}{m}$$ (12-7)

as long as ω^2 is real and positive. If $\omega^2 < 0$, then the displacement ξ increases exponentially and the equilibrium is unstable. The solution $\omega^2 = V''(x_0)/m = 0$ represents an inflection point that many books refer to as a condition of neutral stability. Here we take the point of view that stability (like pregnancy) either obtains or does not; so we shall consider the case $\omega^2 = 0$ to be unstable.

If a frictional dissipation is introduced into the system, ω^2 and ω become complex and a damped oscillation about the equilibrium point characterizes stable systems. It is also possible to construct systems with a positive energy feedback in which energy is supplied to the system at a rate proportional to the velocity $\dot{\xi}$. This situation also results in a complex frequency ω. This case is called *overstability*, a misleading term, since the exponentially increasing amplitude of the oscillation leads to runaway solutions. We shall regard overstable solutions as a form of instability.

EXERCISE

12-1.1. A particle moves in a potential field $V(x) = a - bx^2$, where a and b are constants. Investigate the stability of any equilibria that exist and calculate the frequency of stable oscillations (if any) and the growth rates of instabilities (if any).

12-2 Stability of Circular Orbits—the
Effective Potential

A circular orbit is a possible equilibrium state for a particle moving under the influence of any central force if the force center is located at the center of the orbit. In some cases, circular orbits in which the force and curvature centers are not coincident are also possible (see problem 12-2.1). In the present analysis, we shall restrict our attention to those cases for which the two centers coincide.

In place of the attractive central force $F(r)$ we shall write our equations in terms of the quantity $g(r) = -F(r)/m$; thus $g(r)$ is a positive quantity. The equilibrium condition for a circular orbit of radius a can then be written

$$\frac{u^2}{a} = g(a), \tag{12-8}$$

where u is the linear velocity of the particle in its orbit. We wish to obtain the criteria for stability of this orbit to radial perturbations.

First, we write the radial component of the equation of motion

$$m(\ddot{r} - r\dot{\theta}^2) = -mg(r). \tag{12-9}$$

We shall not eliminate the time from this equation as we did in Section 4-2. Instead we shall use the conservation of angular momentum in a central force field, $J = mr^2\dot{\theta}$, to eliminate $\dot{\theta}$. Thus Eq. (12-9) becomes

$$\ddot{r} - \frac{J^2}{m^2 r^3} = -g(r). \tag{12-10}$$

In order to write the equation of motion for a particle initially traveling in a circle of radius a, subject to a small radial displacement ξ, we shall use

$$r = a + \xi, \tag{12-11}$$

where a is the constant radius of the circular orbit and $\xi \ll a$. Substituting Eq. (12-11) into Eq. (12-10) yields

$$\ddot{\xi} - \frac{J^2}{m^2 a^3 [1 + (\xi/a)]^3} = -g(r). \tag{12-12}$$

This equation is linearized by incorporating our restriction of small displacement. We expand the left-hand side of Eq. (12-12) in a power series in ξ/a, using the binomial expansion

$$(1 + x)^n = 1 + nx + n(n - 1)\frac{x^2}{2!} + n(n - 1)(n - 2)\frac{x^3}{3!} + \cdots \tag{12-13}$$

and the right-hand side by a Taylor expansion of $g(r)$ about the point $a = r$.

To first order in small quantities, we obtain

$$\ddot{\xi} - \frac{J^2}{m^2 a^3}\left(1 - \frac{3\xi}{a}\right) = -g(a) - \xi g'(a). \tag{12-14}$$

From the equilibrium relation (12-8) we have

$$\frac{u^2}{a} = \frac{(a\dot\theta)^2}{a} = \frac{J^2}{m^2 a^3} = g(a). \tag{12-15}$$

Eliminating J^2 between this equation and the linearized equation of motion, Eq. (12-14), we have

$$\ddot{\xi} + \left[\frac{3g(a)}{a} + g'(a)\right]\xi = 0. \tag{12-16}$$

The solution of this equation will be oscillatory only if the quantity in brackets is positive. Thus the criterion for stability is

$$\omega^2 = \frac{3g(a)}{a} + g'(a) > 0. \tag{12-17}$$

If we take the force law to be of the form

$$g(r) = \frac{k}{r^n}, \tag{12-18}$$

where k and n are constants, Eq. (12-17) can be shown to give

$$\frac{g(a)}{a}(3 - n) > 0. \tag{12-19}$$

Thus, for central forces of the form $-k/r^n$, only those cases for which $n < 3$ will provide stable circular orbits. For an inverse cube law ($n = 3$), it would be necessary to go to higher-order terms in Eq. (12-14) in order to determine the stability.

Comparing Eqs. (12-6) and (12-16), we should expect the square bracket in the latter to correspond to the second derivative of some scalar potential evaluated at the point of equilibrium. Recalling the concept of an effective potential for central force motion in the radial direction introduced in Section 4-2,

$$V_{\text{eff}} = \frac{J^2}{2mr^2} + V(r), \tag{12-20}$$

it can be shown that the criterion for stability is given by the minimization of V_{eff}—that is,

$$V''_{\text{eff}} > 0. \tag{12-21}$$

In the case of circular motion, the centrifugal potential $J^2/2mr^2$ is simply equal to the total kinetic energy of the system, and so $V_{\text{eff}} = E = T + V$.

The guiding principle in defining an effective potential energy for the purpose of examining the stability of a mechanical system lies in the answer to the question: Where are the reservoirs of free energy available to drive an instability? In complex systems generally it is necessary to take thermodynamics into account.

As an interesting example, consider the problem of stellar structure. Here the total energy of the star includes thermal and gravitational terms. In order for the star to be gravitationally bound, the total energy must be less than zero,

$$E < 0. \tag{12-22}$$

The condition for an equilibrium to exist is

$$\delta V_{\text{eff}} = \delta E = 0, \tag{12-23}$$

which can be shown to yield

$$\frac{dp}{dr} = -\frac{\rho(r)\gamma}{r^2} \int_0^r 4\pi r^2 \rho(r) \, dr, \tag{12-24}$$

where γ is the gravitational constant. This equation describes the balance between thermal pressure and self-gravitation.

The equilibrium determined by $\delta E = 0$ may be stable or unstable. The criterion for stability

$$\delta^2 E > 0 \tag{12-25}$$

is found to yield

$$\tfrac{2}{3}\alpha^2 \int_0^V (\gamma - \tfrac{4}{3}) p \, d\tau > 0, \tag{12-26}$$

where α is a scale factor and γ is the ratio of specific heats, $\gamma = c_p/c_v$, which is 5/3 for a monatomic gas.

The interpretation of Eqs. (12-22) and (12-26) allows us to place limits on stellar size. Stars with masses less than 0.1 solar masses do not produce enough gravitation potential to ignite the thermo-nuclear fires arrive at an equilibrium state. In large stars dynamic pressure due to radiation begins to play a role in the pressure balance. For radiation pressure, the value of γ is 4/3. Thus Eq. (12-26) tells us that when radiation pressure becomes dominant over thermal pressure, an instability is to be expected. This condition sets an upper limit on stellar mass at about 50 solar masses.

EXERCISES

12-2.1. A particle of mass m, under the action of a central force, moves in a circular orbit of radius a passing through the center of force. Show that the force law is an inverse fifth-power law of attraction given by

$$F(r) = -\frac{8a^2 J^2}{mr^5}.$$

12-2.2. A particle moves in a circular orbit of radius a under the action of an attractive force

$$F(r) = -\left(\frac{b}{r^2} + \frac{c}{r^4}\right),$$

where b and c are positive constants and the center of attraction and the center of the orbit coincide. Show that the orbit is stable to small radial perturbation if $a^2 b > c$.

12-2.3. A particle moves in a circular orbit under the action of an attractive force specified as

$$F(r) = -\frac{k}{r^2}\, e^{-r/a}.$$

Show that the orbit is stable if the radius of the orbit $r_0 < a$ and unstable if $r_0 > a$.

12-2.4. Show that an expansion of the type given in Eq. (12-14), taken to second order in ξ, is also inconclusive in determining the stability of circular orbits under an inverse cube force of attraction. What conclusion can be drawn from an examination of the effective potential?

12-3 Stability of Fluid Systems

In a mechanical system with one degree of freedom, such as that depicted in Fig. 12-1, the number of stable and unstable equilibria is approximately equal. So the chance that an arbitrarily selected equilibrium will turn out to be stable is about 50%. In a two-dimensional system four possible situations exist:

1. $\dfrac{\partial^2 V}{\partial x^2} > 0, \qquad \dfrac{\partial^2 V}{\partial y^2} > 0,$

2. $\dfrac{\partial^2 V}{\partial x^2} > 0, \qquad \dfrac{\partial^2 V}{\partial y^2} < 0,$

3. $\dfrac{\partial^2 V}{\partial x^2} < 0, \qquad \dfrac{\partial^2 V}{\partial y^2} > 0,$

4. $\dfrac{\partial^2 V}{\partial x^2} < 0, \qquad \dfrac{\partial^2 V}{\partial y^2} < 0,$

of which only (1) is stable. Thus only 25% of a random collection of two-dimensional equilibria will be stable. In the case of n degrees of freedom, the chance that an equilibrium is stable reduces to 2^{-n}. In a fluid n is a very large number, and the chance that a stable equilibrium will be selected without prior knowledge of general fluid stability criteria is almost nil.

To illustrate a fluid system in equilibrium, consider two incompressible, immiscible fluids having mass densities ρ_1 and ρ_2, respectively, located in a vessel as shown in Fig. 12-2(a). Clearly, the exchange of any two volume elements within one or the other fluid does not affect the equilibrium. The only perturbations of interest are those that affect the fluid interface, such as the ripple shown in Fig. 12-2(b). If $\rho_2 \neq \rho_1$, this ripple will alter the potential energy of the system, since some of fluid 1 has been lowered and an equal volume of fluid 2 has been raised. If $\rho_1 < \rho_2$, the potential energy is increased, whereas if the upper fluid is heavier ($\rho_1 > \rho_2$), the potential energy has decreased. In the former case, the perturbation leads to stable oscillations about the equilibrium; in the latter, the ripple grows exponentially until the upper fluid breaks through and the two fluids change places to establish a stable configuration. The instability that occurs when $\rho_1 > \rho_2$ is known as the *Rayleigh–Taylor instability*.

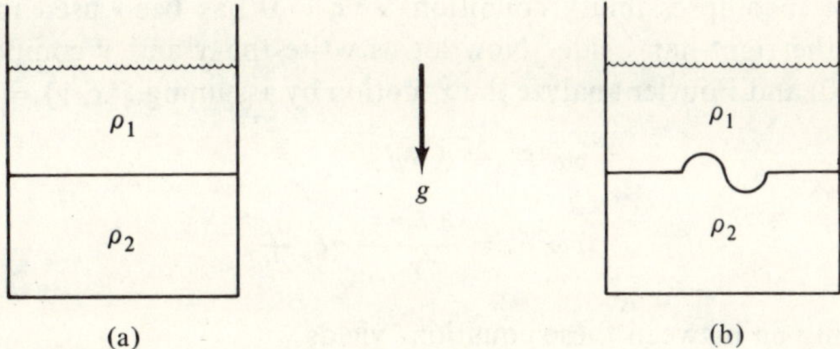

Fig. 12-2

We note that the Dirichlet approach gives us a quick answer to the question of stability. It does not give us the frequency of stable oscillations or the growth rate of the instability, however. For this information we must use the dynamical (Liapunov) approach. The perturbation of the interface between the fluids will be taken to vary sinusoidally in the xy plane as shown in Fig. 12-3. The equation of motion for a perturbed fluid system in static equilibrium to first order in small quantities is

$$\rho\ddot{\xi} = -\nabla\,\delta p + \hat{k}g\,\nabla\cdot(\rho\xi), \tag{12-27}$$

where δp is the increment of pressure arising from the perturbation and

$$g\,\delta\rho = g\,\nabla\cdot(\rho\xi) \tag{12-28}$$

is the corresponding gravitational force. We are interested in perturbations that vary as

$$\xi = \xi(x, y, t) = \xi(x, y)e^{i\omega t}, \tag{12-29}$$

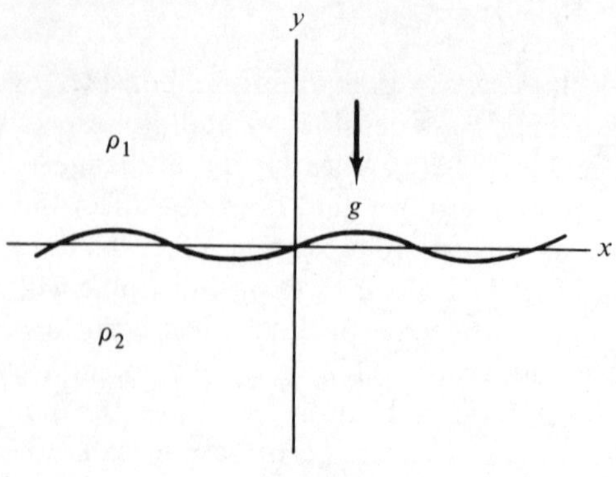

Fig. 12-3

for which Eq. (12-27) becomes

$$\rho\omega^2\boldsymbol{\xi} = \boldsymbol{\nabla}\,\delta p - \hat{k}g\boldsymbol{\xi}\cdot\boldsymbol{\nabla}\rho, \tag{12-30}$$

where the incompressibility condition $\boldsymbol{\nabla}\cdot\boldsymbol{\xi} = 0$ has been used in the last term on the right-hand side. Now let us write the x and y components of Eq. (12-30) and Fourier analyze the x motion by assuming $\xi(x, y) = \xi(y)e^{ikx}$:

$$\rho\omega^2\xi_x = ik\,\delta p, \tag{12-31}$$

$$\rho\omega^2\xi_y = \frac{\partial\,\delta p}{\partial y} - g\xi_y\frac{d\rho}{dy}. \tag{12-32}$$

Eliminating δp between these equations yields

$$\rho\omega^2\xi_y = \frac{\omega^2}{ik}\frac{\partial}{\partial y}(\rho\xi_x) - g\xi_y\frac{d\rho}{dy}. \tag{12-33}$$

Differentiating with respect to x and using the incompressibility condition $\boldsymbol{\nabla}\cdot\boldsymbol{\xi} = 0$ to eliminate ξ_x, we find

$$k^2\left(\rho\omega^2 + g\frac{d\rho}{dy}\right)\xi_y = \omega^2\frac{\partial}{\partial y}\left(\rho\frac{\partial\xi_y}{\partial y}\right). \tag{12-34}$$

For $y \neq 0$, $d\rho/dy$ vanishes and Eq. (12-34) reduces to

$$\frac{\partial^2\xi_y}{\partial y^2} - k^2\xi_y = 0, \tag{12-35}$$

which has as its solutions

$$\begin{aligned}
\xi_y &= \xi_{y0}e^{-ky}e^{ikx}, & y &> 0, \\
\xi_y &= \xi_{y0}e^{ky}e^{ikx}, & y &< 0.
\end{aligned} \tag{12-36}$$

374

Finally, we integrate Eq. (12-34) across the interface at $y = 0$ from $+\epsilon$ to $-\epsilon$, where $\epsilon \to 0$. The result is

$$k^2 g(\rho_1 - \rho_2)\xi_y = \omega^2(-k)\rho_1\xi_y - \omega^2(k)\rho_2\xi_y$$

or

$$\omega^2 = -kg\left(\frac{\rho_1 - \rho_2}{\rho_1 + \rho_2}\right), \tag{12-37}$$

which is the dispersion relation for the displacement ξ. We note that $\omega^2 > 0$ for $\rho_2 > \rho_1$, as we found earlier by the energy method. The case $\rho_1 = \rho_2$ implies $\omega = 0$ and corresponds to a trivial perturbation within a homogeneous liquid. Instability is realized for $\rho_1 > \rho_2$, and the maximum growth rate $\gamma = \pm i\omega$ occurs when $\rho_2 \to 0$:

$$\gamma_{\max} = \sqrt{kg}. \tag{12-38}$$

This situation is realized for the case of a plasma supported in a force field of acceleration g by a magnetic field. The plasma version of the Rayleigh–Taylor instability is known as the *Kruskal–Schwarzschild instability*; the physics is the same.

Another important instability that occurs in fluids takes place when one fluid is in motion across another, assuming that they are immiscible and that the lighter fluid is on top so that the equilibrium is not subject to Rayleigh–Taylor instability. The most common manifestations of this instability, known as the *Kelvin–Helmholtz instability*, are the gravity waves that make the ocean such an interesting place to sail.

We shall assume that the upper fluid of density ρ_1 has a velocity U_1 in the x direction and that the lower fluid of density $\rho_2 > \rho_1$ has a velocity U_2 in the x direction. The xy plane is horizontal and the interface is located at $z = 0$. The equation of motion

$$\rho\,\frac{\partial \mathbf{u}}{\partial t} + \rho(\mathbf{u} \cdot \nabla)\mathbf{u} = -\nabla p - g\rho\hat{\mathbf{k}} \tag{12-39}$$

will have convective terms in its linear form because of the presence of the streaming. If δp is the pressure increment and $\delta\rho$ is the corresponding increment of density that is due to the perturbation of the interface, then the linearized equation of motion, written in components, becomes

$$\rho\,\frac{\partial u_x}{\partial t} + \rho U\,\frac{\partial v_x}{\partial x} + \rho v_z\,\frac{dU}{dz} = -\frac{\partial\,\delta p}{\partial x}, \tag{12-40}$$

$$\rho\,\frac{\partial v_y}{\partial t} + \rho U\,\frac{\partial v_y}{\partial y} = -\frac{\partial\,\delta p}{\partial y}, \tag{12-41}$$

$$\rho\,\frac{\partial v_z}{\partial t} + \rho U\,\frac{\partial v_z}{\partial x} = -\frac{\partial}{\partial z}\,\delta p - g\,\delta\rho, \tag{12-42}$$

where $v = v(v_x, v_y, v_z)$ is the perturbation in velocity space, $|v| \ll |U|$, and the stream velocity U has been specified as an arbitrary function of z, $U(z)$, as has the density $\rho(z)$. The continuity equation and the incompressibility conditions are

$$\frac{\partial\,\delta\rho}{\partial t} + U\frac{\partial\,\delta\rho}{\partial x} = -v_z\frac{d\rho}{dz} \tag{12-43}$$

and

$$\frac{\partial v_x}{\partial x} + \frac{\partial v_y}{\partial y} + \frac{\partial v_z}{\partial z} = 0. \tag{12-44}$$

We shall look for an oscillatory solution for the perturbation of the form

$$v = v_0 \exp i(k_x x + k_y y + \omega t). \tag{12-45}$$

In terms of this assumption, Eqs. (12-40) through (12-44) become

$$i\rho(\omega + k_x U)v_x + \rho v_z\frac{dU}{dz} = -ik_x\,\delta p, \tag{12-46}$$

$$i\rho(\omega + k_x U)v_y = -ik_y\,\delta p, \tag{12-47}$$

$$i\rho(\omega + k_x U)v_z = -\frac{d\,\delta p}{dz} - g\,\delta\rho, \tag{12-48}$$

$$i(\omega + k_x U)\,\delta\rho = -v_z\frac{d\rho}{dz}, \tag{12-49}$$

$$i(k_x v_x + k_y v_y) = -\frac{dv_z}{dz}. \tag{12-50}$$

We multiply Eqs. (12-46) and (12-47) by $-ik_x$ and $-ik_y$, respectively, and add, using Eq. (12-50). The result is

$$i\rho(\omega + k_x U)\frac{dv_z}{dz} - i\rho k_x v_z\frac{dU}{dz} = -k^2\,\delta p, \tag{12-51}$$

where $k^2 \equiv k_x^2 + k_y^2$. Combining Eqs. (12-48) and (12-49), we find

$$i\rho(\omega + k_x U)v_z = -\frac{d\,\delta p}{dz} - ig\frac{d\rho}{dz}\frac{v_z}{\omega + k_x U}. \tag{12-52}$$

Eliminating δp between these two equations gives

$$\frac{d}{dz}\left[\rho(\omega + k_x U)\frac{dv_z}{dz} - \rho k_x v_z\frac{dU}{dz}\right] - k^2\rho(\omega + k_x U)v_z$$

$$= gk^2\left(\frac{v_z}{\omega + k_x U}\right)\frac{d\rho}{dz}. \tag{12-53}$$

If we next require that

$$\rho(z) = \begin{cases} \rho_1, & z > 0, \\ \rho_2, & z < 0, \end{cases} \tag{12-54}$$

and

$$U(z) = \begin{cases} U_1, & z > 0, \\ U_2, & z < 0, \end{cases} \tag{12-55}$$

then Eq. (12-53) evaluated for $|z| > 0$ reduces to

$$\rho(\omega + k_x U)\left(\frac{d^2}{dz^2} - k^2\right)v_z = 0 \tag{12-56}$$

or, since $\rho(\omega + k_x U) \neq 0$, to

$$\left(\frac{d^2}{dz^2} - k^2\right)v_z = 0. \tag{12-57}$$

Prior to integrating this equation, we must establish the boundary condition on v_z at the interface. The displacement of an arbitrary interface δz owing to the perturbation is given by

$$\frac{d\,\delta z}{dt} = \frac{\partial\,\delta z}{\partial t} + U\frac{\partial}{\partial x}\,\delta z = v_z \tag{12-58}$$

or in terms of Fourier components

$$i(\omega + k_x U)\,\delta z = v_z. \tag{12-59}$$

If we apply this equation at the interface and integrate over z from $+\epsilon$ to $-\epsilon$ as $\epsilon \to 0$, we see that the quantity $v_z/(\omega + k_x U)$ must be continuous across the interface between the two fluids. So we write the solutions to Eq. (12-57) as

$$\begin{aligned} v_{z1} &= C(\omega + k_x U_1)e^{-kz}, & z > 0, \\ v_{z2} &= C(\omega + k_x U_2)e^{kz}, & z < 0. \end{aligned} \tag{12-60}$$

Finally, we integrate Eq. (12-53) across the interface, applying Eqs. (12-60), and find

$$\rho_1(\omega + k_x U_1)^2 + \rho_2(\omega + k_x U_2)^2 = gk(\rho_2 - \rho_1). \tag{12-61}$$

Setting

$$\alpha_1 \equiv \frac{\rho_1}{\rho_1 + \rho_2}, \qquad \alpha_2 \equiv \frac{\rho_2}{\rho_1 + \rho_2} \tag{12-62}$$

and expanding the quadratic expressions, Eq. (12-61) becomes

$$\omega^2 + 2k_x(\alpha_1 U_1 + \alpha_2 U_2)\omega + k_x^2(\alpha_1 U_1^2 + \alpha_2 U_2^2) - gk(\alpha_2 - \alpha_1) = 0, \tag{12-63}$$

which has as its roots

$$\omega = -k_x(\alpha_1 U_1 + \alpha_2 U_2) \pm [gk(\alpha_2 - \alpha_1) - k_x^2\alpha_1\alpha_2(U_2 - U_1)^2]^{1/2}. \tag{12-64}$$

Several interesting conclusions can be drawn from this dispersion relation. When $k_x = 0$,

$$\omega = \pm \sqrt{gk(\alpha_2 - \alpha_1)}; \qquad (12\text{-}65)$$

hence perturbations transverse to the x direction do not "see" the streaming and are stable. In every other direction instability occurs when

$$k_x^2 \alpha_1 \alpha_2 (U_2 - U_1)^2 > gk(\alpha_2 - \alpha_1). \qquad (12\text{-}66)$$

This can be reexpressed in a more useful form as

$$k > \frac{g(\alpha_2 - \alpha_1)}{\alpha_1 \alpha_2 (U_2 - U_1)^2 \cos^2 \phi}, \qquad (12\text{-}67)$$

where ϕ is the angle between the vectors $\boldsymbol{k}$ and $\boldsymbol{U}$. According to Eq. (12-67), for a given $U_2 - U_1$, instability arises for the least wave number when $\boldsymbol{k} \parallel \boldsymbol{U}$; then

$$k_{\min} = \frac{g(\alpha_2 - \alpha_1)}{\alpha_1 \alpha_2 (U_2 - U_1)^2} \qquad (12\text{-}68)$$

and instability occurs for $k > k_{\min}$.

Both for the Kelvin–Helmholtz and Rayleigh–Taylor instabilities, the dispersion relations are modified by the inclusion of surface tension and viscosity. In particular, the Kelvin–Helmholtz instability will be suppressed by surface tension if

$$(U_2 - U_1)^2 < \frac{2}{\alpha_1 \alpha_2} \left[\frac{Tg(\alpha_2 - \alpha_1)}{\rho_1 + \rho_2} \right]^{1/2}, \qquad (12\text{-}69)$$

where T is the surface tension of the lower fluid. For air over sea water, $\alpha_1 = 0.00126$, $\rho_2 = 1.02$ g cm^{-3}, $T = 74$ dyn cm^{-1}. Thus Eq. (12-69) predicts stability for

$$|U_2 - U_1| < 650 \text{ cm sec}^{-1}, \qquad (12\text{-}70)$$

or ~ 12.5 knots. The onset of instability will manifest itself as gravity waves with a wavelength of 1.71 cm moving at a (deep water) speed of 0.82 cm sec^{-1}. Munk has verified these results experimentally by showing that although the surface of the sea is ruffled at lesser wind speeds, owing to viscous effects, the number of breaking crests (white caps) suddenly increases as the wind speed reaches 12.5 knots.

EXERCISE

12-3.1. Prove that the first-order gravitational force density is given by

$$\boldsymbol{f} = \nabla \cdot (\rho \boldsymbol{\xi}) g \hat{\boldsymbol{k}}.$$

RELATIVITY THEORY

Newtonian mechanics is invariant under transformation between reference frames in uniform relative motion with respect to one another. Such reference frames are said to be inertial. As we have seen in Chapter 6, a noninertial frame may also be taken as the frame of reference at the expense of having to patch up the theory by the inclusion of various fictitious forces.

The transformation from one inertial frame to another moving in the x direction with speed v is

$$
\begin{array}{ll}
x' = x - vt & x = x' + vt' \\
y' = y & y = y' \\
z' = z & z = z' \\
t' = t & t = t'.
\end{array}
\tag{13-1}
$$

In these (Galilean) transformation relations that accompany Newtonian mechanics, there is no preferred reference frame; all inertial frames serve equally well.

The other great theory of classical physics, the electromagnetic theory of Maxwell, is a wave theory and predicts a definite speed for the propagation of the waves, $c \approx 3 \cdot 10^8$ m sec^{-1}. The obvious question is: With respect to what frame of reference is c measured? In answer, a so-called *aether* frame was postulated. The aether serves the medium in which the electromagnetic waves propagate. In order to respond to the very high frequencies typical of light waves (much less x rays or γ rays), the aether must be quite rigid. At the same time its viscosity must be vanishingly small so that the planetary orbits are unaffected, as observation shows them to be, by drag.

In a landmark experiment carried out in Cleveland in 1887 A. A. Michelson and E. W. Morley sought to measure the speed of the earth through the aether by means of an optical device known as a Michelson interferometer. In this experiment a beam of light is split into two beams. These beams are reflected in the direction of the aether stream and at right angles to it and are then compared interferometrically in order to detect a difference in the propagation speeds of the two beam components.[1] A null result was obtained; that is, the experiment implied that the propagation speed of the light was independent of the motion of the source.

13-1 The Basis of Special Relativity

In 1905 Albert Einstein published a paper entitled "On the Electrodynamics of Moving Bodies."[2] In it Einstein offered two postulates: (1) *Absolute* uniform motion is inherently undetectable; that is, any physical experiment must give the same result no matter which inertial frame is used for reference. There is no absolute (aether) frame. (2) The speed of propagation of electromagnetic waves c is a fixed universal constant and is independent of the motion of the source or observer.

At the time Einstein wrote these postulates he was unaware of the Michelson–Morley experiment or the attempts by Fitzgerald, Poincaré, and Lorentz to explain the null result of the experiment.

Clearly, the second postulate is inconsistent with our "experience," conditioned as it is to Galilean coordinate transformations. Taking the derivative of the top line of Eq. (13-1) with respect to $t = t'$, we find

$$u_x = u_x' + v; \qquad (13\text{-}2)$$

that is, the velocity as perceived by an observer in the unprimed frame is the sum of the velocity v of the primed frame and the velocity u_x' with respect to the primed frame. For example, suppose that we consider an observer O' on a flatcar moving toward an observer O with fixed speed v as shown in Fig. 13-1. The flatcar is also equipped with a loudspeaker that is emitting sound waves. If the speed of sound with respect to still air is v_s and no wind is blowing from the point of view of O, then O will measure the speed of sound to be v_s and O' will measure $v_s - v$ because of the apparent wind speed $-v$ in his reference frame.

Then imagine the identical setup of observers and moving flatcar except

---

[1] For details of the Michelson–Morley experiment, see Joseph Norwood, *Twentieth Century Physics*. Englewood Cliffs: Prentice-Hall, Inc., 1976, pp. 13–18.

[2] *Ann. d. Physik*, **17**, 891 (1905). In this same volume Einstein also published his theories of Brownian motion and photoelectric effect, the latter resulting in the award of the Nobel Prize in 1921.

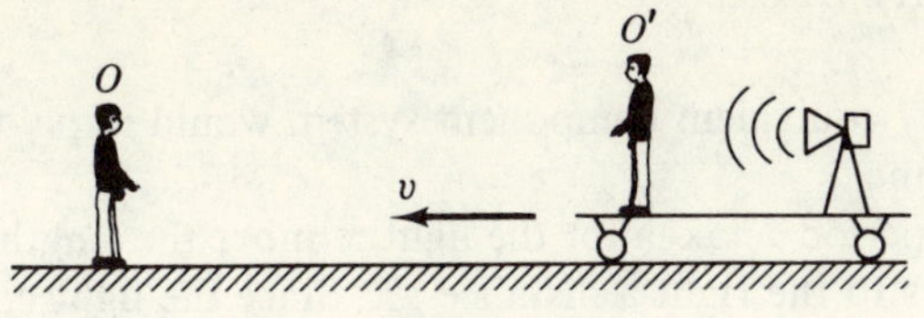

Fig. 13-1

that the loudspeaker has been replaced by a spotlight or some other source of electromagnetic waves. According to Einstein's second postulate, observers O and O' will both measure the wave propagation speed to be c. Neither observer notes any effect on the propagation velocity as a result of differences in the "aether wind strength."

In order to examine the results of Einstein's postulates, we must devise a simple and trustworthy clock. We shall take advantage of the constancy of the speed of light and use as our clock a box of length L_0 with a light source in one end and a detector (for instance, a photomultiplier tube) in the other. If we observe the action of this clock from a frame of reference at rest with respect to the clock, then one tick of the clock—that is, the time required for light to travel a distance L_0—is $T_0 = L_0/c$.

Suppose that the observer is in motion to the left with speed $-v$ or, equivalently, that the clock moves to the right with speed v. What is the period of a tick to such a moving observer? The situation is illustrated in Fig. 13-2. The invariance of all lengths like L_0, perpendicular to the line of

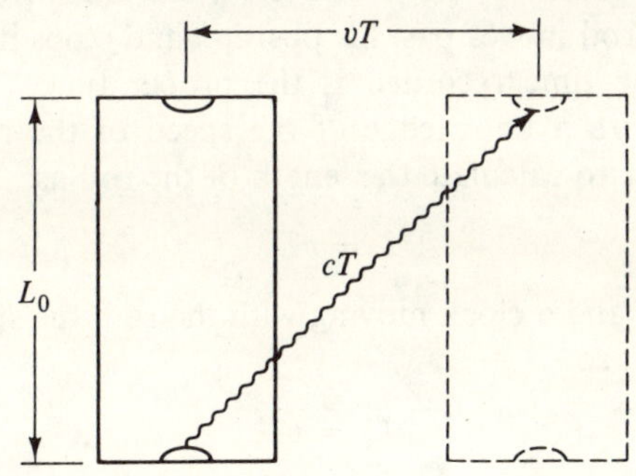

Fig. 13-2

motion, is provided by Einstein's first postulate. The argument is based on the fact that the observation time for both ends of a vertical rod moving horizontally is defined identically in both the moving and the stationary frames. Consequently, both observers can measure the length of the rod. If they did not agree that the length measured by all observers moving uniformly in the horizontal plane is equal to the proper length L_0, then the asymmetry in an apparently equivalent component system would imply the existence of an absolute frame.

In the time period T taken for the light to move the length of the box, the box itself moves to the right a distance vT. Since the light covers a distance cT in this time, we find

$$L_0^2 + v^2T^2 = c^2T^2$$

or
$$T = \frac{T_0}{\sqrt{1 - (v/c)^2}} = \frac{T_0}{\sqrt{1 - \beta^2}} = \gamma T_0 \geq T_0, \qquad (13\text{-}3)$$

where $\beta \equiv v/c$ and $\gamma = (1 - \beta^2)^{-1/2}$ are introduced as a notational convenience.

The time period separating events occurring at the same point in a reference frame, as measured by a clock at rest in the frame, is known as *proper time*. All time period measurements in other inertial frames moving with respect to the observer will be longer by a factor γ. This time dilation effect, also predicted by Lorentz in his electromagnetic theory, applies for all types of clocks, mechanical or biological, and is a direct result of the finite and invariant speed of light.

Let us examine the spatial consequences of Einstein's postulates. Consider a rod moving in a direction parallel to its length at speed v. An observer moving with the rod would measure its length as L_0 (the proper length). A stationary (in the laboratory frame) observer starts his stopwatch when the leading end of the rod moves past his position and stops it when the trailing end goes past. The time recorded is the proper time T_0. This measured period, together with a knowledge of the speed of the rod v, enables the laboratory observer to calculate the length of the rod as

$$L = vT_0. \qquad (13\text{-}4)$$

Using a meterstick and a clock moving with the rod, the stationary observer measures L_0 and T; so

$$L_0 = vT. \qquad (13\text{-}5)$$

Eliminating v between Eqs. (13-4) and (13-5) and using Eq. (13-3), we find

$$L = \frac{L_0}{\gamma} \leq L_0; \qquad (13\text{-}6)$$

that is, moving objects appear to contract in the direction of their motion by a factor γ.

The time dilation and length contraction phenomena are complementary. The time dilation can be directly verified experimentally. For instance, a particular subatomic particle, if created at rest, is known to have a fixed decay time τ_0. This same particle, when found among the fragments resulting from the impact of a high-energy beam on a target, has a lifetime $\tau > \tau_0$ as a result of its velocity. The relationship between τ and τ_0 can be shown to agree with $\tau = \gamma\tau_0$ to within experimental error. Even though the length contraction phenomenon cannot be measured directly, this fact does not cast doubt on its validity. Suppose, for example, that our subatomic particle is created at the top of the earth's atmosphere in a cosmic-ray shower. It is observed to fall a distance L_0 and to live a lifetime $\tau = \gamma\tau_0$. From the particle's point of view the distance that it falls is $L = L_0/\gamma$ and its lifetime is τ_0.

EXERCISE

13-1.1. A meterstick moves past a stationary observer at a speed such that its length is calculated to be 0.5 m. How long does the meterstick take to pass the stationary observer?

13-2 The Lorentz Transformations

Herman Minkowski introduced mathematical refinements to relativity theory, chief among which was the notion of visualizing relativity in terms of coordinate transformations in a four-dimensional space (Minkowski space).

Since spatial dimensions transverse to the direction of motion are invariant, we can deal with a two-dimensional space: t, x. Imagine that an observer aims his flashlight at a mirror located a distance x away. At time t_1 the light is switched on and at some later time t_2 the reflected light beam is seen. If the observer remained motionless during the period $t_2 - t_1$, his trajectory (or *world line*) in our two-dimensional Minkowski space is the vertical time axis, $x = 0$ (see Fig. 13-3). Points in such a diagram are called *events*. The event

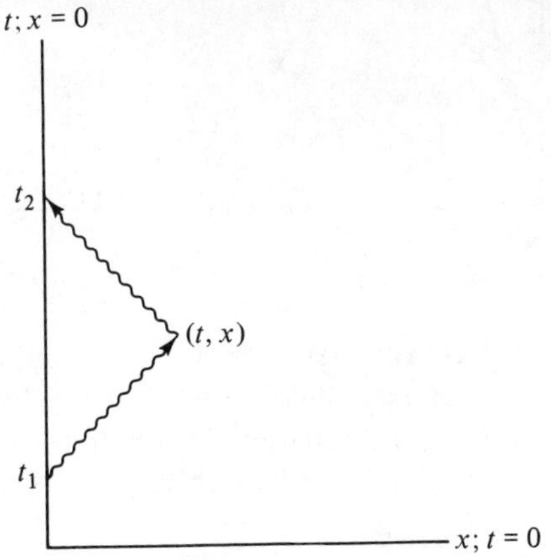

Fig. 13-3

given by the reflection of the leading edge of the light pulse from the flashlight can be calculated as a function of t_1 and t_2. Obviously, the coordinate t is simply the mean of t_2 and t_1:

$$t = \frac{t_1 + t_2}{2}. \tag{13-7}$$

The equation of the outbound light beam is $x = c(t - t_1)$ and that of the reflected beam is $x = c(t_2 - t)$. Eliminating t between these equations or between either of them and Eq. (13-7), we find

$$x = \frac{c(t_2 - t_1)}{2}. \tag{13-8}$$

We next introduce a second observer moving in the x direction at a speed v. Assuming that the stationary (unprimed) and moving (primed) observers coincide in position at $x = t = 0$, we add the world line of the second observer as shown in Fig. 13-4.

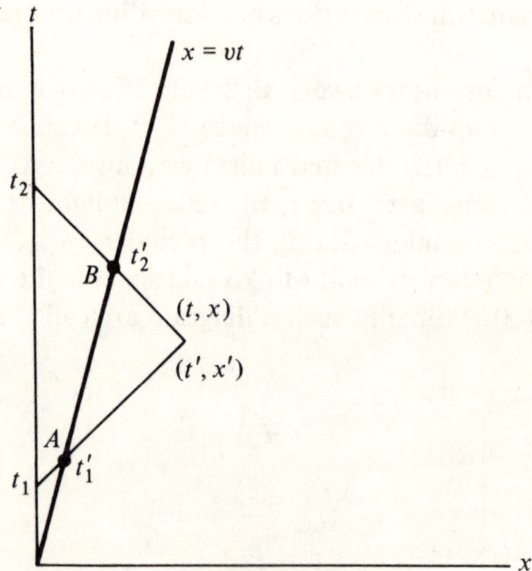

Fig. 13-4

We wish to find the coordinates t', x' for the event, using the moving observer's instruments. First, we calculate the coordinate of point A as seen by the stationary observer. The equations of the lines intersecting at A are $x = vt$ and $x = c(t - t_1)$. Thus, eliminating x,

$$t = \frac{ct_1}{c - v}. \tag{13-9}$$

In order to obtain t_1', the coordinate of A as measured in the moving frame, it is necessary to multiply the value t given in Eq. (13-9) by $1/\gamma$:

$$t_1' = \frac{ct_1}{c - v}\sqrt{1 - \frac{v^2}{c^2}} = t_1\sqrt{\frac{1 + \beta}{1 - \beta}}. \tag{13-10}$$

We do the same for point B and find

$$t_2' = t_2\sqrt{\frac{1 - \beta}{1 + \beta}}. \tag{13-11}$$

Using Eq. (13-7),

$$t' = \frac{t_1' + t_2'}{2} = \frac{1}{2}\left(t_1\sqrt{\frac{1 + \beta}{1 - \beta}} + t_2\sqrt{\frac{1 - \beta}{1 + \beta}}\right). \tag{13-12}$$

Since we are looking for transformation relations in the form $t'(x, t)$ and $x'(x, t)$, we use Eqs. (13-7) and (13-8) to eliminate t_1 and t_2 from Eq. (13-12). We find

$$t' = \frac{1}{2}\left[\left(t - \frac{x}{c}\right)\sqrt{\frac{1 + \beta}{1 - \beta}} + \left(t + \frac{x}{c}\right)\sqrt{\frac{1 - \beta}{1 + \beta}}\right]$$

$$= \frac{[t - (x/c)](1 + \beta) + [t + (x/c)](1 - \beta)}{2\sqrt{1 - \beta^2}}$$

$$= \frac{t - (x\beta/c^2)}{\sqrt{1 - \beta^2}} = \gamma\left(t - \frac{xv}{c^2}\right). \tag{13-13}$$

Similarly, using $x' = c(t_2' - t_1')/2$ and Eqs. (13-10) to (13-12), we obtain

$$x' = \gamma(x - vt). \tag{13-14}$$

These equations, together with $y' = y$ and $z' = z$, constitute the Lorentz transformation relations for relative motion in the x direction. In the limit as $\beta = v/c \to 0$, Eqs. (13-13) and (13-14) reduce to $t' = t$ and $x' = x - vt$, their Galilean counterparts, as we might expect.

EXERCISES

13-2.1. A pair of clocks is synchronized. One of them is then set into motion at a velocity v. After a time t has passed on the stationary clock, how far out of synchronization are the two clocks?

13-2.2. The earth moves in its orbit at a speed of 30 km sec^{-1}. The nearest star (excluding the sun) is about 4 light years away (a light year is the distance traveled by light in one year). By how much time does the earth's speed affect our judgment concerning the simultaneity of events here and there?

13-2.3. Two clocks, one located at the North Pole and the other on the Equator, are synchronized at $t = 0$. Show that the time difference in the clocks after a century will amount to about 0.0038 sec.

13-3 The Addition of Velocities

The quantities x', y', z', t' are the spatial coordinates and time measured by using clocks and metersticks moving in the x direction at speed v. If we were located in the moving (primed) system and made measurements with instruments located in the laboratory frame (receding at speed $-v$ in the negative x direction), the appropriate set of transformations would be the inverse Lorentz transformations with v replaced by $-v$ and the primed and unprimed quantities interchanged:

$$x = \gamma(x' + vt'),$$

$$y = y',$$

$$z = z', \tag{13-15}$$

$$t = \gamma\left(t' + \frac{x'v}{c^2}\right).$$

Note that γ, depending on v only quadratically, remains unchanged.
The differentials of Eq. (13-15) are

$$dx = \gamma(dx' + v\,dt'),$$

$$dy = dy',$$

$$dz = dz', \tag{13-16}$$

$$dt = \gamma\left(dt' + \frac{v\,dx'}{c^2}\right);$$

thus the velocities are

$$u_x = \frac{dx}{dt} = \frac{(dx' + v\,dt')\gamma}{[dt' + (v\,dx'/c^2)]\gamma}$$

$$= \frac{(dx'/dt') + v}{1 + (v/c^2)(dx'/dt')} = \frac{u_x' + v}{1 + (u_x'v/c^2)}, \tag{13-17}$$

$$u_y = \frac{dy}{dt} = \frac{u_y'}{\gamma[1 + (u_x'v/c^2)]}, \tag{13-18}$$

$$u_z = \frac{dz}{dt} = \frac{u_z'}{\gamma[1 + (u_x'v/c^2)]}. \tag{13-19}$$

It can readily be seen that when $v/c \to 0$, Eqs. (13-17) to (13-19) reduce to their Galilean counterparts:

$$u_x = u_x' + v, \qquad u_y = u_y', \qquad u_z = u_z'. \tag{13-20}$$

It is interesting to examine a couple of special cases. First, let $u_x' = c$ and $u_y' = u_z' = 0$. Then using Eq. (13-17), we find

$$u_x = \frac{c + v}{1 + (cv/c^2)} = c. \tag{13-21}$$

In other words, the speed of light in a given direction is constant and equal to c for all observers moving in that direction at *any* speed v.

What about observers moving at right angles to the light beam? To test this situation, let $u_x' = u_1' = 0$ and $u_z' = c$. Equations (13-17) and (13-18) then yield

$$u_x = v$$

and

$$u_y = c\sqrt{1 - \frac{v^2}{c^2}};$$

so

$$u = \sqrt{u_x^2 + u_y^2} = \sqrt{v^2 + c^2 - v^2} = c, \tag{13-22}$$

and we see that the Lorentz transformation equations are consistent with Einstein's assertion that the speed of light is the same for all observers in uniform motion in any direction with respect to the direction in which the light propagates.

EXERCISES

13-3.1. A particle is observed to be moving in the primed frame with a velocity vector $u = 0.577c(\hat{i} + \hat{j} + \hat{k})$. What is the velocity vector as measured by an observer moving at a speed $v = 0.9c$ along the negative x axis?

13-3.2. In 1881 Fizeau carried out an experiment that demonstrated that light is "dragged along" by a moving medium in which it propagates. Fizeau's equation for the speed of the light parallel to the stream velocity v of a water flow is

$$u = \frac{c}{n} + \alpha v,$$

where n is the index of refraction of the water ($n = 1.333$) and α is the drag coefficient for which Fizeau found a value of 0.434.

Use the relativistic formula for addition of velocities to first order in v/c to show that Fizeau's drag coefficient is given by

$$\alpha = 1 - \frac{1}{n^2} = 1 - (1.333)^{-2} = 0.437,$$

in reasonable agreement with Fizeau's measured value.

13-4 The Minkowski Diagram

We wish to construct a diagram as an aid in visualizing relativity problems. In Fig. 13-5 we have plotted Eqs. (13-13) and (13-14) for $t'(t, x)$ and $x'(t, x)$. We also include the *light line*, $x = ct$, as a 45° line, owing to the fact that the diagram is scaled such that $c = 1$. The resulting diagram is a superposition of two coordinate axes—the orthogonal x, t system and its Lorentz transform, the x', t' system, which is not orthogonal.

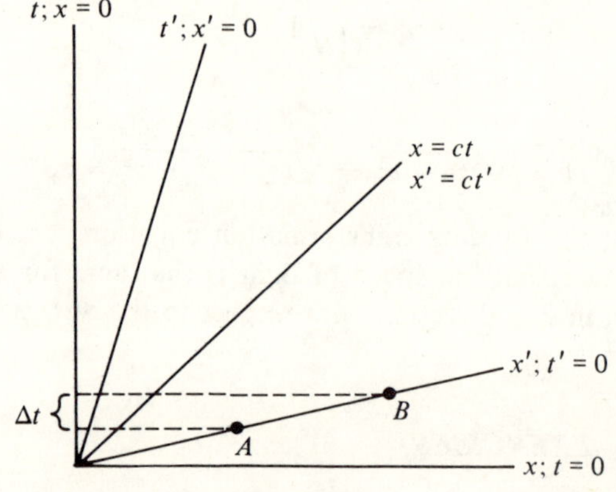

Fig. 13-5

An immediate result of this nonorthogonality of the primed axes is revealed by considering the two events A and B located on the $t' = 0$ line. These events are simultaneous in the primed system; to the stationary observer, however, the event at A appears to occur at a time Δt before the event at B, owing to the finite propagation speed of light.

Let us examine another question. Suppose that two observers are in relative motion at a speed v along the x axis. Since each observer is in motion with respect to the other, each will say that the other's clock runs more slowly than his own. How can each one be right? The situation is illustrated in Fig. 13-6. At point A, $t' = t_0/\gamma < t_0$. Since simultaneity occurs along lines parallel to the x' axis from the point of view of the primed observer, he

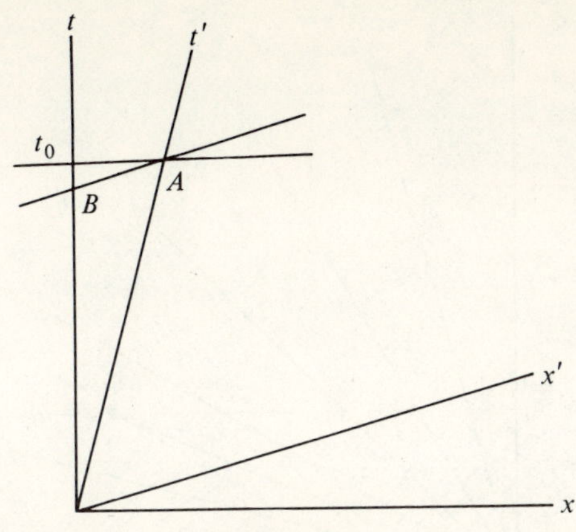

Fig. 13-6

will make his comparison at point B—that is, the intersection with the t axis. The equation of the line AB is

$$\frac{t_0}{\gamma} = \left(t - \frac{x\beta}{c}\right)\gamma$$

or
$$t - t_0 = \frac{v}{c^2}(x - vt_0). \tag{13-23}$$

At the point of intersection with the t axis ($x = 0$), we find

$$t - t_0 = -\frac{v^2}{c^2}t_0$$

or
$$t = t_0(1 - \beta^2) = \frac{t'}{\gamma} < t' \tag{13-24}$$

and the result is indeed symmetric.

A well-known consequence of the relativity theory is that material objects cannot exceed the speed of light. This fact is explained in most textbooks from an energy point of view. There is another more subtle reason for this limiting velocity, however.

In Fig. 13-7 we have superimposed two Minkowski diagrams by including two moving frames—the primed system moving at a speed v_1 along the x axis and the double-primed system moving at a speed $v_2 > v_1$ along the x axis. If the stationary observer were to launch a projectile in the x direction at a speed $u > c$, then the projections of the event onto the t and t' axes are positive, but the projection onto the t'' axis is negative. This situation implies that the double-primed observer would detect the projectile *before* it was launched. This result represents a rather serious break in the causal chain.

389

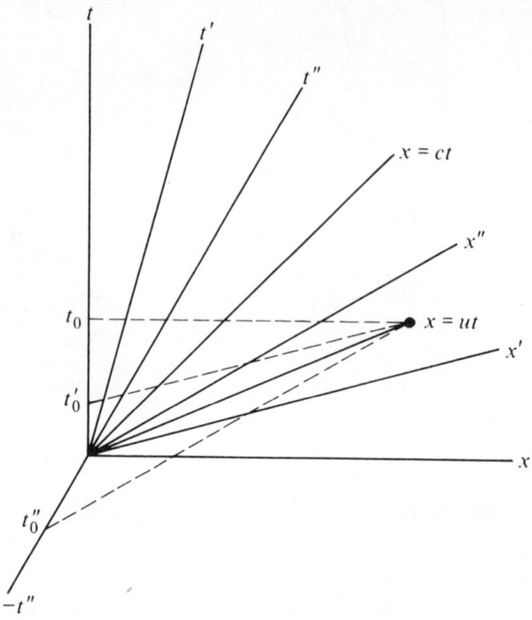

Fig. 13-7 Minkowski diagram for a superluminal event.

Knowledge of the future, together with the reasonable assumption of free will, suggests that the future might be changed by such an observer in order to prevent the launch of the projectile. From a mathematical point of view, it is equivalent to a multivalued solution which classical physics does not permit. For this reason, we assume that no detectable bodies can move faster than light.

EXERCISES

13-4.1. In Fig. 13-8 we have plotted the light lines of a series of light wavefronts having a separation in time of $t_0 = 1/\nu$ from the point of view of an observer at rest with respect to the wave source. An observer moving at a speed v away from the wave source in the direction of wave propagation measures a wave period $t_0' = 1/\nu'$. Show that the wave frequency ν' measured by the moving observer is given by

$$\nu' = \nu\sqrt{\frac{1-\beta}{1+\beta}}. \tag{13-25}$$

This is the relativistic Doppler effect.

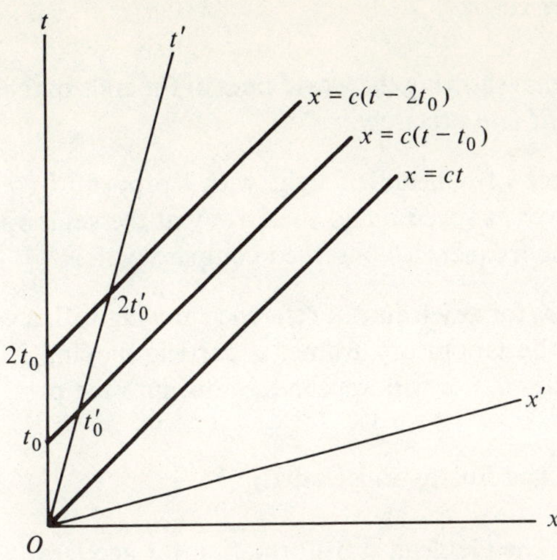

Fig. 13-8 Minkowski diagram for the Doppler effect.

13-4.2. The classical theory of the Doppler effect predicts, for a motionless source and an observer moving away at speed $v_0 \ll c$ ($\beta_0 \ll 1$),

$$\nu' = \nu(1 - \beta_0).$$

If the observer is motionless and the source is moving away at a speed $v_s \ll c$ ($\beta_s \ll 1$), then

$$\nu' = \nu(1 + \beta_s)^{-1}.$$

Show that in the limit of small β these equations agree with each other and with Eq. (13-25).

13-4.3. A stationary clock, synchronized with the observer's clock, is located 10 km distant. The observer views this clock through a telescope. When his own clock reads $t = 0$, what time does he read on the distant clock?

13-4.4. Two events, stationary with respect to each other, are separated in time by a period T_0. What is the maximum value of T_0 such that a frame of reference exists in which the two events are simultaneous?

13-4.5. If an observer takes a photograph, what is the set of events on a Minkowski diagram that is recorded on the film? Answer both for the case of a flash photo and for the case of a long exposure.

13-4.6. Can a fast-moving limousine fit into a garage designed for a minicar? The garage builder says, "Yes, because the limousine has contracted." The driver says, "No, I couldn't fit in when the garage was standing still and now the length of the garage is contracted even more." Resolve this paradox by a

Minkowski diagram showing the world lines of the ends of the car and garage. Define the term *fit into* precisely.

13-4.7. The source of a beam of light with "proper" frequency ν_0 moves toward an observer at speed v and then away at the same speed. How does the average of the frequencies observed compare with ν_0?

13-4.8. Show that, for any frame of reference moving with a velocity $v > c^2/u$ with respect to the laboratory frame, a particle moving in the laboratory frame with velocity $u > c$ will appear to move into the past.

13-5 Momentum and Energy in Relativity

The Galilean transformation for accelerations is simply

$$a'_x = a_x,$$

$$a'_y = a_y, \tag{13-26}$$

$$a'_z = a_z;$$

thus the form of Newton's second law is invariant to a Galilean transformation. Neither $F = ma$ nor the more general form, $F = \dot{p} = d(mu)/dt$, can apply at speeds approaching that of light, however. The Newtonian equations contain no velocity-limiting mechanism; they imply that if the force is applied for sufficiently long times, arbitrarily high speeds can be attained.

In order to unravel the problem, we shall examine the collision of two spheres of equal mass m for which, in Newtonian terms, the total momentum is conserved. In Fig. 13-9 we show the elastic collision of two balls thrown

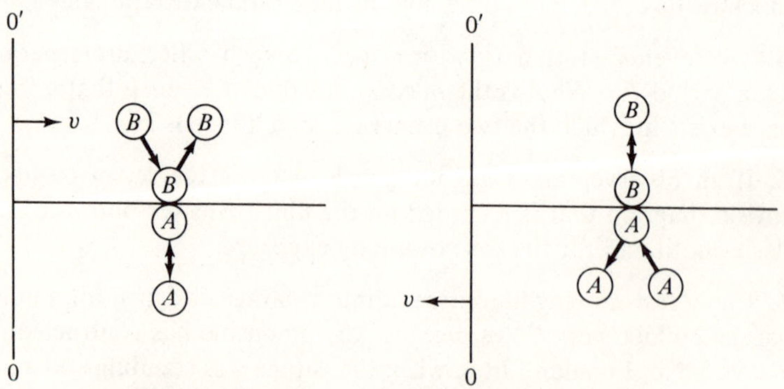

Fig. 13-9 The collision of two balls thrown from two frames O and O' that move with relative speed as viewed from (a) O and (b) O'.

from two frames in uniform motion with respect to each other. The speed of each ball is u_0. The observer in frame O sees as the velocity vector of ball A

$$\boldsymbol{u}_A = (0, 0, u_0), \tag{13-27}$$

whereas the observer in O' sees for ball B

$$\boldsymbol{u}'_B = (0, 0, -u_0). \tag{13-28}$$

What does the observer in O see as the velocity vector of ball B? Using the Lorentz velocity transformation, Eqs. (13-17) to (13-19), we find

$$\boldsymbol{u}_B = \left(v, 0, -\frac{u_0}{\gamma}\right). \tag{13-29}$$

Thus $p_{zA} = mu_0$, whereas $p_{zB} = -mu_0/\gamma$ and the z component of momentum does not vanish! The symmetry between frames O and O' and balls A and B is perfect; the only conclusion is that $\boldsymbol{mu}$ is not conserved in situations where Lorentz transformations are distinguishable from Galilean transformations. If we are to preserve the *concept* of momentum conservation in collisions—that is, the form of Newton's second law—we must define a relativistic momentum $\boldsymbol{p}$ such that

    1. $\boldsymbol{p}$ is conserved in collisions,
    2. $\boldsymbol{p} \to \boldsymbol{mu}$ as $u/c \to 0$.

The definition

$$p = \frac{mu}{\sqrt{1 - (u^2/c^2)}} \tag{13-30}$$

obviously meets the second requirement. As for its conservation in collisions, we have

$$p_{zA} = \frac{mu_0}{\sqrt{1 - (u_0^2/c^2)}}. \tag{13-31}$$

To find p_{zB}, we first note that

$$u_B^2 = u_{xB}^2 + u_{zB}^2 + v^2 + \left(u_0\sqrt{1 - \frac{v^2}{c^2}}\right)^2 = v^2 + u_0^2 - \frac{u_0^2 v^2}{c^2},$$

whence

$$1 - \frac{u_B^2}{c^2} = 1 - \frac{v^2}{c^2} - \frac{u_0^2}{c^2} + \frac{u_0^2 v^2}{c^4} = \left(1 - \frac{v^2}{c^2}\right)\left(1 - \frac{u_0^2}{c^2}\right).$$

Thus we have

$$p_{zB} = \frac{mu_{zB}}{\sqrt{1 - (u_B^2/c^2)}} = -\frac{mu_0\sqrt{1 - (v^2/c^2)}}{\sqrt{1 - (v^2/c^2)}\,\sqrt{1 - (u_0^2/c^2)}}$$

$$= -\frac{mu_0}{\sqrt{1 - (u_0^2/c^2)}} = -p_{zA} \tag{13-32}$$

as required. We shall therefore accept Eq. (13-30) as the definition of linear momentum.

In order to avoid confusion, we wish to introduce the notation

$$\Gamma \equiv \left(1 - \frac{u^2}{c^2}\right)^{-1/2} \tag{13-33}$$

Thus the momentum can be written

$$\boldsymbol{p} = \Gamma m \boldsymbol{u}, \tag{13-34}$$

and Newton's second law written in terms of this momentum

$$\boldsymbol{F} = \frac{d\boldsymbol{p}}{dt} \tag{13-35}$$

can now be used to calculate the relativistic expression for the kinetic energy. Thus

$$T = \int_0^u F\,ds = \int_0^u ds\,\frac{d}{dt}(mu\Gamma) = \int_0^u u\,d(mu\Gamma), \tag{13-36}$$

where integration by parts was carried out in the last step, noting that $u = ds/dt$. It is left as an exercise for the reader to show that

$$d(mu\Gamma) = m\left(1 - \frac{u^2}{c^2}\right)^{-3/2} du; \tag{13-37}$$

so we find for T

$$T = \int_0^u m\left(1 - \frac{u^2}{c^2}\right)^{-3/2} u\,du$$

$$= mc^2\left[\left(1 - \frac{u^2}{c^2}\right)^{-1/2} - 1\right]$$

$$= mc^2(\Gamma - 1). \tag{13-38}$$

It is of interest to examine this result at its limits. In the Newtonian limit where $u \ll c$

$$\Gamma = \left(1 - \frac{u^2}{c^2}\right)^{-1/2} \approx 1 + \frac{1}{2}\frac{u^2}{c^2} + \cdots \tag{13-39}$$

and
$$T \approx mc^2\left(1 + \frac{1}{2}\frac{u^2}{c^2} + \cdots - 1\right) = \tfrac{1}{2}mu^2 \tag{13-40}$$

as expected. In the extreme relativistic limit, $u \to c$ and $T \to \Gamma mc^2$. We shall denote the quantity $mc^2 = T/(\Gamma - 1)$, independent of u, as E_0 and call it the *rest energy*:

$$E_0 = mc^2. \tag{13-41}$$

The total energy E can then be written

$$E = T + E_0 = \Gamma mc^2 = \Gamma E_0. \qquad (13\text{-}42)$$

For many purposes, it is useful to express the total energy in terms of the momentum. It is left as an exercise for the reader to show that

$$E^2 = p^2 c^2 + E_0^2. \qquad (13\text{-}43)$$

Equation (13-41) is probably the most widely known formula of twentieth-century physics. It expresses the equivalence between mass and energy. We note that the coefficient c^2 is very large; that is, a very small mass is equivalent to a *lot* of energy. The conversion of mass to energy was first realized by Enrico Fermi in 1942 in the first fission reactor. Fission reactors today supply a significant percentage of the electricity generated in the developed countries, and fusion reactors seem likely to become operational by the turn of the twenty-first century.

In order to illustrate the mass conversion reaction in a simple manner, let us consider a totally inelastic collision between two similar balls as indicated in Fig. 13-10. In the O frame the two balls approach collision at equal speeds

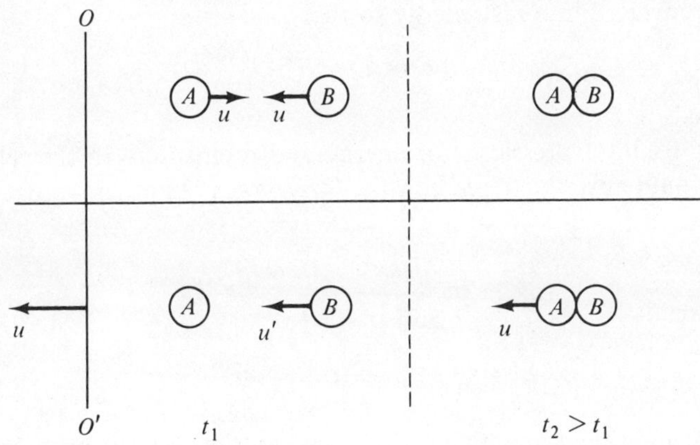

Fig. **13-10** An inelastic collision viewed from two different inertial frames.

and, having collided, do not rebound. The total momentum of the system in frame O is obviously zero. To an observer moving to the right at speed u (the O' system), ball A is at rest and ball B approaches the collision at a speed

$$u' = \frac{2u}{1 + (u^2/c^2)}. \qquad (13\text{-}44)$$

After the collision the two balls move with speed u to the left.

Let m be the mass of each ball before the collision and M be the total mass of the two balls after collision. The initial momentum of the system as measured in O' is

$$p_i' = \frac{mu'}{\sqrt{1 - (u'^2/c^2)}} = \frac{2mu}{1 - (u^2/c^2)} = 2mu\Gamma^2, \qquad (13\text{-}45)$$

and the final momentum of the system is

$$p_f' = \frac{Mu}{\sqrt{1 - (u^2/c^2)}} = Mu\Gamma. \qquad (13\text{-}46)$$

Since momentum is conserved in inelastic collisions as well as elastic ones, $p_i' = p_f'$ or

$$M = 2m\Gamma. \qquad (13\text{-}47)$$

This equation implies an increase in the mass of the system by an amount

$$\Delta m = 2m(\Gamma - 1). \qquad (13\text{-}48)$$

The kinetic energy of the balls as measured in O,

$$T = 2mc^2(\Gamma - 1) = c^2\,\Delta m, \qquad (13\text{-}49)$$

has been converted into rest energy so that

$$E = T + E_0 \qquad (13\text{-}50)$$

is conserved.

Having found expressions for energy and momentum in the relativistic limit, we would now like to see how they transform. In other words, we need to express

$$p' = \frac{mu'}{\sqrt{1 - (u'^2/c^2)}} = mu'\Gamma' \qquad (13\text{-}51)$$

and

$$E' = \frac{mc^2}{\sqrt{1 - (u'^2/c^2)}} = mc^2\Gamma' \qquad (13\text{-}52)$$

in terms of unprimed quantities. To begin, we derive an expression for Γ'. Using the velocity transformations, Eqs. (13-17) to (13-19), we find

$$u^2 = u_x^2 + u_y^2 + u_z^2 = \frac{\gamma^2(u_x' + v)^2 + u_y'^2 + u_z'^2}{\gamma^2[1 + (u_x'v/c^2)]^2} \qquad (13\text{-}53)$$

from which

$$\left(1 - \frac{u^2}{c^2}\right)^{-1/2} = \frac{1 + (u_x'v/c^2)}{\sqrt{1 - (v^2/c^2)}\,\sqrt{1 - (u'^2/c^2)}}$$

or

$$\Gamma = \gamma\left(1 + \frac{u_x'v}{c^2}\right)\Gamma' \qquad (13\text{-}54)$$

and its inverse

$$\Gamma' = \gamma\left(1 - \frac{u_x v}{c^2}\right)\Gamma, \tag{13-55}$$

obtained by priming unprimed quantities (and vice versa) and changing the sign on v. Using Eq. (13-55), we have

$$p'_x = mu'_x\Gamma' = m\left[\frac{u_x - v}{1 - (u_x v/c^2)}\right]\gamma\left(1 - \frac{u_x v}{c^2}\right)\Gamma$$

$$= m\gamma(u_x - v)\Gamma, \tag{13-56}$$

$$p'_y = mu'_y\Gamma' = m\left[\frac{u_y}{\gamma(1 - u_x v/c^2)}\right]\gamma\left(1 - \frac{u_x v}{c^2}\right)\Gamma = mu_y\Gamma, \tag{13-57}$$

$$p'_z = mu_z\Gamma, \tag{13-58}$$

and
$$E' = mc^2\Gamma' = mc^2\gamma\left(1 - \frac{u_x v}{c^2}\right)\Gamma. \tag{13-59}$$

These equations can be written in a more compact form:

$$p'_x = \gamma\left(p_x - \frac{vE}{c^2}\right),$$

$$p'_y = p_y, \tag{13-60}$$

$$p'_z = p_z,$$

$$E' = \gamma(E - vp_x).$$

We see that they have the same form as the space-time transformation equations.

EXERCISES

13-5.1. An astronaut wishes to travel to a star a distance 10^3 light years away and is willing to spend the rest of his life, say 50 years, on the journey. Is this possible, in principle? If so, how fast must he go? If the mass of the spaceship is 10^5 kg, what is the kinetic energy of the ship?

13-5.2. Carry out the steps necessary to establish Eq. (13-37).

13-5.3. Show that Eq. (13-43) follows from Eqs. (13-34) and (13-42).

13-5.4. Calculate the amount of mass that must be turned into energy in order to accelerate a spaceship weighing a million metric tons to escape velocity.

13-5.5. Show that the velocity of a massive object can be written

$$u = \frac{c^2}{E}\, p$$

and its speed as

$$u = \frac{dE}{dp}.$$

13-6 Four-Vectors in Minkowski Space

Suppose that we, following Minkowski, look at relativity in a more geometrical way. We recall that in Euclidean space (the ordinary flat three-dimensional space of our experience) the length of a line segment remains constant no matter in what coordinate system we choose to represent it. This situation can be written

$$s^2 = (\Delta x)^2 + (\Delta y)^2 + (\Delta z)^2 + (\Delta x')^2 + (\Delta y')^2 + (\Delta z')^2$$

$$= \Delta x_i\, \Delta x_i = \Delta x_i'\, \Delta x_i', \qquad i = 1, 2, 3, \tag{13-61}$$

where in the second line we have used the *Einstein summation convention*, which calls for repeated indices to be summed.

In Minkowski space the length of a line is not invariant but depends on the orientation and on the parameter v. There is, however, an invariant quantity on which the geometry can be based—namely, the constancy of the speed of light for all observers, which can be written

$$\text{Invariants} = (\Delta x_1)^2 + (\Delta x_2)^2 + (\Delta x_3)^2 - c^2(\Delta t)^2$$

or $\qquad s^2 = \Delta x_\mu\, \Delta x_\mu = \Delta x_\mu'\, \Delta x_\mu', \qquad \mu = 1, 2, 3, 4, \tag{13-62}$

where $\qquad x_1 = x, \qquad x_2 = y, \qquad x_3 = z, \qquad x_4 = ict. \tag{13-63}$

In order to demonstrate that the "interval" defined in Eq. (13-62) is invariant, we shall drop the two trivial dimensions and represent the equation on a Minkowski diagram. In Fig. 13-11 we have plotted the hyperbola $c^2t^2 - x^2 = c^2t_0^2 = $ invariant. The consequence of this invariance is that all observers take the same time to reach points on this curve starting from the origin—that is, $t_0' = t_0$. To prove it, we eliminate x between the equations $x = vt$ and $c^2t^2 - x^2 = c^2t_0^2$:

$$c^2t^2 - v^2t^2 = c^2t_0^2$$

or $\qquad t = \dfrac{t_0}{\sqrt{1 - (v^2/c^2)}} = \gamma t_0. \tag{13-64}$

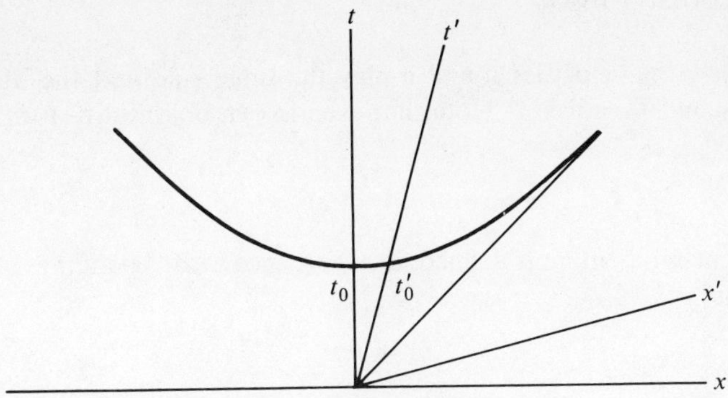

Fig. 13-11 The invariance hyperbola.

Dividing by γ so as to take into account the slower rate of moving clocks, we find

$$t_0' = \frac{t}{\gamma} = t_0, \tag{13-65}$$

which verifies the invariance.

In Minkowski space neither spatial intervals nor time intervals are invariant, unlike the Euclidean space of Newtonian mechanics in which both intervals are invariant; only the combination given in Eq. (13-62) is invariant, owing to the invariance of the speed of light for all observers.

Having defined a four-dimensional length vector x_μ, we can now define a four-velocity by taking the derivative of x_μ with respect to the proper time τ. Thus

$$u_\mu = \frac{dx_\mu}{d\tau} = \frac{dx_\mu}{dt}\frac{dt}{d\tau} = \gamma\frac{dx_\mu}{dt} \tag{13-66}$$

or in components

$$\begin{aligned}
u_1 &= \gamma\dot{x}_1 = \gamma\dot{x}, \\
u_2 &= \gamma\dot{x}_2 = \gamma\dot{y}, \\
u_3 &= \gamma\dot{x}_3 = \gamma\dot{z}, \\
u_4 &= \gamma\dot{x}_4 = i\gamma c.
\end{aligned} \tag{13-67}$$

The basis of our definition of vectors in Euclidean space was their invariant length. If the vectors that we are defining in Minkowski space are to be useful, we must also demonstrate their invariant length. For the velocity four-vector, we find

$$\begin{aligned}
u_\mu u_\mu &= \gamma^2(\dot{x}_1^2 + \dot{x}_2^2 + \dot{x}_3^2 + \dot{x}_4^2) \\
&= \gamma^2(u^2 - c^2) = \frac{u^2 - c^2}{1 - (u^2/c^2)} = -c^2 = \text{inv}.
\end{aligned} \tag{13-68}$$

399

In this case, the velocities u and v play the same role and the distinction between γ and Γ vanishes. Note that even for an object at rest ($u_i = 0$) we still find

$$u_\mu u_\mu = u_4 u_4 = (ic)^2 = -c^2. \tag{13-69}$$

The four-momentum is defined, in accordance with the findings of the last section, as

$$p_\mu = m u_\mu, \tag{13-70}$$

where the first three components have the form

$$p_i = m\gamma \dot{x}_i \tag{13-71}$$

and the fourth component is

$$p_4 = im\gamma c = \frac{iE}{c}. \tag{13-72}$$

It follows directly from Eq. (13-68) that

$$p_\mu p_\mu = -m^2 c^2 = \text{inv.} \tag{13-73}$$

By defining the momentum as we have, Newton's second law has been rendered Lorentz invariant. So we can use this equation to define a force four-vector

$$F_\mu = \frac{dp_\mu}{d\tau} = \frac{d}{d\tau}(m u_\mu). \tag{13-74}$$

The quantity

$$F_\mu u_\mu = u_\mu \frac{d}{d\tau}(m u_\mu) = \frac{d}{d\tau}(\tfrac{1}{2} m u_\mu u_\mu)$$

is zero, since $u_\mu u_\mu = -c^2$ is a constant. Hence

$$F_\mu u_\mu = F_i u_i \gamma^2 + ic\gamma F_4 = 0$$

and we find for the fourth component of the four-force

$$F_4 = \frac{i\gamma}{c} F_i u_i = \frac{i\gamma}{c} \mathbf{F} \cdot \mathbf{u}$$

$$= \frac{i\gamma}{c} \frac{dE}{dt} = \frac{i\gamma}{c} \frac{d}{dt}(\gamma m c^2), \tag{13-75}$$

where we have identified

$$E = \gamma m c^2, \tag{13-76}$$

in agreement with Eq. (13-42). Since

$$p_i = \gamma m u_i, \qquad p_4 = i\gamma mc = \frac{iE}{c}$$

then

$$p_\mu p_\mu = p_i p_i + p_4 p_4 = p^2 - \frac{E^2}{c^2} = -m^2 c^2$$

or
$$E^2 = p^2 c^2 + m^2 c^4, \tag{13-77}$$

in agreement with Eq. (13-43).

Having created such a compact notation, we can now write the Lorentz transform equations for space-time and momentum-energy in four-vector form as

$$\begin{aligned}
x_1' &= \gamma(x_1 + i\beta x_4), \\
x_2' &= x_2, \\
x_3' &= x_3, \\
x_4' &= \gamma(x_4 - i\beta x_1),
\end{aligned} \tag{13-78}$$

and
$$\begin{aligned}
p_1' &= \gamma(p_1 + i\beta p_4), \\
p_2' &= p_2, \\
p_3' &= p_3, \\
p_4' &= \gamma(p_4 - i\beta p_1),
\end{aligned} \tag{13-79}$$

which illustrates the correspondence between x_μ and p_μ.

EXERCISES

13-6.1. Show that the Lorentz space-time transformation Eqs. (13-78) can be written as a matrix equation

$$x_\mu' = \lambda_{\mu\nu} x_\nu, \tag{13-80}$$

where, for relative translation in the x_1 direction at speed $v = \beta c$,

$$\lambda = \begin{pmatrix} \gamma & 0 & 0 & i\beta\gamma \\ 0 & 1 & 0 & 0 \\ 0 & 0 & 1 & 0 \\ -i\beta\gamma & 0 & 0 & \gamma \end{pmatrix}. \tag{13-81}$$

13-6.2. In Fig. 13-12 we see two rotated coordinate frames in Euclidean three-space. The transformation is defined by the familiar orthogonal equations

$$x_1' = x_1 \cos \theta + x_2 \sin \theta,$$

$$x_2' = -x_1 \sin \theta + x_2 \cos \theta.$$

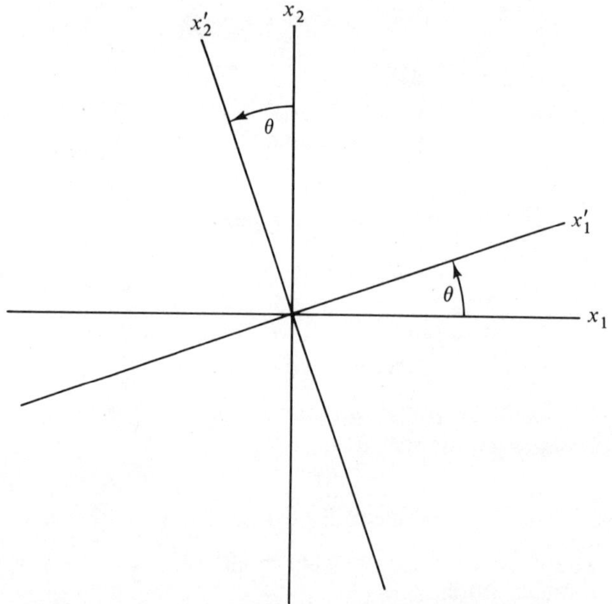

Fig. 13-12 An orthogonal transformation in Euclidean space.

Show that the Lorentz transform shown in Fig. 13-13, where $x = x_1$ and $ict = x_4$, can be written

$$x' = x \cosh \varphi - ct \sinh \varphi$$

$$ct' = -x \sinh \varphi + ct \cosh \varphi$$

and is subject to interpretation as a rotation through the imaginary angle φ in Minkowski space. What is φ as a function of v?

13-6.3. Just as the equation

$$\nabla^2 \phi = \sum_{i=1}^{3} \frac{\partial^2 \phi}{\partial x_i^2} = 0$$

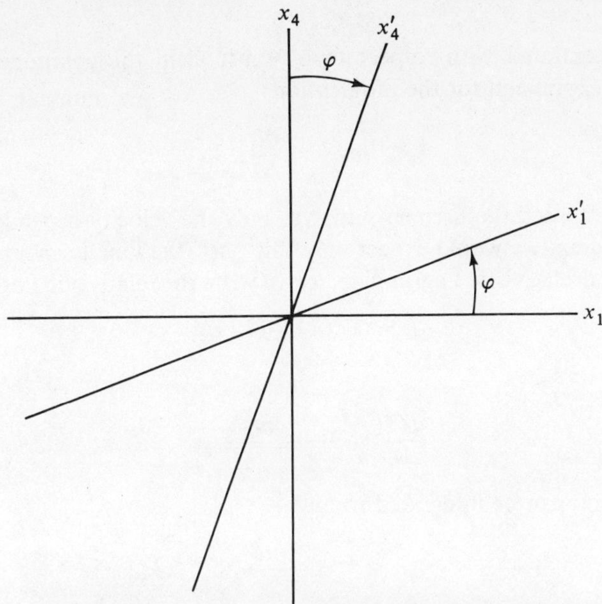

Fig. 13-13 A Lorentz transformation in Minkowski space.

is invariant to Galilean transformation, show that the wave equation that describes the propagation of light in free space

$$\Box^2\phi = \nabla^2\phi - \frac{1}{c^2}\frac{\partial^2\phi}{\partial t^2} = \sum_{\mu=1}^{4}\frac{\partial^2\phi}{\partial x_\mu^2} = 0$$

is invariant under Lorentz transformation but not under Galilean transformation.

13-7 The Relativistic Lagrangian

Our prior discussion of Lagrangian and Hamiltonian dynamics is correct only in the nonrelativistic limit. The extension to a relativistically correct formulation is straightforward, however.

It will be recalled that, for a single particle moving in a velocity-independent potential, we defined the generalized momentum as

$$p_i = \frac{\partial L}{\partial u_i}. \tag{13-82}$$

We shall adopt this definition and require that the *relativistic* Lagrangian,

403

when differentiated with respect to u_i, must yield the components of the relativistic expression for the momentum:

$$mu_i\gamma = \frac{\partial L}{\partial u_i}. \tag{13-83}$$

Since this imposed requirement involves only the velocity-dependent part of the Lagrangian, we would expect only this part (the kinetic energy) to differ from the nonrelativistic Lagrangian. If we write the relativistic Lagrangian as

$$L = T^*(u_i) - V(x_i), \tag{13-84}$$

then

$$\frac{dT^*}{du_i} = \frac{mu_i}{\sqrt{1 - (u^2/c^2)}}. \tag{13-85}$$

This equation can be integrated to find

$$T^* = -\frac{mc^2}{\gamma},$$

plus an integration constant that can be ignored. Thus

$$L = -\frac{mc^2}{\gamma} - V(x_i) \tag{13-86}$$

is found for the relativistic Lagrangian. Note that this expression is not equal to $T - V$, as in the nonrelativistic case. The equations of motion are found by operating on Eq. (13-86) with the Euler–Lagrange equations in the usual way.

The Hamiltonian can be calculated, using Eq. (11-68), as

$$H = \sum_i u_i p_i - L$$

$$= \sum_i \frac{mu_i^2}{\sqrt{1 - (u^2/c^2)}} + mc^2\sqrt{1 - \frac{u^2}{c^2}} + V$$

$$= \frac{mu^2 + mc^2[1 - (u^2/c^2)]}{\sqrt{1 - (u^2/c^2)}} + V$$

$$= \gamma mc^2 + V = T + E_0 + V = E. \tag{13-87}$$

Therefore the relativistic Hamiltonian is equal to the total energy, including the rest energy.

EXERCISE

13-7.1. Show that $T^* = -mc^2\sqrt{1 - (u^2/c^2)}$ is an integral of Eq. (13-84).

13-8 General Relativity

Newton's laws govern the dynamical behavior of matter for observers located in inertial frames of reference. In his 1905 special theory of relativity, Einstein showed that not only mechanical but also electromagnetic laws are invariant for inertial observers. In this theory it is shown that information cannot be propagated at speeds in excess of c, the speed of light, and so instantaneous action at a distance implicit in Newton's theory of gravitation cannot be correct.

Furthermore, we have seen that Newton's laws can be retained in non-inertial frames of reference only by including additional forces having no physically evident source—namely, the centrifugal and Coreolis forces.

Consider two deformable spheres located arbitrarily far from each other so that any distortions due to their mutual gravitational attraction are negligible. One of the spheres is caused to rotate about their common axis. Observers on each of the spheres would see the other sphere rotate with respect to his own reference frame, and each one might be tempted to say that the motion is purely relative. If the two observers measure the shape of their respective spheres, however, one of them will find that his ball is still spherical, whereas the other will discover that his has assumed a flattened spheroidal shape under the action of the centrifugal force. Similarly, both spheres could have been equipped with Foucault pendula operating at their north poles. The pendulum on the undistorted ball will not be observed to precess, whereas that on the flattened one, because of the Coreolis force, will. Newton viewed this sort of thing as evidence that *absolute space* is somehow endowed with a physical property that causes inertial forces to act on bodies accelerated with respect to it. In no other way does absolute space manifest itself. Uniform velocities with respect to absolute space are indistinguishable; otherwise we should have no relativity at all. Note also that absolute space can apparently act on matter via the inertial forces without being acted on in return, an apparent contradiction to Newton's third law.

Two directions might be taken in an attempt to improve our view of inertia. If we choose to take the modern *field* view, we can still consider space to be the seat of the inertial forces; however, we must allow space to be *nonabsolute* and to vary locally as a result of reactions. This was the view adopted by Einstein in formulating his general theory of relativity. For pedagogic reasons, we shall, for the present, explore the notion that the inertial forces have their seat in other bodies and therefore are not fictitious at all. From this point of view, Newton's laws hold in all frames of reference, some of which are subject to inertial forces and some not.

Bishop Berkeley in Ireland suggested 20 years after the publication of Newton's *Principia* that distant matter (the stars) makes a much larger contribution to the inertial force experienced, say, by water in a rotating bucket than does the bucket itself, owing to the greater mass of the former, which overcompensates the distance. Ernst Mach, writing 150 years after Berkeley, made an important extension of this approach to an inertial theory; he suggested that inertial frames are distinguished from noninertial frames by being unaccelerated with respect to the average distribution of matter in the universe. This statement is known as *Mach's principle.*

We would like to pursue this question of inertial theory by using a particle sort of approach to illustrate the continuity with Newtonian concept. In doing so, we shall use arguments first presented by D. W. Sciama. First, we ask ourselves: On what property of a body does the inertial force act? The answer must be the inertial mass by virtue of Newton's second law. Of greater significance is the question: What property of a body causes it to *exert* inertial forces? These same questions also apply to electrical forces and there we know that the action of an electric field and its establishment both depend on charge. Thus charge plays both an active and a passive role, in accordance with Newton's third law requiring that the force exerted by a pair of charges on each other be equal and opposite. For the same reason, we must assume that inertial mass can play an active as well as a passive role; hence the inertial force between two masses m_1 and m_2 must be proportional to their product

$$F \propto m_1 m_2. \tag{13-88}$$

Returning to the electrical case for the moment, suppose that the two charges are in relative motion with respect to each other. Electromagnetic theory tells us that here the force is a function of the velocity and that in the general case of nonuniform motion there is also an acceleration effect. This acceleration-dependent term is simply proportional to the acceleration $\ddot{r}$. Experience shows that inertial forces are exerted on a body only when it is accelerated with respect to the stars; so we should expect position or velocity-dependent terms in the intertial law, if any, to add to zero, presumably because of the symmetry in the distribution of matter in the universe. Acceleration-dependent terms are free of this restriction; hence we should expect that

$$F \propto m_1 m_2 \ddot{r}. \tag{13-89}$$

The acceleration-dependent force between two charges is given by

$$F = \frac{q_1 q_2}{4\pi\epsilon_0 c^2 r^3}\, r \times (\ddot{r} \times r) \tag{13-90}$$

or in scalar terms

$$F \propto \frac{q_1 q_2}{c^2 r}\, \ddot{r}_\perp, \tag{13-91}$$

where $\ddot{r}_\perp$ is the component of acceleration perpendicular to r, the line joining the two charges. Can we also justify postulating a $1/r$ distance-dependence for the inertial force?

Recalling Newton's observation that inertial forces on water in a rotating bucket do not depend on nearby masses (such as the bucket), we know that a long-range force law is needed. Newton's conclusion was strengthened by experiments performed by the brothers Friedländer in 1896 on heavy fly-wheels. An even more sensitive test is provided by the orbital motion of the sun relative to the earth. If the sun made a significant contribution to the inertial forces experienced by a Foucault pendulum, then the precession would have a yearly rather than a daily dependence, which it does not.

If the matter in the universe is considered to be uniformly distributed (a reasonable assumption on a cosmical scale), then successive spherical layers will contain an amount of matter that increases as r^2. So only a force law that varies as $1/r^n$, where $n < 2$, will favor the effect of distant mass over that of nearby mass. If we limit n to being integer for esthetic reasons, we are left with a $1/r$ law, as the electrical analogy suggests.

If we consider the static case and invoke the electrical analogy

$$F = \frac{q_1 q_2}{4\pi\epsilon_0 r^2}, \tag{13-92}$$

we might expect a static inertial interaction to have the form

$$F \propto \frac{m_1 m_2}{r^2}.$$

It is plausible to identify this interaction with Newton's law of gravitation for which the constant of proportionality is γ, the gravitation constant. Using estimates of the mass content of the universe obtained by counting galactic distribution and a radius obtained from Hubble's observation of the expansion rate of the universe, it can be shown that, by replacing $q_1 q_2$ by $m_1 m_2$ and $1/4\pi\epsilon_0$ by γ in Eq. (13-89), a plausible equation is obtained for the inertial forces

$$\boldsymbol{F} = \frac{\gamma m_1 m_2}{c^2 r^3} \boldsymbol{r} \times (\ddot{\boldsymbol{r}} \times \boldsymbol{r}). \tag{13-93}$$

In 1911 Einstein generalized these results in his *principle of equivalence.*

Within a region sufficiently small to be able to neglect effects due to nonuniformity, the laws of physics must be the same in a gravitational field with acceleration g as in a uniformly accelerated coordinate system with acceleration $-g$.

This postulate of Einstein's implies that light is affected by gravity even though it has no mass!

In Fig. 13-14(a) we show two observers displaced by a vertical distance l in a uniform gravitational field of acceleration g. In Fig. 13-14(b) the two observers are located in a noninertial frame with upward acceleration g. According to the principle of equivalence, these situations are equivalent and any physics experiments performed in either frame should give the same answers.

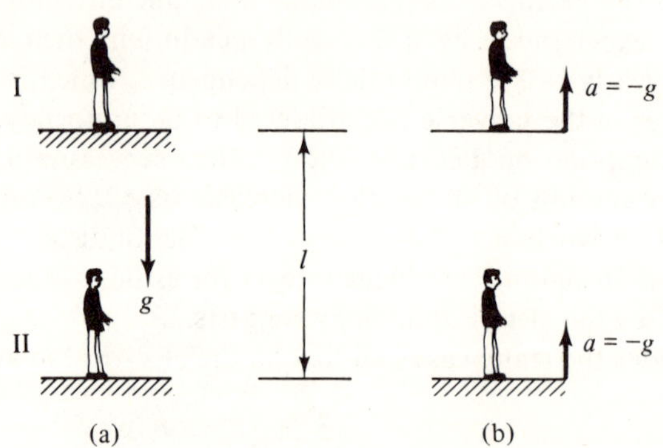

Fig. 13-14 Equivalent noninertial frames.

Suppose that observer I in frame (b) flashes a light of frequency ν toward observer II. The time required for the light to travel from I to II is

$$t = \frac{c}{g}\left(\sqrt{1 + \frac{2gl}{c^2}} - 1\right) \approx \frac{l}{c}. \tag{13-94}$$

If the light source at I has a velocity u at the time of the emission of the pulse, then at the time of arrival at observer II, the second observer will be rising at a velocity

$$u + at \approx u + \frac{gl}{c}. \tag{13-95}$$

The effective relative motion of observers I and II is therefore $v = gl/c$, and so a Doppler shift in the frequency amounting to

$$\nu' = \nu\sqrt{\frac{1 + gl/c^2}{1 - gl/c^2}} \approx \nu\left(1 + \frac{gl}{c^2}\right) \tag{13-96}$$

will be observed by II. If we invoke the principle of equivalence, we see that this result should also be obtained in frame (a). We must then say that the light in "falling" through the gravitational field has been shifted to a higher frequency. This gravitational spectral shift was confirmed by Pound and Snider in 1965 to an accuracy of 1.5%.

Not only is light shifted in its frequency by gravitation, its path can also be deflected. To illustrate, consider a small, windowless room being accelerated in deep space with an acceleration equal to g. We would expect experiments carried out in such a laboratory to have the same result as those obtained by an equivalent laboratory on the surface of the earth. If a beam of light is propagated horizontally in the accelerated laboratory, it will be observed to strike the opposite wall at a point somewhat below the point that would be illuminated by a straight light beam. The apparent curvature is understood to be caused by the upward acceleration of the room and the finite propagation time for the beam. In the terrestrial laboratory, however, the light beam will also be seen to curve and this curvature can only be due to gravity. The curvature of light in the earth's gravitational field is very slight; however, over cosmical distances or in the neighborhood of very dense stars where g is much larger, the curvature of a light beam becomes significant.

This question of the path taken by a beam of light is intimately associated with our perception of space. Thus it would seem that only an empty universe can be Euclidean; in the presence of a gravitational field, space would seem to be curved.

From a mathematical point of view, the geometry of any space can be specified by writing the equation for an invariant interval. For ordinary three-dimensional Euclidean space, we have

$$ds^2 = dx^i\, dx^i, \qquad i = 1, 2, 3. \tag{13-97}$$

In Minkowski space, the space of special relativity

$$ds^2 = g_{\mu\nu}\, dx^\mu\, dx^\nu, \qquad \mu, \nu = 1, 2, 3, 4, \tag{13-98}$$

where
$$g_{\mu\nu} = \begin{pmatrix} 1 & 0 & 0 & 0 \\ 0 & 1 & 0 & 0 \\ 0 & 0 & 1 & 0 \\ 0 & 0 & 0 & -1 \end{pmatrix} \tag{13-99}$$

The quantity $g_{\mu\nu}$ is known as the *metric tensor*.

As we know, a straight line is the shortest distance between two points in Euclidean space. Straight-line paths are also extremal in Minkowski space; they represent, however, the longest path between two events.

In a flat space, the metric tensor can always be put into the form in which its off-diagonal elements are all zero and all diagonal elements have an absolute value of one. The metric tensor for a curved space is more complicated; in its simplest form it contains nonunitary and off-diagonal terms.

Einstein, in his general theory of relativity, deduced a field

$$R_{\mu\nu} - \tfrac{1}{2} g_{\mu\nu} R + \Lambda g_{\mu\nu} = \frac{8\pi\gamma}{c^4}\, T_{\mu\nu}, \tag{13-100}$$

relating a complete mathematical description of the geometry of the space (on the left-hand side of the equation) to the stress-energy-momentum content of the space. The terms are second-rank tensor terms in four dimensions. The equation is highly nonlinear and can only be solved for certain simple cases. The coefficient of the stress-energy-momentum tensor $T_{\mu\nu}$ on the right of Eq. (13-100) has a numerical value in the MKS system of $2.07 \cdot 10^{-45}$. The effects on a terrestrial scale are *very* small.

In 1916 Karl Schwarzschild found an exact solution of Einstein's field equation for the four-space surrounding a spherically symmetrical mass. This metric is

$$ds^2 = \left(1 - \frac{2\gamma M}{c^2 r}\right) dt^2 - r^2(d\theta^2 + \sin^2\theta\, d\phi^2) - \frac{dr^2}{[1 - (2\gamma M/c^2 r)]}, \quad (13\text{-}101)$$

where M is the mass and r is the distance from the center of the mass. This solution permitted the calculation of effects that might be verified experimentally.

The angle of deflection of light from a star that just grazes the solar disk was calculated and verified by experiment to within one part in 500.

The planetary orbits are also slightly affected by general relativity; the perihelion of a planetary orbit is predicted to precess at a very slow rate because of relativity effects. The measured value of 42.6 ± 0.9 sec or arc per century for Mercury, the planet that experiences the longest effect, is regarded to be in satisfactory agreement with the calculated value of 43.0 sec/century.

One of the more interesting phenomena predicted by the general theory of relativity is the *black hole*. We note in examining the Schwarzschild metric, Eq. (13-94), that there exists a singularity for $r = 2\gamma M/c^2$. This value of the radius for a sphere of mass M corresponds to such a degree of compactness that light cannot escape the gravitational field. It can be shown that any object that collapses to this radius must continue to collapse to zero size and infinite density! There are various reasons to believe that the universe may have a sizable fraction of its mass in the form of black holes.

As a result in part of recent advances in high-energy astrophysics and radar astronomy, general relativity and its extension to include electromagnetic phenomena (unified field theory) constitute a rapidly expanding field of research. Thus classical mechanics continues to be refined and is far from being a subject of historical interest only.

EXERCISES

13-8.1. The equation for the gravitational shift in a nonuniform field is

$$\frac{\Delta\nu}{\nu} = \frac{\Delta\phi}{c^2},$$

where $\Delta\phi$ is the change in gravitational potential. Consider the light emitted from the surface of a neutron star having a mass equal to that of the sun and a radius of 10 km. How cool must the surface of this star be in order that the gravitational spectral shift be greater than the Doppler broadening due to thermal motion of the emitting ions?

13-8.2. Calculate the angular deflection of a beam of light projected from one vertical wall to another a distance l away in the presence of a uniform gravitational field of acceleration g.

13-8.3. Write the metric tensor for the surface of a sphere of radius r.

13-8.4. Let T be the period of a clock located in a field-free region of space and T' the corresponding period in the presence of a gravitational field characterized by a uniform acceleration g. By the principle of equivalence, the effect of such a field can be replaced by an acceleration g of the clocks' common frame of reference. We have shown that light radiated by such a clock—for example, a vibrating atom—has its observed frequency shifted to the red by an amount

$$\frac{\nu' - \nu}{\nu} = -\frac{gl}{c^2},$$

where l is the distance from the clock to the observer. Show that, to first order in small quantities, the period of the clock in the gravitational field is

$$T' = T\left(1 + \frac{gl}{c^2}\right).$$

13-8.5. We can measure the circumference of a disk of radius R by placing a large number of small rulers on the circumference and counting them. Suppose that the disk is rotating with a constant angular frequency ω. We can attach a Lorentz frame to each small ruler and make the assumption of uniform motion for short times. Under these assumptions, find the ratio of the circumference of the disk to its diameter from the viewpoint of (a) an observer rotating with the disk and (b) a nonrotating observer. What conclusions do you reach regarding the geometry of the space occupied by the disk in each case?

13-8.6. During its active life a star produces by fusion reactions an amount of energy equivalent to about 7 per mil of its rest energy. Calculate the radius to which a cloud of gas must collapse, starting at infinity, in order to release an amount of energy $0.007Mc^2$. What is the radius if the energy release is Mc^2? What is the significance of this latter radius?

REFERENCES

BOHM, D., *Special Theory of Relativity.* New York: W. A. Benjamin, Inc., 1965.

BONDI, H., *Relativity and Common Sense.* Garden City, N.Y.: Doubleday & Co., Inc., 1964.

BORN, M., *Einstein's Theory of Relativity.* New York: Dover Publications, Inc., 1962.

EDDINGTON, SIR ARTHUR, *Space, Time, and Gravitation.* New York: Harper & Row, 1959.

EINSTEIN, A., *The Meaning of Relativity.* Princeton, N.J.: Princeton University Press, 1970.

KACSER, C., *Introduction to the Special Theory of Relativity.* Englewood Cliffs, N.J.: Prentice-Hall, Inc., 1967.

LANCZOS, C., *Albert Einstein and the Cosmic World Order.* New York: Interscience Publishers, Inc., 1965.

NORWOOD, J., *Twentieth Century Physics.* Englewood Cliffs, N.J.: Prentice-Hall, Inc., 1976. Chapters 3 and 4.

PANOFSKY, W. K. H., and M. PHILLIPS, *Electricity and Magnetism.* Reading, Mass.: Addison-Wesley Publishing Co., Inc., 1955. Chapters 15 to 17.

"Resource Letter SRT-1 on Special Relativity Theory," *Am. J. Phys.*, No. 30 (1962), p. 462.

RUFFINI, REMO, and JOHN A. WHEELER, "Introducing the Black Hole," *Physics Today*, **xxiv**, No. 1 (1971), 30.

SCIAMA, D. W., *Physical Foundations of General Relativity.* Garden City, N.Y.: Doubleday & Company, Inc., 1969.

"Special Relativity Theory" reprints published for the American Association of Physics Teachers, American Institute of Physics, New York, 1962.

TAYLOR, E. F., and J. A. WHEELER, *Spacetime Physics.* San Francisco: W. H. Freeman and Co., 1966.

TAYLOR, J. G., *Black Holes.* New York: Avon Books, 1975.

INDEX